Barron's Regents Exams and Answers

Algebra II

Gary M. Rubinstein
B.A. Mathematics
Tufts University
M.S. Computer Science
University of Colorado

Barron's Educational Series, Inc.

Copyright © 2018, 2017 by Barron's Educational Series, Inc.

All rights reserved.
No part of this book may be reproduced or distributed in any form or by any means without the written permission of the copyright owner.

All inquiries should be addressed to:
Barron's Educational Series, Inc.
250 Wireless Boulevard
Hauppauge, NY 11788
www.barronseduc.com

ISBN: 978-1-4380-0843-1
ISSN: 2472-9175

PRINTED IN THE UNITED STATES OF AMERICA
9 8 7 6 5 4 3 2 1

Contents

Preface   **v**

How to Use This Book   **1**

Test-Taking Tips and Strategies   **3**

Some Key Algebra II Facts and Skills   **17**

  1. Polynomial Expressions and Equations   17
  2. Rational Expressions and Equations   36
  3. Exponential and Logarithmic Expressions and Equations   52
  4. Radical Expressions and Equations   65
  5. Trigonometric Expressions and Equations   78
  6. Systems of Equations   97
  7. Functions   112
  8. Sequences   121
  9. Probability   130
 10. Normal Distribution   142
 11. Inferential Statistics   155

Glossary of Terms   **169**

Regents Examinations, Answers, and Self-Analysis Charts 179

June 2016 Exam.................................181
August 2016 Exam..............................237
June 2017 Exam.................................293
August 2017 Exam..............................347

Preface

In 2009 New York State adopted the Common Core Standards in order to qualify for President Obama's Race To The Top initiative. The Common Core math is more difficult than the previous math curriculum, and the new state tests, including the Common Core Algebra II Regents, are more difficult as well.

Certain topics that had been in the Algebra II curriculum for decades have been removed for not being rigorous enough. Other topics have been added with the goal of making 21st-century American students more career and college ready than their predecessors.

The main topics that have been cut from the curriculum are permutations, combinations, Bernoulli trials, binomial expansion, and the majority of the trigonometric identities like the sine sum, cosine sum, sine difference, cosine difference, and the different double and half angle formulas. Other topics have been moved into earlier grades like the Law of Sines and the Law of Cosines. More than half the Trigonometry that had once been part of Algebra II is no longer part of the course.

Other topics have been added to fill the gaps left by those topics now considered obsolete. Primarily these are topics often taught in AP statistics as part of inferential statistics.

Aside from the change in topics, there is a change in the style of questions. There is a lot more need now for students to think more deeply about the topics as questions are intentionally phrased in a less straightforward way than they had been in the past.

Preface

Conquering the Algebra II test, something that was never an easy feat beforehand, has gotten that much more difficult and will require more test preparation than before. Getting this book is a great first step toward that goal. In addition to a review of all the topics that can appear on this test, there are nearly 1,000 practice questions of various levels of difficulty. The book can serve as a review or even as a way to learn the material for the first time. Teachers can also use this book to guide their pacing and to focus on the types of questions that are most likely to appear on the test while spending less focus on complicated aspects of the Common Core curriculum that are unlikely to be on the test.

The Common Core is part of a grand plan that is intended to propel our country to the top of the international rankings in math and reading. Good luck. We are all counting on you!

Gary Rubinstein
Math Teacher
2016

How to Use This Book

This book is designed to help you get the most out of your review for the new Regents exam in Algebra II (Common Core). Use this book to improve your understanding of the Algebra II topics and improve your grade.

TEST-TAKING TIPS

The first section in this book contains test-taking tips and strategies to help prepare you for the Algebra II Regents exam. This information is valuable, so be sure to read it carefully and refer to it during your study time. Remember: no single problem-solving strategy works for all problems—you should have a toolbox of strategies to pick from as you're facing unfamiliar or difficult problems on the test.

PRACTICE WITH KEY ALGEBRA II FACTS AND SKILLS

The second section in this book provides you with key Algebra II facts, useful skills, and practice problems with solutions. It provides you with a quick and easy way to refresh the skills you learned in class.

2   **How to Use This Book**

REGENTS EXAMS AND ANSWERS

The final section of the book contains actual Algebra II Regents. These exams and thorough answer explanations are probably the most useful tool for your review, as they let you know what's most important. By answering the questions on these exams, you will be able to identify your strengths and weaknesses and then concentrate on the areas in which you may need more study.

Remember, the answer explanations in this book are more than just simple solutions to the problems—they contain facts and explanations that are crucial to success in the Algebra II course and on the Regents exam. Careful review of these answers will increase your chances of doing well.

SELF-ANALYSIS CHARTS

Each of the Algebra II Regents exams ends with a Self-Analysis Chart. This chart will further help you identify weaknesses and direct your study efforts where needed. In addition, the chart classifies the questions on each exam into an organized set of topic groups. This feature will also help you to locate other questions on the same topic in the other Algebra II exam.

IMPORTANT TERMS TO KNOW

The terms that are listed in the glossary are the ones that have appeared most frequently on past Algebra II/Trigonometry and the most recent Algebra II (Common Core) exams. All terms and their definitions are conveniently organized for a quick reference.

Test-Taking Tips and Strategies

Knowing the material is only part of the battle in acing the new Algebra II Regents exam. Things like improper management of time, careless errors, and struggling with the calculator can cost valuable points. This section contains some test-taking strategies to help you perform your best on test day.

TIP 1

Time Management

SUGGESTIONS

- *Don't rush.* The Algebra II Regents exam is three hours long. While you are officially allowed to leave after 90 minutes, you really should stay until the end of the exam. Just as it wouldn't be wise to come to the test an hour late, it is almost as bad to leave a test an hour early.
- *Do the test twice.* The best way to protect against careless errors is to do the entire test twice and compare the answers you got the first time to the answers you got the second time. For any answers that don't agree, do a "tie breaker" third attempt. Redoing the test and comparing answers is much more effec-

tive than simply looking over your work. Students tend to miss careless errors when looking over their work. By redoing the questions, you are less likely to make the same mistake.
- *Bring a watch.* What will happen if the clock is broken? Without knowing how much time is left, you might rush and make careless errors. Yes, the proctor will probably write the time elapsed on the board and update it every so often, but it's better to be safe than sorry.

The TI-84 graphing calculator has a built in clock. Press the [MODE] to see it. If the time is not right, go to SET CLOCK and set it correctly. The TI-Nspire does not have a built-in clock.

 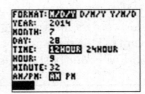

TIP 2

Know How to Get Partial Credit

SUGGESTIONS

- *Know the structure of the exam.* The Algebra Regents exam has 37 questions. The first 24 of those questions are multiple-choice worth two points each. There is no partial credit if you make a mistake on one of those questions. Even the smallest careless error, like missing a negative sign, will result in no credit for that question. Parts II, III, and IV are free-response questions with no multiple-choice. Besides giving a numerical answer, you may be asked to explain your reasoning. Part II

Test-Taking Tips and Strategies   5

has eight free-response questions worth two points each. The smallest careless error will cause you to lose one point, which is half the value of the question. Part III has four free-response questions worth four points each. These questions generally have multiple parts. Part IV has one free-response question worth six points. This question will have multiple parts.

- *Explain your reasoning.* When a free-response question asks you to "Justify your answer," "Explain your answer," or "Explain how you determined your answer," the grader is expecting a few clearly written sentences. For these, you don't want to write too little since the grader needs to see that you understand why you did the different steps you did to solve the equation. You also don't want to write too much because if anything you write is not accurate, points can be deducted.

Here is an example followed by two solutions. The first would not get full credit, but the second would.

Example 1

Use algebra to solve for x in the equation $\frac{2}{3}x + 1 = 11$. Justify your steps.

Solution 1 (partial credit):

$\frac{2}{3}x + 1 = 11$ $-1 = -1$ $\frac{2}{3}x = 10$ $x = 15$	I used algebra to get the x by itself. The answer was $x = 15$.

6   Test-Taking Tips and Strategies

Solution 2 (full credit):

$\dfrac{2}{3}x + 1 = 11$ $-1 = -1$ $\dfrac{2}{3}x = 10$ $\dfrac{3}{2} \cdot \dfrac{2}{3}x = \dfrac{3}{2} \cdot 10$ $1x = 15$ $x = 15$	I used the subtraction property of equality to eliminate the $+1$ from the left-hand side. Then to make it so the x had a 1 in front of it, I used the multiplication property of equality and multiplied both sides of the equation by the reciprocal of $\dfrac{2}{3}$, which is $\dfrac{3}{2}$. Then since $1 \cdot x = x$, the left-hand side of the equation just became x and the right-hand side became 15.

- *Computational errors vs. conceptual errors*

   In the Part III and Part IV questions, the graders are instructed to take off one point for a "computational error" but half credit for a "conceptual error." This is the difference between these two types of errors.

   If a four-point question was $x - 1 = 2$ and a student did it like this,

$$x - 1 = 2$$
$$\underline{+1 = +1}$$
$$x = 4$$

the student would lose one point out of 4 because there was one computational error since $2 + 1 = 3$ and not 4.

   Had the student done it like this,

$$x - 1 = 2$$
$$\underline{-1 = -1}$$
$$x = 1$$

the student would lose half credit, or 2 points, since this error was conceptual. The student thought that to eliminate the -1, he should subtract 1 from both sides of the equation.

   Either error might just be careless, but the conceptual error is the one that gets the harsher deduction.

TIP 3

Know Your Calculator

SUGGESTIONS

- *Which calculator should you use?* The two calculators used for this book are the TI-84 and the TI-Nspire. Both are very powerful. The TI-84 is somewhat easier to use for the functions needed for this test. The TI-Nspire has more features for courses in the future. The choice is up to you. This author prefers the TI-84 for the Algebra Regents. Graphing calculators come with manuals that are as thick as the book you are holding. There are also plenty of video tutorials online for learning how to use advanced features of the calculator. To become an expert user, watch the online tutorials or read the manual.
- *Clearing the memory.* You may be asked at the beginning of the test to clear the memory of your calculator. When practicing for the test, you should clear the memory too so you are practicing under test-taking conditions.

This is how you clear the memory.

For the TI-84:

Press [2ND] and then [+] to get to the MEMORY menu. Then press [7] for Reset.

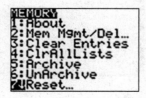

8 Test-Taking Tips and Strategies

Use the arrows to go to [ALL] for All Memory. Then press [1].

Press [2] for Reset.

The calculator will be reset as if in brand new condition. The one setting that you may need to change is to turn the diagnostics on if you need to calculate the correlation coefficient.

For the TI-Nspire:

The TI-Nspire must be set to Press-to-Test mode when taking the Algebra Regents. Turn the calculator off by pressing [ctrl] and [home]. Press and hold [esc] and then press [home].

Test-Taking Tips and Strategies 9

While in Press-to-Test mode, certain features will be deactivated. A small green light will blink on the calculator so a proctor can verify the calculator is in Press-to-Test mode.

To exit Press-to-Test mode, use a USB cable to connect the calculator to another TI-Nspire. Then from the home screen on the calculator in Press-to-Test mode, press [doc], [9], and select Exit Press-to-Test.

- *Use parentheses*
The calculator always uses the order of operations where multiplication and division happen before addition and subtraction. Sometimes, though, you may want the calculator to do the operations in a different order.

Suppose at the end of a quadratic equation, you have to round $x = \frac{-1+\sqrt{5}}{2}$ to the nearest hundredth. If you enter (–) (1) (+) (2ND) (x^2) (5) (/) (2), it displays

 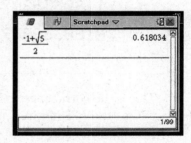

which is not the correct answer.

One reason is that for the TI-84 there needs to be a closing parenthesis (or on the TI-Nspire, press [right arrow] to move out from under the radical sign) after the 5 in the square root symbol. Without it, it calculated $-1+\sqrt{\frac{5}{2}}$. More needs to be done, though, since

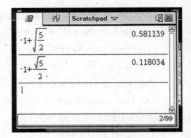

still is not correct. This is the solution to $-1+\sqrt{\dfrac{5}{2}}$.

To get this correct, there also needs to be parentheses around the entire numerator, $-1 + \sqrt{5}$.

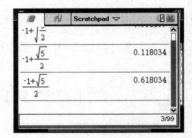

This is the correct answer.

On the TI-Nspire, fractions can also be done with [templates].

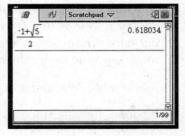

Test-Taking Tips and Strategies 11

- *Using the ANS feature*
  The last number calculated with the calculator is stored in something called the ANS variable. This ANS variable will appear if you start an expression with a +, −, ×, or ÷. When an answer has a lot of digits in it, this saves time and is also more accurate.

  If for some step in a problem you need to calculate the decimal equivalent of $\frac{1}{7}$, it will look like this on the TI-84:

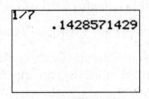

For the TI-Nspire, if you try the same thing, it leaves the answer as $\frac{1}{7}$. To get the decimal approximation, press [ctrl] and [enter] instead of just [enter].

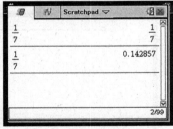

Now if you want to multiply this by 3, just press [×], and the calculator will display "Ans*"; press [3] and [enter].

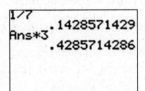

12 Test-Taking Tips and Strategies

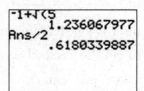

The ANS variable can also help you do calculations in stages. To calculate $x = \dfrac{-1+\sqrt{5}}{2}$ without using so many parentheses as before, it can be done by first calculating $-1 + \sqrt{5}$ and then pressing [÷] and [2] and Ans will appear automatically.

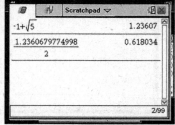

The ANS variable can also be accessed by pressing [2ND] and [–] at the bottom right of the calculator. If after calculating the decimal equivalent of $\dfrac{1}{7}$ you wanted to subtract $\dfrac{1}{7}$ from 5, for the TI-84 press [5], [–], [2ND], [ANS], and [ENTER]. For the TI-Nspire press [5], [–], [ctrl], [ans], and [enter].

Test-Taking Tips and Strategies 13

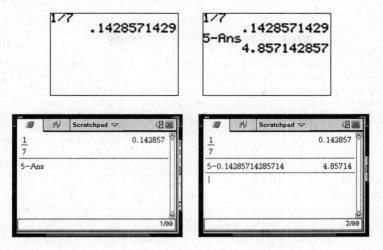

TIP 4

Use the Reference Sheet

SUGGESTIONS

- In the back of the Algebra II Regents booklet is a reference sheet that contains 17 conversion facts, such as inches to centimeters and quarts to pints, and also 17 formulas. Many of these conversion facts and formulas will not be needed for an individual test, but the quadratic formula and the arithmetic sequence formula are the two that will come in the handiest.

14   Test-Taking Tips and Strategies

High School Math Reference Sheet

1 inch = 2.54 centimeters	1 kilometer = 0.62 mile	1 cup = 8 fluid ounces
1 meter = 39.37 inches	1 pound = 16 ounces	1 pint = 2 cups
1 mile = 5280 feet	1 pound = 0.454 kilogram	1 quart = 2 pints
1 mile = 1760 yards	1 kilogram = 2.2 pounds	1 gallon = 4 quarts
1 mile = 1.609 kilometers	1 ton = 2000 pounds	1 gallon = 3.785 liters
		1 liter = 0.264 gallon
		1 liter = 1000 cubic centimeters

Triangle	$A = \frac{1}{2}bh$	Pythagorean Theorem	$a^2 + b^2 = c^2$
Parallelogram	$A = bh$	Quadratic Formula	$x = \dfrac{-b \pm \sqrt{b^2 - 4ac}}{2a}$
Circle	$A = \pi r^2$	Arithmetic Sequence	$a_n = a_1 + (n - 1)d$
Circle	$C = \pi d$ or $C = 2\pi r$	Geometric Sequence	$a_n = a_1 r^{n-1}$
General Prisms	$V = Bh$	Geometric Series	$S_n = \dfrac{a_1 - a_1 r^n}{1 - r}$ where $r \neq 1$
Cylinder	$V = \pi r^2 h$	Radians	1 radian = $\dfrac{180}{\pi}$ degrees
Sphere	$V = \frac{4}{3}\pi r^3$	Degrees	1 degree = $\dfrac{\pi}{180}$ radians
Cone	$V = \frac{1}{3}\pi r^2 h$	Exponential Growth/Decay	$A = A_0 e^{k(t - t_0)} + B_0$
Pyramid	$V = \frac{1}{3}Bh$		

Test-Taking Tips and Strategies 15

HOW MANY POINTS DO YOU NEED TO PASS?

The Algebra II Regents exam is scored out of a possible 86 points. Unlike most tests given in the year by your teacher, the score is not then turned into a percent out of 86. Instead each test has a conversion sheet that varies from year to year. For the June 2016 test, the conversion sheet looked like this.

Raw Score	Scale Score	Raw Score	Scale Score	Raw Score	Scale Score
86	100	57	82	28	68
85	99	56	82	27	67
84	98	55	81	26	66
83	97	54	81	25	65
82	97	53	81	24	64
81	96	52	80	23	63
80	95	51	80	22	61
79	94	50	80	21	60
78	94	49	79	20	58
77	93	48	79	19	55
76	92	47	79	18	54
75	91	46	78	17	53
74	91	45	78	16	51
73	90	44	77	15	49
72	89	43	77	14	47
71	89	42	77	13	44
70	88	41	76	12	42
69	88	40	76	11	39
68	87	39	76	10	37
67	87	38	75	9	34
66	86	37	75	8	31
65	86	36	74	7	27
64	85	35	73	6	24
63	84	34	73	5	20
62	84	33	72	4	17
61	84	32	72	3	13
60	83	31	71	2	9
59	83	30	70	1	4
58	82	29	69	0	0

Test-Taking Tips and Strategies

On this test, 30 points became a 70, 57 points became an 82, and 73 points became a 90. This means that for this examination a student who got 30 out of 86, which is just 35% of the possible points, would get a 73 on this exam. 57 out of 86 is 66%, but this scaled to an 86. 64 out of 86, however, is actually 85% and became an 85. So in the past there has been a curve on the exam for lower scores, though the scaling is not released until after the exam.

17

Some Key Algebra II Facts and Skills

1. POLYNOMIAL EXPRESSIONS AND EQUATIONS

1.1 POLYNOMIAL ARITHMETIC

A **polynomial** is an expression that combines numbers and variables raised to powers. $3x^2 + 5x + 1$ is an example of a polynomial.

The *terms* of this polynomial are $3x^2$, $5x$, and 1. Each term has a **coefficient**, which is the number in front of the term. In the term $3x^2$, the coefficient is 3.

Since the highest exponent in this polynomial is 2, we say that the **degree** of this polynomial is 2. Polynomials that have a degree of 2 are sometimes called **quadratic** polynomials. Polynomials that have a degree of 3 are sometimes called **cubic** polynomials. This polynomial has 3 terms. Polynomials with 3 terms are sometimes called **trinomials**. Polynomials with 2 terms are sometimes called **binomials**.

Adding and Subtracting Polynomials

When terms have the same variable and the same exponent, they are called **like terms**. Like terms can be easily added or subtracted.

$$5x^2 + 3x^2 = 8x^2$$
$$5x^2 - 3x^2 = 2x^2$$

18   **Some Key Algebra II Facts and Skills**

Polynomials can be added or subtracted by combining their like terms.

$$(2x^2 + 5x - 3) + (3x^2 - 2x + 5) = 5x^2 + 3x + 2$$
$$(2x^2 + 5x - 3) - (3x^2 - 2x + 5) = -1x^2 + 7x - 8$$

Multiplying Polynomials

To multiply a polynomial by a constant, multiply that constant by each of the coefficients of each of the terms.

$$5(3x^2 + 2x - 1) = 15x^2 + 10x - 5$$

- To multiply a polynomial by another polynomial, each term in the polynomial on the left must be multiplied by each term in the polynomial on the right, and all those products should then be added together.

$$(2x + 3)(4x - 5) = 2x \cdot 4x + 2x(-5) + 3 \cdot 4x + 3(-5)$$
$$= 8x^2 - 10x + 12x - 15 = 8x^2 + 2x - 15$$

In this example, both polynomials were binomials. When multiplying two binomials, this process is often called **FOIL** since the four combinations represent the **F**irsts, **O**uters, **I**nners, and **L**asts.

Dividing Polynomials

Polynomial division is like long division.

Example

Find the quotient and remainder of $(6x^2 + 13x + 8) \div (2x + 1)$.

Solution:

- **Step 1**: Write the problem as you would for long division.

$$2x + 1 \overline{)6x^2 + 13x + 8}$$

Then, at each step, figure out what the first term of the divisor needs to be multiplied by.

Polynomial Expressions and Equations   19

- **Step 2**: Determine what $2x$ must be multiplied by to make it $6x^2$, and then multiply $2x + 1$ by that and subtract and bring down the next term.

$$2x+1\overline{)6x^2+13x+8} \quad \begin{array}{r} 3x \\ \hline \end{array}$$
$$- \underline{(6x^2+3x)}$$
$$10x+8$$

- **Step 3**: Determine what $2x$ must be multiplied by to make it $10x$, and then multiply $2x + 1$ by that and subtract. There will be no other terms to bring down.

$$2x+1\overline{)6x^2+13x+8} \quad \begin{array}{r} 3x+5 \\ \hline \end{array}$$
$$- \underline{(6x^2+3x)}$$
$$10x+8$$
$$- \underline{10x+5}$$
$$3$$

- **Step 4**: The solution is $3x + 5$ and the remainder is 3. You can write $3x + 5$ R3 or as a fractional remainder $3x + 5 + \dfrac{3}{2x+1}$.

20   **Some Key Algebra II Facts and Skills**

1.2 FACTORING POLYNOMIALS

Greatest Common Factor Factoring
If all the terms of a polynomial have a common factor, that factor can be "factored out."

Example
Factor $6x^3 + 15x^2 + 21x$.

Solution:
Since each coefficient is divisible by 3 and each variable part is divisible by x, the common factor is $3x$. When it is factored out, it becomes $3x(2x^2 + 5x + 7)$.

Difference of Perfect Square Factoring
A polynomial of the form $x^2 - a^2$ can be factored into $(x - a)(x + a)$.

Example
Factor $x^2 - 9$.

Solution:
Since this can be written as $x^2 - 3^2$, it can be factored into

$$(x - 3)(x + 3)$$

Perfect Square Trinomial Factoring
A polynomial of the form $x^2 + 2ax + a^2$ can be factored into

$$(x + a)(x + a) = (x + a)^2$$

Example
Factor $x^2 + 10x + 25$.

Solution:
Since this can be written as $x^2 + 2 \cdot 5 \cdot x + 5^2$, it can be factored into $(x + 5)^2$. In general, if the square of half the coefficient of the x is equal to the constant term, this factoring pattern can be used. Since $\left(\dfrac{10}{2}\right)^2 = 25$, this factoring pattern applies.

Polynomial Expressions and Equations 21

General Trinomial Factoring

A polynomial of the form $x^2 + bx + c$ can be factored into

$$(x + m)(x + n)$$

if m and n can be found so that $m + n = b$ and $m \cdot n = c$.

Example

Factor $x^2 + 5x + 6$.

Solution:

Since there are two numbers 2 and 3 such that $2 + 3 = 5$ and $2 \cdot 3 = 6$, this can be factored into $(x + 2)(x + 3)$.

Factor by Grouping

Certain cubic expressions can be factored by a two-step process called **grouping**.

Example

Factor $2x^3 + 4x^2 + 6x + 12$.

Solution:

- **Step 1:** Find a common factor for the first two terms, and factor it out of them: $2x^2(x + 2) + 6x + 12$

- **Step 2:** Find a common factor for the last two terms, and factor it out of them: $2x^2(x + 2) + 6(x + 2)$

- **Step 3:** If these two parts have a common factor, as they do in this case with $(x + 2)$, factor that out of both terms:

$$(x + 2)(2x^2 + 6)$$

Notice that $2x^2 + 6$ can be further factored as $2(x^2 + 3)$. The final answer is $2(x + 2)(x^2 + 3)$.

22   Some Key Algebra II Facts and Skills

1.3 THE REMAINDER THEOREM AND THE FACTOR THEOREM

There is a short cut for finding the remainder when a polynomial is divided by a binomial of the form $x - a$.

MATH FACTS

The remainder when a polynomial is divided by $x - a$ will always be the same as the value of the polynomial when $x = a$ is substituted into it.

Example

What is the remainder of $(x^2 + 5x - 8) \div (x - 3)$?

Solution:

According to the Remainder Theorem, the remainder will be equal to $3^2 + 5 \cdot 3 - 8 = 9 + 15 - 8 = 16$.

MATH FACTS

There is a shortcut for determining if $x - a$ is a factor of another polynomial. If the value of the polynomial is zero when $x = a$ is substituted into it, then $x - a$ is a factor.

Example

Is $x - 3$ a factor of $x^3 + 2x^2 - 12x - 9$?

Solution:

Since $3^3 + 2 \cdot 3^2 - 12 \cdot 3 - 9 = 27 + 18 - 36 - 9 = 0$, $x - 3$ is a factor of $x^3 + 2x^2 - 12x - 9$.

Polynomial Expressions and Equations   23

1.4 POLYNOMIAL EQUATIONS

A **polynomial equation** has a polynomial on either one or both sides of the equal sign. The simplest type of polynomial equation has a zero on one side of the equal sign and a polynomial that can be factored on the other side.

Example

Solve for all values of x if $x^2 + 7x + 10 = 0$.

Solution:

$$x^2 + 7x + 10 = 0$$
$$(x + 2)(x + 5) = 0$$

For a product to be zero, at least one of the factors must be zero so either $x + 2 = 0$ or $x + 5 = 0$.

$$
\begin{array}{ccc}
x + 2 = 0 & \text{or} & x + 5 = 0 \\
\underline{-2 = -2} & & \underline{-5 = -5} \\
x = -2 & \text{or} & x = -5
\end{array}
$$

The solution set is $\{-2, -5\}$.

Solving Quadratic Equations When the Polynomial Cannot Be Factored

An equation like $x^2 + 6x + 7 = 0$ cannot be solved like the previous example since the polynomial cannot be factored. When this happens, there are two methods for solving it.

Method 1: Completing the Square

- **Step 1:** Eliminate the constant from the right side by subtracting 7 from both sides.

$$
\begin{array}{r}
x^2 + 6x + 7 = 0 \\
\underline{-7 = -7} \\
x^2 + 6x = -7
\end{array}
$$

Some Key Algebra II Facts and Skills

- **Step 2:** "Complete" the square by adding the square of half the coefficient on the x to both sides of the equation. For this example, add $\left(\dfrac{6}{2}\right)^2 = 9$ to both sides.

$$x^2 + 6x = -7$$
$$\underline{+9 = +9}$$
$$x^2 + 6x + 9 = 2$$

- **Step 3:** The left-side polynomial is now a perfect square polynomial, which can be factored to $(x + 3)^2$. Take the square root of both sides and eliminate the constant from the left-hand side.

$$x^2 + 6x + 9 = 2$$
$$(x+3)^2 = 2$$
$$\sqrt{(x+3)^2} = \pm\sqrt{2}$$
$$x + 3 = \pm\sqrt{2}$$
$$\underline{-3 = -3}$$
$$x = -3 \pm \sqrt{2}$$

Method 2: The Quadratic Formula

To solve equations of the form $ax^2 + bx + c = 0$, use the formula $x = \dfrac{-b \pm \sqrt{b^2 - 4ac}}{2a}$. For this example, $a = 1$, $b = 6$, and $c = 7$.

$$x = \frac{-6 \pm \sqrt{6^2 - 4 \cdot 1 \cdot 7}}{2 \cdot 1}$$
$$= \frac{-6 \pm \sqrt{36 - 28}}{2}$$
$$= \frac{-6 \pm \sqrt{8}}{2}$$
$$= \frac{-6 \pm 2\sqrt{2}}{2}$$
$$= -3 \pm \sqrt{2}$$

Polynomial Expressions and Equations 25

1.5 GRAPHS OF EQUATIC EXPRESSIONS

The graph of $y = ax^2 + bx + c$ will be a parabola. If $a > 0$, the parabola will open "upward" like the letter u. If $a < 0$, the parabola will open "downward" like the letter n.

The graph of $y = x^2 + 6x + 8$ looks like this:

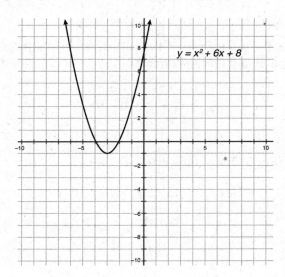

- Four important points on the parabola are the vertex, the y-intercept, and the x-intercepts. There can be 0, 1, or 2 x-intercepts.

The y-intercept is at $(0, c)$, which is $(0, 8)$ for $y = x^2 + 6x + 8$.

The x-intercept(s) can be found by solving, if possible, the equation $0 = x^2 + 6x + 8$. This might require factoring, completing the square, or using the quadratic formula.

$$0 = x^2 + 6x + 8$$
$$0 = (x + 2)(x + 4)$$

$$\begin{array}{rcl} x + 2 = 0 & \quad\text{or}\quad & x + 4 = 0 \\ \underline{-2 = -2} & & \underline{-4 = -4} \\ x = -2 & \quad\text{or}\quad & x = -4 \end{array}$$

The x-intercepts are $(-2, 0)$ and $(-4, 0)$.

The vertex is the low point if the parabola looks like a u and the high point if it looks like an n. The x-coordinate of the vertex is always $-\frac{b}{2a}$. For this example, that is

$$-\frac{6}{2 \cdot 1} = -\frac{6}{2} = -3$$

To get the y-intercept of the vertex, substitute the x-value you just calculated into the equation.

$$y = x^2 + 6x + 8$$
$$y = (-3)^2 + 6(-3) + 8$$
$$y = 9 - 18 + 8$$
$$y = -1$$

The coordinates of the vertex are $(-3, -1)$.

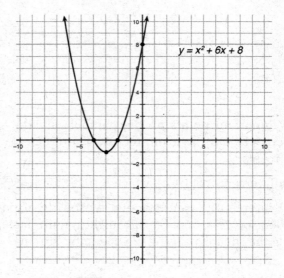

- If an equation has the form $y = a(x - h)^2 + k$, the graph will also be a parabola. If $a > 0$, it looks like a u. Otherwise, it looks like an n. This form is called *vertex form* since the numbers in the place of h and k will be the vertex of the parabola.

Polynomial Expressions and Equations

Example
What is the vertex of the graph of $y = 1(x + 3)^2 - 1$?

Solution:
Since this can be written as $y = 1(x - (-3))^2 + (-1)$, the h is -3 and the k is -1 so the vertex is at $(-3, -1)$.

Focus and Directrix of a Parabola
- Every parabola also has associated with it an invisible horizontal line called the **directrix** and an invisible point called the **focus**.

If the equation is $y = ax^2 + bx + c$ or $y = a(x - h)^2 + k$, the focus is $\frac{1}{4a}$ units above the vertex, and the directrix is $\frac{1}{4a}$ units below the vertex.

For $y = x^2 + 6x + 8$, the vertex is $(-3, -1)$ and $\frac{1}{4a} = \frac{1}{4 \cdot 1} = \frac{1}{4}$ so the focus is $\frac{1}{4}$ units above the vertex at $\left(-3, -\frac{3}{4}\right)$ and the directrix is $y = -1\frac{1}{4}$.

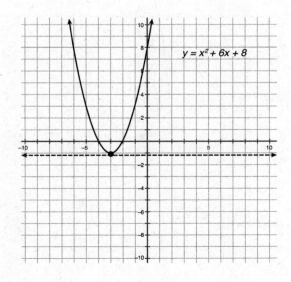

Graphs of Higher-Degree Polynomials

For degree greater than 2, the graph of the polynomial resembles a snake with the number of turns usually one less than the degree.

- The graph of the cubic $y = (x - 1)(x + 1)(x - 2)$ has three x-intercepts because $0 = (x - 1)(x + 1)(x - 2)$ has three solutions.

$$
\begin{array}{ccccc}
x - 1 = 0 & \text{or} & x + 1 = 0 & \text{or} & x - 2 = 0 \\
x = 1 & \text{or} & x = -1 & \text{or} & x = 2
\end{array}
$$

The three x-intercepts are $(1, 0)$, $(-1, 0)$, and $(2, 0)$.

The graph of $y = (x - 1)(x + 1)(x - 2)$ looks like this

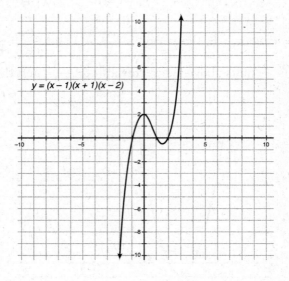

Polynomial Expressions and Equations

A **double root** happens when two of the x-intercepts are located at the same point.

In the equation $y=(x-2)(x-2)(x+1)$, the three x-intercepts are (2, 0), (2, 0), and (−1, 0). In the graph, the curve "bounces off" the x-axis at (2, 0), the location of the double root.

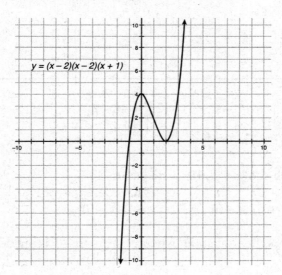

Practice Exercises

1. What is $(5x^2 - 3x + 8) - (2x^2 + 4x - 2)$?
   (1) $3x^2 - 7x + 10$
   (2) $3x^2 + x + 10$
   (3) $3x^2 - 7x + 6$
   (4) $3x^2 + x + 6$

2. Which shows $x^2 - 12x + 36$ factored?
   (1) $(x + 6)^2$
   (2) $(x + 6)(x - 6)$
   (3) $(x - 1)(x - 36)$
   (4) $(x - 6)^2$

3. What is the remainder when $x^3 + x^2 - 9x - 5$ is divided by $x - 3$?
   (1) 4
   (2) 5
   (3) 6
   (4) 7

4. Which of the following is a factor of $x^3 + 3x^2 - 10x - 24$?
   (1) $(x - 1)$
   (2) $(x - 2)$
   (3) $(x - 3)$
   (4) $(x - 4)$

5. What are the solutions to $x^2 - 6x + 4 = 0$?
   (1) $x = 5 \pm \sqrt{3}$
   (2) $x = 3 \pm \sqrt{2}$
   (3) $x = 3 \pm \sqrt{5}$
   (4) $x = 2 \pm \sqrt{5}$

Polynomial Expressions and Equations   31

6. Which is an equation of a parabola with its vertex at $(3, -1)$?
   (1) $y = (x - 3)^2 + 1$ (3) $y = (x + 3)^2 + 1$
   (2) $y = (x - 3)^2 - 1$ (4) $y = (x + 3)^2 - 1$

7. What is the vertex of the parabola whose equation is
   $y = x^2 + 4x - 12$?
   (1) $(-2, -16)$ (3) $(2, 0)$
   (2) $(-2, -24)$ (4) $(2, -16)$

8. What is the equation of the directrix and the coordinates of the
   focus of the parabola $y = 3x^2$?

   (1) $y = -12$ and $(0, 12)$ (3) $y = -\dfrac{1}{12}$ and $\left(0, \dfrac{1}{12}\right)$

   (2) $y = -\dfrac{1}{4}$ and $\left(0, \dfrac{1}{4}\right)$ (4) $y = \dfrac{1}{12}$ and $\left(0, -\dfrac{1}{12}\right)$

9. What are the x-intercepts and y-intercept of
   $y = (x - 2)(x + 4)(x + 5)$?
   (1) $(-2, 0), (4, 0), (5, 0), (0, -40)$
   (2) $(-2, 0), (4, 0), (5, 0), (0, 40)$
   (3) $(2, 0), (-4, 0), (-5, 0), (0, -40)$
   (4) $(2, 0), (-4, 0), (-5, 0), (0, 40)$

10. If a and b are positive integers, which could be the graph of $y = (x - b)(x^2 + 2ax + a^2)$?

(1)

(3)

(2)

(4)

Polynomial Expressions and Equations

Solutions

1. $(5x^2 - 3x + 8) - (2x^2 + 4x - 2)$
   $5x^2 - 3x + 8 - 2x^2 - 4x + 2$
   $3x^2 - 7x + 10$

   The correct choice is **(1)**.

2. Because the square of half the b is equal to the constant term, c, $\left(\dfrac{-12}{2}\right)^2 = 36$, this is a perfect square trinomial and can be factored into $\left(x + \dfrac{b}{2}\right)^2 = \left(x + \dfrac{-12}{2}\right)^2 = (x - 6)^2$.

   The correct choice is **(4)**.

3. The Remainder Theorem says that you can get the remainder when dividing by $x - a$ by substituting a for x into the expression you are dividing into.

   $$3^3 + 3^2 - 9(3) - 5 = 27 + 9 - 27 - 5 = 4$$

   The correct choice is **(1)**.

4. To check if $(x - a)$ is a factor of a polynomial, substitute a for x into the polynomial, and if it evaluates to 0, $(x - a)$ is a factor. Testing all the choices, this only happens for choice (3).

   $$3^3 + 3 \cdot 3^2 - 10(3) - 24 = 27 + 27 - 30 - 24 = 0$$

   The correct choice is **(3)**.

Some Key Algebra II Facts and Skills

5. Using the quadratic formula,

$$x = \frac{-(-6) \pm \sqrt{(-6)^2 - 4(1)(4)}}{2(1)}$$

$$= \frac{6 \pm \sqrt{36 - 16}}{2}$$

$$= \frac{6 \pm \sqrt{20}}{2}$$

$$= \frac{6 \pm 2\sqrt{5}}{2}$$

$$= 3 \pm \sqrt{5}$$

The correct choice is **(3)**.

6. When the equation is in vertex form, $y = (x - h)^2 + k$, the vertex of the parabola will be (h, k). If the vertex is $(3, -1)$, the equation is $y = (x - 3)^2 + (-1) = (x - 3)^2 - 1$.

The correct choice is **(2)**.

7. The x-coordinate of the vertex can be found with the formula

$$x = \frac{-b}{2a} = \frac{-4}{2(1)} = -2$$

The y-coordinate can be found by substituting the x-coordinate into the equation and solving for y.

$$y = (-2)^2 + 4(-2) - 12 = 4 - 8 - 12 = -16$$

The vertex is $(-2, -16)$.

The correct choice is **(1)**.

Polynomial Expressions and Equations   35

8. The focus is $\dfrac{1}{4a}$ units above the vertex. When the equation for a parabola is of the form $y = ax^2$, the vertex is $(0, 0)$. Since $a = 3$, $\dfrac{1}{4a} = \dfrac{1}{4(3)} = \dfrac{1}{12}$. So the focus is $\left(0,\, 0+\dfrac{1}{12}\right) = \left(0,\, \dfrac{1}{12}\right)$.

The directrix is a horizontal line $\dfrac{1}{4a}$ units below the vertex. So the directrix is $y = 0 - \dfrac{1}{12} = -\dfrac{1}{12}$.

The correct choice is **(3)**.

9. To find the y-intercept, substitute 0 into the equation

$$y = (0 - 2)(0 + 4)(0 + 5) = (-2)(4)(5) = -40$$

so the y-intercept is $(0, -40)$.

To find the x-intercepts, find the solutions to the equation

$$0 = (x - 2)(x + 4)(x + 5)$$

To get a product of zero, one of the factors must be 0 so either $x - 2 = 0$, $x + 4 = 0$, or $x + 5 = 0$. The solutions to these three equations are 2, -4, and -5 so the x-intercepts are $(2, 0)$, $(-4, 0)$, and $(-5, 0)$.

The correct choice is **(3)**.

10. The second factor is a perfect square trinomial so this can be factored to $y = (x - b)(x + a)(x + a) = (x - b)(x + a)^2$. This means that there is a root at $x = b$ to the right of the y-axis, and a double root at $x = -a$ to the left of the y-axis. At a double root, the curve bounces off the x-axis. This happens in choice (1).

The correct choice is **(1)**.

2. RATIONAL EXPRESSIONS AND EQUATIONS

2.1 ARITHMETIC WITH RATIONAL EXPRESSIONS

- A **rational expression** involves a fraction that has a polynomial in its denominator.

  $\dfrac{1}{x^2+5x+6}$ is a rational expression.

  A rational expression can have a polynomial in both the denominator and the numerator, such as the rational expression $\dfrac{2x+6}{x^2+5x+6}$.

Reducing Rational Expressions

To reduce a rational expression, factor the numerator and the denominator and "cancel out" any common factors.

Example

Reduce the expression $\dfrac{2x+6}{x^2+5x+6}$.

Solution:

$$\frac{2x+6}{x^2+5x+6} = \frac{2\cancel{(x+3)}}{(x+2)\cancel{(x+3)}} = \frac{2}{x+2}$$

A Warning About 'Canceling Out'

When the numerator has two or more terms separated by a + or a −, you can only 'cancel out' something in the denominator if each of the terms in the numerator shares the common factor.

So

$$\frac{2(x+5)+5(x+5)}{(x+2)(x+5)} = \frac{2\cancel{(x+5)}+5\cancel{(x+5)}}{(x+2)\cancel{(x+5)}} = \frac{2+5}{x+2} = \frac{7}{x+2}$$

Rational Expressions and Equations 37

But you cannot do

$$\frac{2(x+3)+5}{(x+2)(x+3)} = \frac{2\cancel{(x+3)}+5}{(x+2)\cancel{(x+3)}} = \frac{2+5}{x+2} = \frac{7}{x+2}$$

since the 5 did not get divided by the $(x+3)$.

Multiplying Rational Expressions

To multiply two rational expressions, first factor all polynomials in the numerators and denominators. You can then cancel out any factors in either of the denominators with any matching factors in any of the numerators.

Example

Multiply $\dfrac{2x+6}{5x-10} \cdot \dfrac{x^2-4}{x^2+5x+6} = \dfrac{2\cancel{(x+3)}}{5\cancel{(x-2)}} \cdot \dfrac{\cancel{(x-2)}(x+2)}{(x+2)\cancel{(x+3)}} = \dfrac{2}{5}$

Dividing Rational Expressions

Divide by multiplying the expression on the left by the reciprocal of the expression on the right. If the division is represented as a fraction, multiply the numerator by the reciprocal of the denominator.

Example

$$\frac{5}{x+3} \div \frac{2}{x^2+5x+6} = \frac{5}{x+3} \cdot \frac{x^2+5x+6}{2}$$

$$= \frac{5}{\cancel{x+3}} \cdot \frac{(x+2)\cancel{(x+3)}}{2}$$

$$= \frac{5(x+2)}{2}$$

38   **Some Key Algebra II Facts and Skills**

Adding Rational Expressions

If two rational expressions have the same denominator, their sum has that same denominator, and the numerator is the sum of the two numerators.

Example

$$\frac{5x}{x+3}+\frac{2}{x+3}=\frac{5x+2}{x+3}$$

- If two rational expressions have different denominators that have no common factor, the numerator and denominator of each rational expression must first be multiplied by the denominator of the other rational expression. The new rational expressions will have a common denominator and can then be added.

Example

What is $\dfrac{5}{x+2}+\dfrac{3}{x+5}$?

Solution:

$$\frac{5}{x+2}+\frac{3}{x+5}=\frac{5(x+5)}{(x+2)(x+5)}+\frac{3(x+2)}{(x+2)(x+5)}=\frac{5(x+5)+3(x+2)}{(x+2)(x+5)}$$

$$=\frac{5x+25+3x+6}{(x+2)(x+5)}=\frac{8x+31}{(x+2)(x+5)}$$

- If two rational expressions have different denominators that *do* have a common factor, the common denominator will have each of the factors of each of the denominators in it, but just one of the common factors.

Example

What is $\dfrac{5}{(x+2)(x+3)}+\dfrac{3}{(x+2)(x+5)}$?

Rational Expressions and Equations

39

Solution:

Because both denominators have a factor of $(x + 2)$, the common denominator will be $(x + 2)(x + 3)(x + 5)$.

$$\frac{5}{(x+2)(x+3)} + \frac{3}{(x+2)(x+5)}$$

$$= \frac{5(x+5)}{(x+2)(x+3)(x+5)} + \frac{3(x+3)}{(x+2)(x+3)(x+5)}$$

$$= \frac{5(x+5)+3(x+3)}{(x+2)(x+3)(x+5)} = \frac{5x+25+3x+9}{(x+2)(x+3)(x+5)}$$

$$= \frac{8x+34}{(x+2)(x+3)(x+5)}$$

Subtracting Rational Expressions

Subtracting is just like adding, but you need to be more careful when subtracting since people sometimes improperly distribute the negative sign through the right-hand expression.

Example

What is $\dfrac{5}{x+2} - \dfrac{3}{x+5}$?

Solution:

$$\frac{5}{x+2} - \frac{3}{x+5} = \frac{5(x+5)}{(x+2)(x+5)} - \frac{3(x+2)}{(x+2)(x+5)} = \frac{5(x+5)-3(x+2)}{(x+2)(x+5)}$$

$$= \frac{5x+25-3x-6}{(x+2)(x+5)} = \frac{2x+19}{(x+2)(x+5)}$$

A common error is to have a +6 instead of a –6 in the second to last step because of not distributing the –3 through.

40   Some Key Algebra II Facts and Skills

2.2 RATIONAL EQUATIONS

A **rational equation** involves one or more rational expressions, like $\dfrac{2}{x+2} + \dfrac{3}{x+5} = \dfrac{7x+10}{(x+2)(x+5)}$.

- **Step 1**: Combine the two rational expressions using the methods from the previous section.

$$\frac{2}{x+2} + \frac{3}{x+5} = \frac{7x+10}{(x+2)(x+5)}$$

$$\frac{2(x+5)}{(x+2)(x+5)} + \frac{3(x+2)}{(x+2)(x+5)} = \frac{7x+10}{(x+2)(x+5)}$$

$$\frac{2(x+5) + 3(x+2)}{(x+2)(x+5)} = \frac{7x+10}{(x+2)(x+5)}$$

$$\frac{2x+10+3x+6}{(x+2)(x+5)} = \frac{7x+10}{(x+2)(x+5)}$$

$$\frac{5x+16}{(x+2)(x+5)} = \frac{7x+10}{(x+2)(x+5)}$$

- **Step 2**: If the denominators are different, cross multiply. If the denominators are the same, as they are in this example, they can be ignored.

$$5x+16 = 7x+10$$
$$-5x = -5x$$
$$\overline{16 = 2x+10}$$
$$-10 = -10$$
$$\overline{6 = 2x}$$
$$\frac{6}{2} = \frac{2x}{2}$$
$$3 = x$$

- **Step 3**: Check your answer. Sometimes this process produces "fake" answers so the only way to know whether or not to reject any potential solutions is to substitute back into the *original* equation and see if the solution makes that equation true.

    Checking $x = 3$:

    $$\frac{2}{3+2} + \frac{3}{3+5} = \frac{7 \cdot 3 + 10}{(3+2)(3+5)}$$

    $$\frac{2}{5} + \frac{3}{8} = \frac{31}{40}$$

    $$\frac{16}{40} + \frac{15}{40} = \frac{31}{40}$$

    $$\frac{31}{40} = \frac{31}{40}$$

Since this is true, $x = 3$ is the solution to the original equation.

2.3 GRAPHS OF RATIONAL FUNCTIONS

The graph of a rational function like $y = \frac{2x-5}{x-3}$ has some properties that graphs of polynomial functions do not.

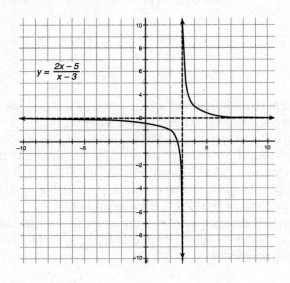

An asymptote of a graph is a line that is not part of the graph but functions like a wall that the graph gets closer and closer to without touching. The dotted lines indicate the location of these invisible asymptotes of the graph.

The graph above has a horizontal asymptote of $y = 2$ and a vertical asymptote of $x = 3$.

The graph of a rational function can have more than one vertical asymptote. Below is the graph of $y = \dfrac{3}{(x-3)(x+1)}$.

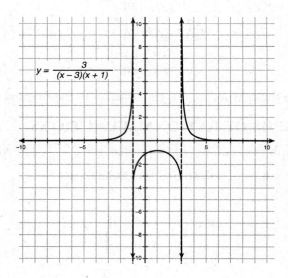

This graph has a horizontal asymptote of $y = 0$ and has two vertical asymptotes, one at $x = 3$ and the other at $x = -1$.

It is possible to determine the equation of the vertical asymptotes by finding the x-value(s) that make the denominator of the rational expression equal to zero. For the above example, the equation $(x - 3)(x + 1) = 0$ has two solutions, $x = 3$ and $x = -1$, which are also the equations of the two vertical asymptotes.

The horizontal asymptote is more complicated to calculate by hand but can easily be found by graphing the rational expression on the graphing calculator.

Practice Exercises

1. What is $\dfrac{2x+6}{x^2-2x-15}$ reduced to simplest terms?

   (1) $\dfrac{6}{x^2-15}$

   (2) $\dfrac{2x}{x^2-15}$

   (3) $\dfrac{2}{x-5}$

   (4) $\dfrac{6}{2x-3}$

2. What is $\dfrac{12-3x}{5x-5} \cdot \dfrac{x^2-1}{x^2-3x-4}$ reduced to simplest terms?

   (1) $\dfrac{3}{5}$

   (2) $\dfrac{12}{5}$

   (3) $-\dfrac{3}{5}$

   (4) $-\dfrac{12}{5}$

Some Key Algebra II Facts and Skills

3. What is $\dfrac{4x+16}{x^2+5x+6} \div \dfrac{x^2-16}{5x+15}$ reduced to simplest terms?

(1) $\dfrac{20}{x+2}$

(2) $\dfrac{20}{x-4}$

(3) $\dfrac{20}{(x+2)(x+3)}$

(4) $\dfrac{20}{(x+2)(x-4)}$

4. What is $\dfrac{4}{(x+2)(x+3)} + \dfrac{3}{(x+2)(x-2)}$?

(1) $\dfrac{7x+1}{(x+2)(x-2)(x+3)}$

(2) $\dfrac{7}{(x+2)(x-2)(x+3)}$

(3) $\dfrac{7x+3}{(x+2)(x-2)(x+3)}$

(4) $\dfrac{7}{(x+2)^2(x-2)(x+3)}$

Rational Expressions and Equations 45

5. What is $\dfrac{x}{x^2+7x+12} - \dfrac{3}{x^2-9}$?

   (1) $\dfrac{x^2-6x+12}{(x+3)(x-3)(x+4)}$

   (2) $\dfrac{x-3}{7x+21}$

   (3) $\dfrac{x^2-6x-12}{(x+3)(x-3)(x+4)}$

   (4) $\dfrac{x-3}{7x+3}$

6. Solve for x.
$$\frac{3}{x+2}+\frac{2}{x-3}=\frac{15}{(x+2)(x-3)}$$

   (1) 2
   (2) 3
   (3) 4
   (4) 5

7. Solve for x.
$$\frac{1}{x-4}+\frac{3}{x-1}=\frac{7}{x^2-5x+4}$$

   (1) 4
   (2) 5
   (3) 6
   (4) 7

Some Key Algebra II Facts and Skills

8. A group of people contribute equal amounts of money to get a $24 gift for a friend. If two more people contributed and they all paid equal amounts, they would each pay $1 less. How many people were there originally?

9. What are the asymptotes of the graph of $R(x) = \dfrac{3}{x+4}$?

   (1) $x = 4, y = 0$
   (2) $x = -4, y = 0$
   (3) $x = 0, y = 4$
   (4) $x = 0, y = -4$

10. Which could be the equation for this graph?

   (1) $R(x) = \dfrac{12}{x+3}$

   (2) $R(x) = \dfrac{x+3}{12}$

   (3) $R(x) = \dfrac{x-3}{12}$

   (4) $R(x) = \dfrac{12}{x-3}$

Rational Expressions and Equations 47

Solutions

1. Factor the numerator and denominator, and cancel the common factor $(x + 3)$.

$$\frac{2x+6}{x^2-2x-15} = \frac{2(x+3)}{(x+3)(x-5)} = \frac{2}{x-5}$$

The correct choice is **(3)**.

2. Factor both numerators and denominators, and cancel the common factors. A factor from the numerator in one of the fractions can be cancelled with a denominator in the other fraction.

$$\frac{12-3x}{5x-5} \cdot \frac{x^2-1}{x^2-3x-4} = \frac{3(4-x)}{5(x-1)} \cdot \frac{(x-1)(x+1)}{(x-4)(x+1)}$$

Cancel the $(x + 1)$s and the $(x - 1)$s. Then, since $(4 - x) = -1(x - 4)$, the $(4 - x)$ in the numerator of the first fraction can be canceled out with the $(x - 4)$ in the second fraction, leaving a -1 in the numerator.

$$\frac{3(4-x)}{5} \cdot \frac{1}{(x-4)} = \frac{3(-1)}{5} = -\frac{3}{5}$$

The correct choice is **(3)**.

3. Divide by multiplying the first expression by the reciprocal of the second.

$$\frac{4x+16}{x^2+5x+6} \cdot \frac{5x+15}{x^2-16} = \frac{4(x+4)}{(x+2)(x+3)} \cdot \frac{5(x+3)}{(x-4)(x+4)}$$

Cancel the $(x + 3)$s and the $(x + 4)$s.

$$\frac{4}{(x+2)} \cdot \frac{5}{(x-4)} = \frac{20}{(x+2)(x-4)}$$

The correct choice is **(4)**.

Some Key Algebra II Facts and Skills

4. Since the two denominators have a common factor of $(x + 2)$, this factor is only used once in the common denominator, $(x + 2)(x - 2)(x + 3)$

$$\frac{4(x-2)}{(x+2)(x-2)(x+3)} + \frac{3(x+3)}{(x+2)(x-2)(x+3)}$$

$$= \frac{4(x-2)+3(x+3)}{(x+2)(x-2)(x+3)}$$

$$= \frac{4x-8+3x+9}{(x+2)(x-2)(x+3)}$$

$$= \frac{7x+1}{(x+2)(x-2)(x+3)}$$

The correct choice is (**1**).

5. First factor the two denominators.

$$\frac{x}{(x+3)(x+4)} - \frac{3}{(x+3)(x-3)}$$

Since the denominators have a common factor of $(x + 3)$, this factor is only used once in the common denominator $(x + 3)(x - 3)(x + 4)$.

$$\frac{x(x-3)}{(x+3)(x-3)(x+4)} - \frac{3(x+4)}{(x+3)(x-3)(x+4)}$$

$$= \frac{x(x-3)-3(x+4)}{(x+3)(x-3)(x+4)}$$

$$= \frac{x^2-3x-3x-12}{(x+3)(x-3)(x+4)}$$

$$= \frac{x^2-6x-12}{(x+3)(x-3)(x+4)}$$

The correct choice is (**3**).

Rational Expressions and Equations   **49**

6. The common denominator for all three terms is $(x + 2)(x - 3)$. Convert all terms to the common denominator, combine the two terms on the left side of the equal sign, and create an equation to find the value of x that makes the two numerators equal. Check your answer by substituting it back into the original equation.

$$\frac{3(x-3)}{(x+2)(x-3)} + \frac{2(x+2)}{(x+2)(x-3)} = \frac{15}{(x+2)(x-3)}$$

$$\frac{3(x-3) + 2(x+2)}{(x+2)(x-3)} = \frac{15}{(x+2)(x-3)}$$

$$\frac{3x-9+2x+4}{(x+2)(x-3)} = \frac{15}{(x+2)(x-3)}$$

$$3x-9+2x+4 = 15$$

$$5x-5 = 15$$

$$5x = 20$$

$$x = 4$$

The correct choice is **(3)**.

7. Factor the denominator of the fraction on the right side of the equal sign into $(x - 4)(x - 1)$. The common denominator for all three terms is $(x - 4)(x - 1)$. Convert all terms to the common denominator, combine the two terms on the left side of the equal sign, and create an equation to find the value of x that makes the two numerators equal. Check your answer by substituting it back into the original equation.

Some Key Algebra II Facts and Skills

$$\frac{1(x-1)}{(x-4)(x-1)} + \frac{3(x-4)}{(x-4)(x-1)} = \frac{7}{(x-4)(x-1)}$$

$$\frac{1(x-1) + 3(x-4)}{(x-4)(x-1)} = \frac{7}{(x-4)(x-1)}$$

$$\frac{x-1+3x-12}{(x-4)(x-1)} = \frac{7}{(x-4)(x-1)}$$

$$x - 1 + 3x - 12 = 7$$

$$4x - 13 = 7$$

$$4x = 20$$

$$x = 5$$

The correct choice is **(2)**.

8. 6 people

If there were originally x people, the amount each person would pay is $\frac{24}{x}$. If two more people were involved, there would be $x + 2$ people so the amount each person would pay is $\frac{24}{x+2}$. Since each person's share would be \$1 less if there were two more people, the equation relating these expressions is $\frac{24}{x} = \frac{24}{x+2} + 1$. This can be solved by making all three terms have a common denominator of $x(x + 2)$ and solving.

$$\frac{24(x+2)}{x(x+2)} = \frac{24x}{x(x+2)} + \frac{x(x+2)}{x(x+2)}$$

$$\frac{24x+48}{x(x+2)} = \frac{24x+x^2+2x}{x(x+2)}$$

$$24x + 48 = 24x + x^2 + 2x$$

$$0 = x^2 + 2x - 48$$

$$0 = (x+8)(x-6)$$

$$x + 8 = 0 \quad \text{or} \quad x - 6 = 0$$

$$x = -8 \quad \text{or} \quad x = 6$$

Rational Expressions and Equations 51

Since the amount each pays must be positive, the -8 is rejected, and the only answer is $x = 6$.

9. One way is to graph the function on the graphing calculator and see the curve approach the x-axis, which has equation $y = 0$, for large x-values and approach the vertical line $x = -4$ for values close to -4.

Even without the graphing calculator, you can find the equation of the vertical asymptote(s) by setting the denominator equal to zero.

$$x + 4 = 0$$
$$x = -4$$

This is the equation of the vertical asymptote and only choice (2) has this as one of the asymptotes.

The correct choice is **(2)**.

10. Since there is a vertical asymptote at $x = 3$ on the graph, there must be a $x - 3$ in the denominator of the rational expression. This only happens in choice (4).

The correct choice is **(4)**.

52   **Some Key Algebra II Facts and Skills**

3. EXPONENTIAL AND LOGARITHMIC EXPRESSIONS AND EQUATIONS

3.1 PROPERTIES OF EXPONENTS

In an exponential expression like $3^5 = 3 \cdot 3 \cdot 3 \cdot 3 \cdot 3 = 243$ the 3 is called the *base* and the 5 is called the **exponent** or the *power*.

To multiply two exponential expressions with the same base like $3^8 \cdot 3^2$, the product has the same base as the factors do and the power of the product will be the sum (and *not* the product) of the powers of the factors.

$$3^8 \cdot 3^2 = 3^{8+2} = 3^{10}$$

USEFUL FORMULA

$$x^a \cdot x^b = x^{a+b}$$

- Division is similar except you subtract the two exponents instead of adding them.

$$3^8 \div 3^2 = 3^{8-2} = 3^6$$

USEFUL FORMULA

$$x^a \div x^b = x^{a-b}$$

$$\frac{x^a}{x^b} = x^{a-b}$$

- To raise an exponential expression to a power, like $(3^8)^2$, multiply the two exponents together.

$$(3^8)^2 = 3^{8 \times 2} = 3^{16}$$

Exponential and Logarithmic Expressions and Equations   53

> ## USEFUL FORMULA
> $$(x^a)^b = x^{ab}$$

- To raise a number or expression to a negative exponent, make a fraction with a numerator of 1 and a denominator with the same expression, but with the negative sign changed to a positive sign.

$$5^{-3} = \frac{1}{5^{+3}} = \frac{1}{125}$$

> ## USEFUL FORMULA
> $$x^{-a} = \frac{1}{x^a}$$

- To raise a number or expression to a fractional exponent, like $8^{\frac{5}{3}}$, take the third root of 8 and raise the answer to the fifth power.

$$8^{\frac{5}{3}} = \left(\sqrt[3]{8}\right)^5 = 2^5 = 32$$

> ## USEFUL FORMULA
> $$x^{\frac{a}{b}} = \left(\sqrt[b]{x}\right)^a$$

Anything but 0 raised to the 0th power is 1, for example $8^0 = 1$.

It is *not* true that $(5 + 3)^2 = 5^2 + 3^2$ since the left side is equal to 64 while the right side is equal to 34.

There is, though, a distributive property of exponents when a single term with multiple factors is raised to a power like $(5 \cdot 3)^2 = 5^2 \cdot 3^2$.

54   **Some Key Algebra II Facts and Skills**

3.2 EXPONENTIAL EQUATIONS

If an equation involves two exponential equations with the same base, the exponents must be equal.

To solve for x in the equation $4^{2x+1} = 4^{3x-5}$, solve the equation $2x + 1 = 3x - 5$ so $x = 6$.

If the bases are different, it is sometimes possible to convert one or both of the expressions so they do have the same base.

Example

Solve for x in $8^{x+4} = 2^{5x-4}$.

Solution:

Notice that 8 can be replaced with 2^3.

$$8^{x+4} = 2^{5x-4}$$
$$2^{3(x+4)} = 2^{5x-4}$$
$$3(x+4) = 5x-4$$
$$3x+12 = 5x-4$$
$$16 = 2x$$
$$8 = x$$

3.3 LOGARITHMS

A logarithmic expression looks like $\log_2 64$. This can be evaluated to a number by thinking "To what power must the number 2 be raised in order to get a value of 64?" Since $2^6 = 64$, $\log_2 64 = 6$.

USEFUL FORMULA

A logarithmic equation like $\log_a b = c$ can be rearranged into an exponential equation $a^c = b$.

Exponential and Logarithmic Expressions and Equations 55

Example
Write the equation $\log_7 2{,}401 = x$ as an exponential equation. Then solve that equation by guess-and-check.

Solution:
The equation is $7^x = 2{,}401$. By testing different positive integers, it is quickly found that $7^4 = 2{,}401$ so the solution is $x = 4$.

The TI-84 and the TI-Nspire calculators have logarithm (or, for short, log) functions, which can be used to solve logarithmic equations.

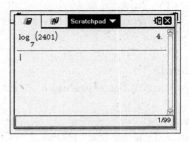

Example
Convert the equation $5^x = 700$ into a logarithmic equation, and then use the graphing calculator to approximate the answer to the nearest hundredth.

Solution:
The equation is $\log_5 700 = x$. Using the log function of your calculator, $x = 4.07$.

If, in an exponential equation, there is a number multiplied by the exponential expression, divide both sides of the equation by that number before converting the equation to a log equation.

56   **Some Key Algebra II Facts and Skills**

Example
Solve for x in the equation $800 = 200 \cdot 1.04^x$.

Solution:
Do not multiply 200 and 1.04. Instead first divide both sides of the equation by 200 to get $4 = 1.04^x$. This can then be solved with the graphing calculator as $\log_{1.04} 4 = 35.35$.

- The number e is a mathematical constant that is equal to approximately 2.72. Many exponential equations have a base of e. Since e is such a popular base, there is a special key on the calculator, the LN key, which can be used to calculate log base e.

Example
Solve for x in the equation $e^x = 150$.

Solution:
First rewrite as $\log_e 150 = x$. This can be solved with the graphing calculator by either doing $\log_e 150 = 5.01$ or, even shorter, $\ln 150 = 5.01$.

3.4 GRAPHS OF EXPONENTIAL AND LOGARITHMIC FUNCTIONS

The graph of $y = 2^x$ looks like a playground slide going up to the right. When the base of the exponent is less than 1, like in $y = \left(\frac{1}{2}\right)^x$, it looks like a playground slide going down to the right.

Exponential and Logarithmic Expressions and Equations

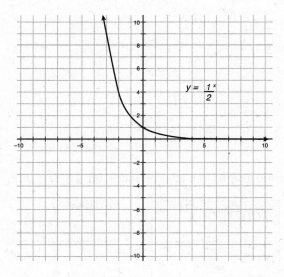

Some Key Algebra II Facts and Skills

Since any number except zero raised to the 0th power is 1, exponential graphs of the form $y = b^x$ have a y-intercept of $(0, 1)$. They will also have a horizontal asymptote of $y = 0$ since raising a number greater than 1 to a very large negative power becomes a fraction with a very large denominator. Fractions with very large denominators are very close to zero.

The graph of a logarithmic function like $y = \log_2 x$ is a reflection of the graph of $y = 2^x$. This graph has an x-intercept of $(1, 0)$ and a vertical asymptote at $x = 0$.

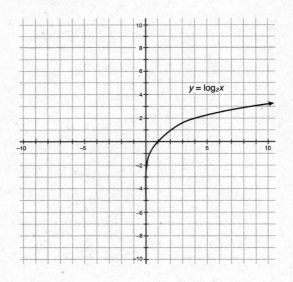

Logarithmic graphs can also be made on the graphing calculator.

3.5 MODELING REAL-WORLD SCENARIOS WITH EXPONENTIAL EQUATIONS

Many real-world problems involve things that grow or decay in a way that can be modeled with exponential equations.

Example
The population of a country can be modeled by the equation $P = 50(1.02)^T$, where T is the number of years since 2010 and P is the population in millions. In what year will the population reach 75 million?

Solution:

$$75 = 50(1.02)^T$$
$$\frac{75}{50} = \frac{50(1.02)^T}{50}$$
$$1.5 = 1.02^T$$
$$T = \log_{1.02} 1.5 = 20.5$$

It will reach the population of 75 million 20.5 years after 2010, or sometime between 2030 and 2031.

Practice Exercises

1. What is $a^{\frac{3}{4}}$?

   (1) $\sqrt[3]{a^4}$
   (3) $a^{-\frac{4}{3}}$

   (2) $a^{-\frac{3}{4}}$
   (4) $\left(\sqrt[4]{a}^{-3}\right)$

2. Simplify $(2x^3y^4)^3$.

   (1) $8x^{27}y^{64}$
   (3) $8x^9y^{12}$

   (2) $6x^9y^{12}$
   (4) $6x^{27}y^{64}$

3. Solve $4^x = 256$.

   (1) 4
   (3) 16

   (2) 8
   (4) 64

4. Solve $8^{x+4} = 2^{x+4}$.

   (1) 2
   (3) 1

   (2) −2
   (4) −4

5. The equation $m^x = n$ is equivalent to which of the following?

   (1) $\log_n m = x$
   (3) $\log_m n = x$

   (2) $\log_m x = n$
   (4) $\log_n x = m$

Exponential and Logarithmic Expressions and Equations 61

6. What is the solution to x to the *nearest* tenth in $3 \cdot 4^x = 270$?
   (1) 3.0   (3) 3.2
   (2) 3.1   (4) 3.3

7. If $x = \log_7 7^{123456789}$, what is the value of x?

8. The population of West Algebra can be modeled by the equation $P = 30 \cdot 1.04^T$ where T is the number of years since 2000 and P is the population in millions. How many million people will there be in 2020?
   (1) 63.7   (3) 65.7
   (2) 64.7   (4) 66.7

9. The temperature of a cup of herbal tea can be modeled by the equation $T = 90 \cdot 0.7^M + 75$ where T is the temperature and M is the number of minutes since the tea was taken off the stove. How hot will the tea be 15 minutes after it is taken off the stove?
   (1) 75°   (3) 79°
   (2) 77°   (4) 81°

10. Which equation is equivalent to the equation $y = 50(1.07)^{12x}$?
    (1) $y = 50(2.1)^x$ (3) $y = 50(1.3)^{6x}$
    (2) $y = 50(1.5)^{2x}$ (4) $y = 53.5^{12x}$

62   Some Key Algebra II Facts and Skills

Solutions

1. The denominator of the fraction goes outside the radical sign and the numerator becomes the exponent that the radical gets raised to.

$$a^{\frac{3}{4}} = (\sqrt[4]{a})^3$$

The correct choice is **(4)**.

2. Each factor in the parentheses gets raised to the 3rd power. When something raised to a power is then raised to another power, the base does not change but the exponent becomes the product of the two exponents.

$$(2x^3y^4)^3 = 2^3(x^3)^3(y^4)^3 = 8x^9y^{12}$$

The correct choice is **(3)**.

3. Because this is a multiple-choice question, you can test the four choices and find that for choice (1), $4^4 = 256$.

The correct choice is **(1)**.

4. If possible, change both expressions so they have the same base. Since $8 = 2^3$, this equation can be written as

$$(2^3)^{x+4} = 2^{x+4}$$
$$2^{3(x+4)} = 2^{x+4}$$

If the bases are equal and the expressions are equal, the exponents must be equal. This lead to a new equation.

$$3(x + 4) = x + 4$$
$$3x + 12 = x + 4$$
$$2x = -8$$
$$x = -4$$

The correct choice is **(4)**.

Exponential and Logarithmic Expressions and Equations   63

5. If $m^x = n$, then by the definition of logs, $\log_m n = x$.

The correct choice is **(3)**.

6. First eliminate the 3 by dividing both sides by 3.

$$\frac{3 \cdot 4^x}{3} = \frac{270}{3}$$
$$4^x = 90$$

Rearrange this exponential equation into a log equation, and solve with the log function of the graphing calculator. For the TI-84, press [MATH] [A]. For the TI-Nspire, press [log].

$$\log_4 90 = x$$

$x = 3.2459$, which, rounded to the nearest tenth, is 3.2.

The correct choice is **(3)**.

7. 123456789

This can be rearranged to the exponential equation

$$7^x = 7^{123456789}$$
$$x = 123456789$$

8. Substitute $T = 20$ into the equation, and calculate with your calculator.

$$P = 30 \cdot 1.04^{20}$$
$$P = 65.7$$

The correct choice is **(3)**.

64　　　**Some Key Algebra II Facts and Skills**

9. Substitute $M = 15$ into the equation, and calculate with your calculator.

$$T = 90 \cdot 0.7^{15} + 75$$
$$T = 75°$$

The correct choice is **(1)**.

10. The rule for raising a power to a power is $(x^a)^b = x^{ab}$. This rule also works in reverse $x^{ab} = (x^a)^b$. So $(1.07)^{12x} = (1.07^{12})^x$. Calculate $1.07^{12} = 2.25$ so one equivalent equation is $y = 50(2.25)^x$. Unfortunately, this is not one of the choices. But if you do this process for each of the choices you will see that choice (2), $y = 50(1.5)^{2x} = 50(1.5^2)^x = 50(2.25)^x$ is the correct answer.

The correct choice is **(2)**.

Radical Expressions and Equations 65

4. RADICAL EXPRESSIONS AND EQUATIONS

4.1 SIMPLIFYING RADICALS

The square root of a perfect square, like $\sqrt{25}$, will be a positive integer. In this case, $\sqrt{25} = 25$ because $5^2 = 25$.

- The square root of something that is not a perfect square, like $\sqrt{27}$, will be an irrational number. In this case $\sqrt{27}$ will be between 5 and 6 because $5^2 = 25$ and $6^2 = 36$ and 27 is between 25 and 36. Using a calculator you can find that $\sqrt{27} \approx 5.196$.

- Radical expressions like $\sqrt{7} \cdot \sqrt{11}$ can be multiplied by multiplying the numbers inside the radicals together, $\sqrt{7} \cdot \sqrt{11} = \sqrt{7 \cdot 11} = \sqrt{77}$. This process can also be reversed when simplifying the square root of any composite number like $\sqrt{8} = \sqrt{2 \cdot 4} = \sqrt{2} \cdot \sqrt{4}$.

If the number inside the square root sign has a factor that is a perfect square, the expression can be simplified. Since $27 = 9 \cdot 3$ and 9 is a perfect square, $\sqrt{27} = \sqrt{9 \cdot 3} = \sqrt{9} \cdot \sqrt{3} = 3 \cdot \sqrt{3} = 3\sqrt{3}$.

Two radical expressions can be added or subtracted if the number inside the radical is the same, so $7\sqrt{2} + 3\sqrt{2} = 10\sqrt{2}$ and $7\sqrt{2} - 3\sqrt{2} = 4\sqrt{2}$.

66 Some Key Algebra II Facts and Skills

4.2 IMAGINARY AND COMPLEX NUMBERS

The number i is defined as $\sqrt{-1}$.

USEFUL FACTS

The first four powers of i are:
$$i^1 = i, i^2 = -1, i^3 = i^2 \cdot i^1 = -1 \cdot i = -i,$$
$$\text{and } i^4 = i^2 \cdot i^2 = (-1) \cdot (-1) = 1$$

Anytime a negative number is inside a square root sign it can be rewritten with an i. Because of this definition, $i^2 = \left(\sqrt{-1}\right)^2 = -1$.

$$\sqrt{-9} = \sqrt{9} \cdot \sqrt{-1} = 3i$$

A number of the form ai, where a is a real number, is known as an **imaginary number**.

- Imaginary numbers can be added or subtracted

$$5i + 2i = 7i$$
$$5i - 2i = 3i$$

- Imaginary numbers can be multiplied. Whenever an i^2 appears in a solution, replace it with a -1.

$$5i \cdot 2i = 10i^2 = 10 \cdot (-1) = -10$$

A number of the form $a + bi$, where a and b are real numbers, is called a **complex number**. $5 + 3i$ is a complex number.

Complex numbers can be added and subtracted

$$(5 + 3i) + (2 + 7i) = 7 + 10i$$
$$(5 + 3i) - (2 + 7i) = 3 - 4i$$

Radical Expressions and Equations 67

- Complex numbers can be multiplied. Whenever an i^2 appears in a solution, replace it with a -1.

$$(5 + 3i)(2 + 7i) = 10 + 35i + 6i + 21i^2 = 10 + 41i - 21 = -11 + 41i$$

Sometimes complex numbers appear when solving quadratic equations with the quadratic formula.

Example
Solve for all solutions in the equation $x^2 - 4x + 13 = 0$.

Solution:
Using the quadratic equation, $a = 1$, $b = -4$, and $c = 13$.

$$x = \frac{-(-4) \pm \sqrt{(-4)^2 - 4(1)(13)}}{2(1)} = \frac{4 \pm \sqrt{16 - 52}}{2} = \frac{4 \pm \sqrt{-36}}{2}$$

Since there is a negative number inside the radical sign, the solutions will be complex numbers.

$$x = \frac{4 \pm 6i}{2} = 2 \pm 3i$$

There are two complex solutions $x = 2 + 3i$ or $x = 2 - 3i$.

- Complex numbers can be graphed as points on **the complex plane**.

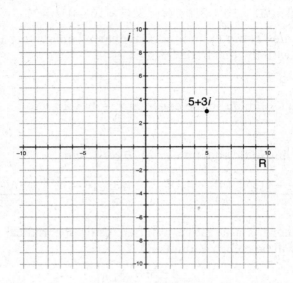

USEFUL FORMULA

The absolute value of a complex number $a + bi$ is the distance of the point that represents that number on the complex plane from the point that represents the number $0 + 0i$. The formula is $|a + bi| = \sqrt{a^2 + b^2}$.

The absolute value of $5 + 3i$ is $\sqrt{5^2 + 3^2} = \sqrt{25 + 9} = \sqrt{34}$.

Radical Expressions and Equations 69

4.3 RADICAL EQUATIONS

If an equation has one radical sign, isolate it and then square both sides of the equation to get a new equation. The solutions to that new equation may be solutions to the original equation.

Example

Solve for all values of x that satisfy the equation $\sqrt{x} - x = -6$.

Solution:

- **Step One:** Isolate the radical.

$$\sqrt{x} - x = -6$$
$$\underline{+x = +x}$$
$$\sqrt{x} = x - 6$$

- **Step Two:** Square both sides.

$$\left(\sqrt{x}\right)^2 = (x - 6)^2$$
$$x = x^2 - 12x + 36$$
$$0 = x^2 - 13x + 36$$
$$0 = (x - 4)(x - 9)$$

$$x - 4 = 0 \quad \text{or} \quad x - 9 = 0$$
$$x = 4 \quad \text{or} \quad x = 9$$

- **Step Three:** Substitute answers back into the original equation to check if some solutions should be rejected.

$$\sqrt{4} - 4 = 2 - 4 = -2$$
$$\sqrt{9} - 9 = 3 - 9 = -6$$

The 4 must be rejected since it does not satisfy the original equation so the only solution is $x = 9$.

Some Key Algebra II Facts and Skills

Solving an equation that involves two radical terms like $\sqrt{x+1} + \sqrt{x-4} = 5$ is a lengthy process. To solve it, first isolate one of the radical terms, square both sides, and continue with the problem by isolating the other radical term.

$$\sqrt{x+1} + \sqrt{x-4} = 5$$
$$\underline{-\sqrt{x-4} = -\sqrt{x-4}}$$
$$\sqrt{x-4} = 5 - \sqrt{x-4}$$
$$\left(\sqrt{x+1}\right)^2 = \left(5 - \sqrt{x-4}\right)^2$$
$$x+1 = 25 - 10\sqrt{x-4} + x - 4$$
$$\underline{-x = -x}$$
$$1 = 21 - 10\sqrt{x-4}$$
$$\underline{-21 = -21}$$
$$-20 = -10\sqrt{x-4}$$
$$\frac{-20}{-10} = \frac{-10\sqrt{x-4}}{-10}$$
$$2 = \sqrt{x-4}$$
$$2^2 = \left(\sqrt{x-4}\right)^2$$
$$4 = x - 4$$
$$\underline{+4 = +4}$$
$$8 = x$$

Check to see if the answer should be rejected.

$$\sqrt{8+1} + \sqrt{8-4} = \sqrt{9} + \sqrt{4} = 3 + 2 = 5$$

The solution is $x = 8$.

4.4 GRAPHS OF RADICAL FUNCTIONS

The graph of $y = \sqrt{x}$ is half of a sideways parabola.

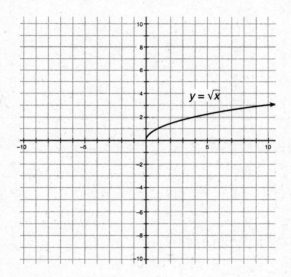

Practice Exercises

1. What is $\sqrt{12} + \sqrt{75}$?

   (1) $7\sqrt{3}$ (3) $7\sqrt{6}$

   (2) $\sqrt{87}$ (4) $3\sqrt{7}$

2. What is $2\sqrt{3} \cdot 5\sqrt{3}$?

   (1) 30   (3) $\sqrt{30}$

   (2) $10\sqrt{3}$ (4) $7\sqrt{3}$

3. What is i^{66}?

   (1) -1 (3) $-i$

   (2) 1 (4) i

4. What is $(3 + 5i) + (2 - 7i)$?

   (1) $5 + 2i$ (3) $-5 - 2i$

   (2) $-5 + 2i$ (4) $5 - 2i$

5. Solve the quadratic equation $x^2 - 6x + 34 = 0$.

   (1) $3 \pm 5i$ (3) $13, -7$

   (2) $-3 \pm 5i$ (4) $-13, 7$

Radical Expressions and Equations

6. What is $|12 - 5i|$?

    (1) $\sqrt{119}$ (3) 13

    (2) 12   (4) 7

7. What value(s) of x make the equation $\sqrt{x} + x = 20$ true?

    (1) 14   (3) 16

    (2) 25   (4) 16, 25

8. Find all solutions to the equation $\sqrt{x+3} + x = 9$.

    (1) 13, 6   (3) 6

    (2) 13   (4) 4

9. Solve for x.

$$\sqrt{x-3} + \sqrt{x+9} = 6$$

    (1) 6   (3) 8

    (2) 7   (4) 9

10. Solve for x.

$$5\sqrt{x-4} + 3 = 18$$

74   **Some Key Algebra II Facts and Skills**

Solutions

1. Simplify each radical and, if they have the same number under the radical after simplifying, combine them.

$$\sqrt{12} + \sqrt{75} = \sqrt{4 \cdot 3} + \sqrt{25 \cdot 3}$$
$$= \sqrt{4}\sqrt{3} + \sqrt{25}\sqrt{3}$$
$$= 2\sqrt{3} + 5\sqrt{3}$$
$$= 7\sqrt{3}$$

The correct choice is (**1**).

2. $2\sqrt{3} \cdot 5\sqrt{3} = 2 \cdot 5 \cdot \sqrt{3} \cdot \sqrt{3} = 10 \cdot 3 = 30$

The correct choice is (**1**).

3. Since $i^4 = 1$, $i^{64} = (i^4)^{16} = 1^{16} = 1$. So i raised to any multiple of 4 will be 1. 66 is not a multiple of 4, but it can be written as $64 + 2$ so $i^{66} = i^{64 + 2} = i^{64} \cdot i^2 = 1 \cdot i^2 = -1$.

The correct choice is (**1**).

4. Combine like terms

$(3 + 5i) + (2 - 7i) = 3 + 5i + 2 - 7i = 3 + 2 + 5i - 7i = 5 - 2i$

The correct choice is (**4**).

Radical Expressions and Equations 75

5. Using the quadratic formula with $a = 1$, $b = -6$, and $c = 34$,

$$x = \frac{-(-6) \pm \sqrt{(-6)^2 - 4(1)(34)}}{2(1)}$$

$$= \frac{6 \pm \sqrt{36 - 136}}{2}$$

$$= \frac{6 \pm \sqrt{-100}}{2}$$

$$= \frac{6 \pm 10i}{2}$$

$$= 3 \pm 5i$$

The correct choice is (**1**).

6. The absolute value of the complex number $a + bi$ is

$$|a + bi| = \sqrt{a^2 + b^2}$$

So $|12 - 5i| = \sqrt{12^2 + (-5)^2} = \sqrt{144 + 25} = \sqrt{169} = 13$.
The correct choice is (**3**).

7. Isolate the radial expression, and square both sides.

$$\sqrt{x} + x = 20$$

$$\underline{-x = -x}$$

$$\sqrt{x} = -x + 20$$

$$\left(\sqrt{x}\right)^2 = (-x + 20)^2$$

$$x = x^2 - 40x + 400$$

$$\underline{-x = -x}$$

$$0 = x^2 - 41x + 400$$

$$0 = (x - 16)(x - 25)$$

$$x - 16 = 0 \quad \text{or} \quad x - 25 = 0$$

$$x = 16 \quad \text{or} \quad x = 25$$

76　　　**Some Key Algebra II Facts and Skills**

Test each of these in the original equation to make sure that an extraneous solution was not created when both sides were squared.

For $x = 16$, $\sqrt{16} + 16 = 4 + 16 = 20$.

For $x = 25$, $\sqrt{20} + 25 = 5 + 25 = 30 \neq 20$.

The only solution is $x = 16$

The correct choice is **(3)**.

8. Isolate the radical expression, square both sides, and solve the resulting equation.

$$\sqrt{x+3} + x = 9$$
$$\underline{\phantom{\sqrt{x+3}} -x = -x }$$
$$\sqrt{x+3} = -x+9$$
$$\left(\sqrt{x+3}\right)^2 = (-x+9)^2$$
$$x+3 = x^2 - 18x + 81$$
$$0 = x^2 - 19x + 78$$
$$0 = (x-13)(x-6)$$

$$x - 13 = 0 \quad \text{or} \quad x - 6 = 0$$
$$x = 13 \quad \text{or} \quad x = 6$$

Check both of these answers to see if they satisfy the original equation.

For $x = 13$, $\sqrt{13+3} + 13 = \sqrt{16} + 13 = 4 + 13 = 17 \neq 9$.

For $x = 6$, $\sqrt{6+3} + 6 = \sqrt{9} + 6 = 3 + 6 = 9$.

The only solution is $x = 6$.

The correct choice is **(3)**.

Radical Expressions and Equations   77

9. Isolate one radical; square both sides. Then isolate the other radical and square both sides. Solve the resulting equation, and check your answers for extraneous roots.

$$\sqrt{x-3}+\sqrt{x+9}=6$$
$$\underline{\quad -\sqrt{x+9}=-\sqrt{x+9}\quad}$$
$$\sqrt{x-3}=6-\sqrt{x+9}$$
$$\left(\sqrt{x-3}\right)^2=\left(6-\sqrt{x+9}\right)^2$$
$$x-3=36-12\sqrt{x+9}+x+9$$
$$x-3=45-12\sqrt{x+9}+x$$
$$\underline{-45-x=-45-x\qquad\qquad}$$
$$-48=-12\sqrt{x+9}$$
$$\frac{-48}{-12}=\frac{-12\sqrt{x+9}}{-12}$$
$$4=\sqrt{x+9}$$
$$4^2=\left(\sqrt{x+9}\right)^2$$
$$16=x+9$$
$$\underline{-9=-9\quad}$$
$$7=x$$

Check $x=7$ in the original equation,

$$\sqrt{7-3}+\sqrt{7+9}=\sqrt{4}+\sqrt{16}=2+4=6$$

The correct choice is **(2)**.

10.
$$5\sqrt{(x-4)}+3=18$$
$$\underline{\qquad\quad -3=-3}$$
$$\frac{5\sqrt{(x-4)}}{5}=\frac{15}{5}$$
$$\sqrt{(x-4)}=3$$
$$\left(\sqrt{(x-4)}\right)^2=3^2$$
$$x-4=9$$
$$\underline{\quad +4=+4}$$
$$x=13$$

5. TRIGONOMETRIC EXPRESSIONS AND EQUATIONS

5.1 UNIT CIRCLE TRIGONOMETRY

A **unit circle** is a circle centered at (0, 0) that has a radius of one unit.

If P is a point on the unit circle and point A is at (1, 0), then $\angle AOP$ is said to be in **standard position**. In the diagram below m$\angle AOP$ = 27°.

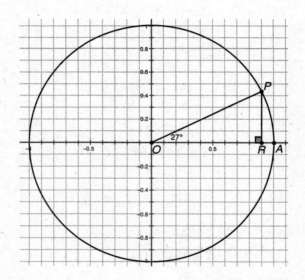

Because triangle OPR is a right triangle, the x-coordinate and y-coordinate of point P can be found with trigonometry. The x-coordinate will always equal $\cos(\angle AOP)$ and the y-coordinate will always equal $\sin(\angle AOP)$. In the following diagram, the coordinates of P are ($\cos 27°$, $\sin 27°$) = (0.89, 0.45)

Trigonometric Expressions and Equations

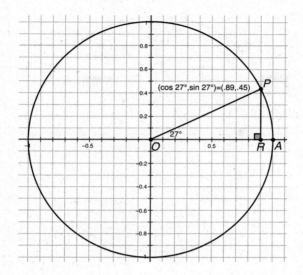

When point P is in quadrant II, the x-coordinate is negative so $\cos(\angle AOP) < 0$.

When point P is in quadrant III, both the x-coordinate and the y-coordinate are negative so $\cos(\angle AOP) < 0$ and $\sin(\angle AOP) < 0$.

When point P is in quadrant IV, the y-coordinate is negative so $\sin(\angle AOP)$ is negative.

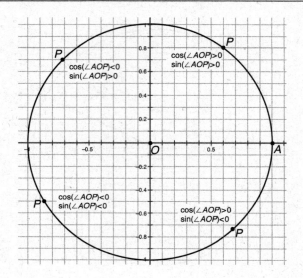

- If the x-coordinate of a point P on the unit circle is known, the $\sin^{-1}$ function on the calculator can be used as a step in finding the measure of $\angle AOP$. If the y-coordinate of a point P on the unit circle is known, the $\cos^{-1}$ function on the calculator can be used as a step in finding the measure of $\angle AOP$.

Example
If the coordinates of P on the unit circle are (0.48, 0.87), what is m$\angle AOP$?

Solution:
P is in quadrant I so m$\angle AOP$ is between 0° and 90°. Since the x-coordinate of the point is 0.48, $\angle AOP = \cos^{-1}(0.48) = 61°$. You could also use the y-coordinate to find $\angle AOP = \sin^{-1}(0.48) = 39°$.

Trigonometric Expressions and Equations

- If point P is not in quadrant I, you find the measure of the *reference angle* by taking either the $\sin^{-1}(|y\text{-coordinate}|)$ or $\cos^{-1}(|x\text{-coordinate}|)$. $\angle AOP$ can be calculated by following one of the four rules.

Quadrant of P	$\angle AOP$
I	reference angle
II	180° − reference angle
III	180° + reference angle
IV	360° − reference angle

Example

If $\sin(\angle AOP) = -\dfrac{4}{5}$ and $\cos(\angle AOP) < 0$, what is $m\angle AOP$?

Solution:

Since the x-coordinate is the cosine, point P has a negative x-coordinate. Since the y-coordinate is the sine, point P has a negative y-coordinate. This means that point P is in quadrant III.

The reference angle is $\sin^{-1}\left(\left|-\dfrac{4}{5}\right|\right) = \sin^{-1}\left(\dfrac{4}{5}\right) = 53.1°$. In quadrant III, $\angle AOP = 180° + \text{reference angle} = 180° + 53.1° = 223.1°$.

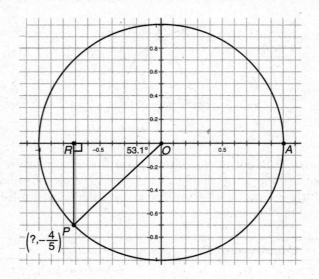

5.2 RADIAN MEASURE

Just as there are different units for measuring lengths, like inches and meters, there are different units for measuring angles. The most common unit for measuring angles is the degree. There are 360 degrees in a circle.

A **radian** is a unit of measurement that is much larger than a degree. A radian is approximately 57.3 degrees.

USEFUL MATH FACTS

To convert radians to degrees, multiply by $\dfrac{180°}{\pi}$.

To convert degrees to radians, multiply by $\dfrac{\pi}{180°}$.

Trigonometric Expressions and Equations 83

Example

Convert $\frac{\pi}{3}$ radians to degrees.

Solution:

$\frac{\pi}{3}$ radians is equal to $\frac{\pi}{3} \cdot \frac{180°}{\pi} = 60°$.

USEFUL MATH FACTS

Some common radian to degree conversions are:

Degrees	Radians
360°	2π
180°	π
90°	$\frac{\pi}{2}$
60°	$\frac{\pi}{3}$
45°	$\frac{\pi}{4}$
30°	$\frac{\pi}{6}$

5.3 GRAPHS OF THE SINE AND COSINE FUNCTIONS

The graph of $y = \sin x$ looks like this.

The graph of $y = \cos x$ looks like this

Curves that look like these are known as **sinusoidal curves** and have a **midline**, an **amplitude**, and a **period**. The midline is the invisible horizontal line passing through the middle of the curve. The amplitude is the distance between the midline and either a high point or a low point on the curve. The period is the horizontal distance between two consecutive high points or low points.

For $y = \sin x$ and $y = \cos x$, the midlines are $y = 0$, the amplitudes are 1, and the periods are 360° (or 2π radians).

Trigonometric Expressions and Equations

MATH FACTS

The graph of an equation of the form $y = A\sin(Bx) + D$ or $y = A\cos(Bx) + D$ has a midline of $y = D$, an amplitude of A, and a period of $\frac{360°}{B}$ (or $\frac{2\pi}{B}$ radians). If A is negative, the curve is "upside down."

Example

What is the midline, amplitude, and period of the curve graphed below with equation $y = 3\sin(2x) - 1$?

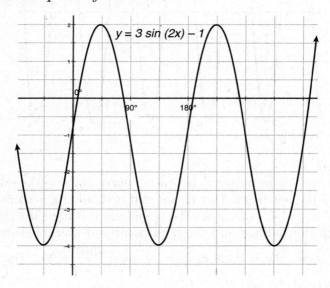

Solution:
The midline is $y = -1$, the amplitude is 3, and the period is $\frac{360°}{2} = 180°$.

Example

Below is one cycle of a sinusoidal curve. What could be the equation of this curve?

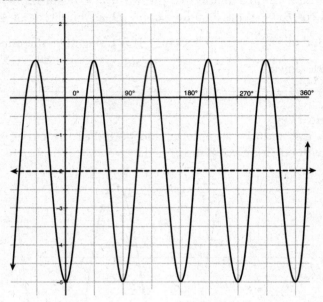

Solution:
This resembles an upside down cosine curve because the "starting point" is the low point. This means that A should be negative. The midline is $y = -2$ so $D = -2$. The amplitude is 3 since the distance between the midline and the high point is 3 units. Because the period is 90°, the B value must be 4 since $\frac{360°}{4} = 90°$. The equation is $y = -3\cos(4x) - 2$.

Trigonometric Expressions and Equations

5.4 TRIGONOMETRY EQUATIONS

An equation that involves one of the trigonometric functions, like sine or cosine, is called a trigonometric equation. Trigonometric equations often have more than one solution. The simplest way to find the solutions to a trigonometric equation is to use a graphing calculator.

Example

Use your graphing calculator to find all solutions between $0° \leq x < 360°$ that satisfy the equation $\sin x = -0.94$.

Solution:

Graph $y = \sin x$ and $y = -0.94$ on the same set of axes. Use the Zoom Trig command to set a useful window. Then use the intersect feature of the calculator to find the x-coordinates of the intersection points that are between 0 and 360.

The two intersection points are $(250, -0.94)$ and $(290, -0.94)$ so the solutions are $x = 250$ and $x = 290$.

5.5 MODELING REAL-WORLD SCENARIOS WITH TRIG FUNCTIONS

Certain real-world scenarios can be modeled with a trigonometric equation. The graphing calculator can be used to solve this equation.

Example

The radius of this Ferris wheel is 10 feet, and the bottom of the Ferris wheel is 3 feet above the ground. The amount of time it takes to make a complete revolution around the Ferris wheel is 100 seconds.

When Evelyn gets on the Ferris wheel, she is exactly 3 feet above the ground. 50 seconds after starting, she will be 23 feet above the ground, at the peak of the Ferris wheel. 100 seconds after starting Evelyn will be back to her starting position, again exactly 3 feet above the ground.

The graph of Evelyn's height above the ground will look like this.

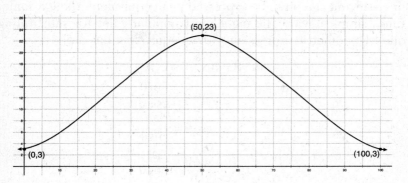

After how many seconds will she be 18 feet above the ground?

Solution:
When the unit on the x-axis is not degrees, as in this case where it is seconds, assume everything is in radians.

The equation is $h = -10 \cos\left(\frac{\pi}{50}t\right) + 13$. The B value was calculated by using the formula period $= \frac{2\pi}{B}$. The period is given as 100 so $100 = \frac{2\pi}{B}$, $100B = 2\pi$, $B = \frac{2\pi}{100} = \frac{\pi}{50}$.

Graphing $y = -10 \cos\left(\frac{\pi}{50}x\right) + 13$ and $y = 18$ on the graphing calculator and finding the x-coordinates of the intersection points gets solutions of $x = 33$ and $x = 67$ seconds.

90   Some Key Algebra II Facts and Skills

5.6 TRIGONOMETRIC IDENTITIES

An expression involving sine, cosine, tangent, cosecant, secant, or cotangent can often be expressed in an equivalent, though sometimes simpler, way. The eight most popular rules for substituting one trigonometric expression with an equivalent one are:

$$\sin^2 \theta + \cos^2 \theta = 1$$

$$\tan \theta = \frac{\sin \theta}{\cos \theta}$$

$$\csc \theta = \frac{1}{\sin \theta}$$

$$\sec \theta = \frac{1}{\cos \theta}$$

$$\cot \theta = \frac{1}{\tan \theta}$$

$$1 + \cot^2 \theta = \csc^2 \theta$$

$$\tan^2 \theta + 1 = \sec^2 \theta$$

- These identities can be used to check if more complicated expressions involving trigonometric functions are equivalent.

Example
Show that $\sin x \tan x + \cos x = \sec x$.

Solution:

$$\sin x \tan x + \cos x$$

$$= \sin x \frac{\sin x}{\cos x} + \cos x$$

$$= \frac{\sin^2 x}{\cos x} + \frac{\cos x}{1}$$

$$= \frac{\sin^2 x}{\cos x} + \frac{\cos^2 x}{\cos x}$$

$$= \frac{\sin^2 x + \cos^2 x}{\cos x}$$

$$= \frac{1}{\cos x}$$

$$= \sec x$$

Practice Exercises

1. In unit circle O, the coordinates of point B are $(-0.91, -0.41)$. What is the measure of $\angle AOB$?

    (1) 211°
    (2) 237°
    (3) 246°
    (4) 204°

2. Convert 240° to radians.

    (1) $\dfrac{4}{3}$
    (2) $\dfrac{5\pi}{6}$
    (3) $\dfrac{4\pi}{3}$
    (4) $\dfrac{5\pi}{4}$

3. Convert $\frac{\pi}{4}$ radians to degrees.
   (1) 90° (3) 30°
   (2) 60° (4) 45°

4. What is the graph of $y = -3 \sin(x) + 1$?

(1)

(3)

(2)

(4)

5. What is the y-coordinate of one of the maximum points of $y = 2\cos(x) - 1$?
   (1) 3 (3) 1
   (2) 2 (4) 0

6. What is the period, in degrees, of the graph of $y = 4\sin\left(\frac{2}{3}x\right) + 1$?
   (1) 540° (3) 270°
   (2) 360° (4) 240°

Trigonometric Expressions and Equations

7. What is the period, in radians, of the graph of $y = 4\cos\left(\dfrac{\pi}{10}x\right) - 3$?

   (1) 5 (3) 15
   (2) 10 (4) 20

8. Graphically solve the equation $2\cos(3x) + 3 = 4$ for $0 \le x < \dfrac{2\pi}{3}$.

   (1) 0.25 and 1.65 (3) 0.35 and 1.75
   (2) 0.30 and 1.70 (4) 0.40 and 1.80

9. The height above ground of a Ferris wheel car can be modeled with the equation $h = -20\cos\left(\dfrac{\pi}{15}t\right) + 24$ where h is the height in feet and t is the time in seconds.
   (a) How many seconds does it take for the Ferris wheel to make a complete revolution?
   (b) What is the maximum height of the Ferris wheel?
   (c) Sketch a graph that shows how the height in the air of a Ferris wheel car relates to the amount of time since the beginning of the ride?

10. $\cos^2\theta - \sin^2\theta$ is equal to which of the following?
    (1) $2\cos^2\theta - 1$ (3) -1
    (2) $2\cos^2\theta + 1$ (4) 1

Some Key Algebra II Facts and Skills

Solutions

1. Point B is in quadrant III so the angle is between $180°$ and $270°$. One way to get the solution is to work out

$$180° + \sin^{-1} 0.41 = 204°$$

The correct choice is (**4**).

2. To convert degrees to radians, multiply the number of degrees by $\dfrac{\pi}{180°}$.

$$240° \cdot \frac{\pi}{180°} = \frac{4\pi}{3}$$

The correct choice is (**3**).

3. To convert radians to degrees, multiply the number of radians by $\dfrac{180°}{\pi}$.

$$\frac{\pi}{4} \cdot \frac{180°}{\pi} = 45°$$

The correct choice is (**4**).

4. All the choices have an amplitude of 3. Because of the $+1$, the midline of the solution must be $y = 1$. This eliminates choices (1) and (2). Since the coefficient of the $\sin(x)$ is negative, the sine curve must start by going "down" to the left. This eliminates choice (3).

The correct choice is (**4**).

5. The -1 at the end of the equation means the midline is $y = -1$. The 2 in front of the $\cos(x)$ means the amplitude of the curve is 2 so the maximum points are 2 above the midline or $-1 + 2 = 1$.

The correct choice is (**3**).

Trigonometric Expressions and Equations   95

6. The formula for period, in degrees, is $\dfrac{360°}{B}$ where B is the coefficient of the x-term. For this example, that is

$$\dfrac{360°}{\tfrac{2}{3}} = 360° \cdot \dfrac{3}{2} = 540°$$

The correct choice is **(1)**.

7. The formula for period, in degrees, is $\dfrac{2\pi}{B}$ where B is the coefficient of the x-term. For this example, that is

$$\dfrac{2\pi}{\tfrac{\pi}{10}} = 2\pi \cdot \dfrac{10}{\pi} = 20$$

The correct choice is **(4)**.

8. Graph $y = 4$ and $y = 2\cos(3x) + 3$ together on the graphing calculator (in radian mode) and find the x-coordinates of the first two intersection points to the right of the y-axis. The intersection points are $(0.35, 4)$ and $(1.75, 4)$.

The correct choice is **(3)**.

9. (a) The length of time that it takes to make one revolution is the period of the curve. This can be calculated with the formula $\dfrac{2\pi}{B}$ where B is the coefficient of t. For this example,

this is $\dfrac{2\pi}{\tfrac{\pi}{15}} = 2\pi \cdot \dfrac{15}{\pi} = 30$ so the time for one revolution is

30 seconds.

(b) Because of the +24, the midline of the curve is at $y = 24$. The coefficient of the cosine is −20 so the high point is 20 above the midline and $24 + 20 = 44$ feet.

(c) The graph starts at (0, 44) and goes down to (15, 4) and then back up to (30, 44).

10. One of the most important trig identities is $\sin^2\theta + \cos^2\theta = 1$. This can be rearranged as $\sin^2\theta = 1 - \cos^2\theta$. Replace the $\sin^2\theta$ with $1 - \cos^2\theta$ in the given expression to get

$$\cos^2\theta - (1 - \cos^2\theta) = \cos^2\theta - 1 + \cos^2\theta = 2\cos^2\theta - 1$$

The correct choice is **(1)**.

Systems of Equations 97

6. SYSTEMS OF EQUATIONS

6.1 TWO EQUATIONS IN TWO UNKNOWNS

A system of two equations in two unknowns is two equations that each have two (or more) variables. A solution to a system of equations is an ordered pair that makes both equations true.

- An example of a system of two equations in two unknowns is

$$x + y = 10$$
$$x - y = 4$$

This system has as its only solution $(7, 3)$ since $7 + 3 = 10$ and $7 - 3 = 4$.

There are two main ways to solve systems of equations with algebra.

The Substitution Method

If one of the variables is already isolated in one of the equations, it is possible to simplify the two equations with two variables into just one equation with one variable.

Example

$$y = 3x - 2$$
$$x + 2y = 24$$

Solution:

Since the y is isolated in the first equation, the y in the second equation can be replaced with $(3x - 2)$.

$$x + 2(3x - 2) = 24$$
$$x + 6x - 4 = 24$$
$$7x - 4 = 24$$
$$\underline{+4 = +4}$$
$$7x = 28$$
$$\frac{7x}{7} = \frac{28}{7}$$
$$x = 4$$

98   **Some Key Algebra II Facts and Skills**

- To get the other number in the ordered pair, substitute the $x = 4$ into the first equation.

$$y = 3 \cdot 4 - 2 = 12 - 2 = 10$$

- The ordered pair that satisfies both equations is $(4, 10)$.

The Elimination Method for Two Equations in Two Unknowns

If the two equations can be combined in a way that *eliminates* one of the variables, the remaining equation can be solved for one part of the ordered pair. Sometimes it requires multiplying both sides of one or both equations by some constant.

Example
Solve the system of equations.

$$2x + 6y = 4$$
$$5x + 3y = 34$$

Solution:
The coefficient of the y in the first equation, 6, is a multiple of the coefficient of the y in the second equation, 3. For the equation that has the smaller coefficient, multiply both sides of the equation by the number that would make that coefficient the opposite of the one from the other equation. For this example, multiply both sides of the second equation by -2.

$$2x + 6y = 4$$
$$-2(5x + 3y) = -2(34)$$

$$2x + 6y = 4$$
$$+ \; \underline{-10x - 6y = -68}$$
$$-8x = -64$$

$$\frac{-8x}{-8} = \frac{-64}{-8}$$

$$x = 8$$

Systems of Equations

$$2 \cdot 8 + 6y = 4$$
$$16 + 6y = 4$$
$$\underline{-16 = -16}$$
$$6y = -12$$
$$\frac{6y}{6} = \frac{-12}{6}$$
$$y = -2$$

The solution is $(8, -2)$.

Example
Solve the system of equations.

$$3x + 4y = 42$$
$$5x - 6y = 32$$

Solution:
For the x-terms, the 5 is not a multiple of 3. For the y-terms, the -6 is not a multiple of the 4. When this happens, both equations must be changed.

You can choose whether you want to change the equation so that either the x-variables will be eliminated or the y-variables will be eliminated.

To eliminate the y-variables, find the least common multiple of the two coefficients. In this case, the least common multiple of 4 and 6 is 12. The goal is to convert one of the y-coefficients into a $+12$ and the other into a -12. This can be done by multiplying both sides of the first equation by $+3$ and both sides of the second equation by $+2$.

$$3(3x + 4y) = 3(42)$$
$$2(5x - 6y) = 2(32)$$

$$9x + 12y = 126$$
$$\underline{+ \quad 10x - 12y = 64}$$
$$19x = 190$$
$$\frac{19x}{19} = \frac{190}{19}$$
$$x = 10$$

Some Key Algebra II Facts and Skills

Substitute 10 for x into either of the original equations.

$$3 \cdot 10 + 4y = 42$$
$$30 + 4y = 42$$
$$\underline{-30 = -30}$$
$$4y = 12$$
$$\frac{4y}{4} = \frac{12}{4}$$
$$y = 3$$

The solution is (10, 3).

Infinite Solutions or No Solutions

While doing the systems of equations process, sometimes all the variables get canceled out and you have an equation that has only numbers. If the numbers on either side of the equal sign are not equal, the system has no solution. If the numbers on either side of the equal sign are equal, there are an infinite number of solutions to the system.

Example

How many solutions does this system of equations have?

$$2x + 3y = -10$$
$$-2x - 3y = 12$$

(1) No solutions
(2) One solution
(3) Two solutions
(4) More than two solutions

Solution:

When the two equations are added together it becomes 0 = 2, which is never true. There are no solutions to the original system of equations. The correct choice is (1).

Systems of Equations 101

6.2 THE ELIMINATION METHOD FOR THREE EQUATIONS IN THREE UNKNOWNS

When there are three equations and three variables, the process for solving the system of equations is much longer. It still involves multiplying the equations by constants with the goal of "eliminating" the variables until there is just one variable left.

Example

$$-3x - 7y - 2z = -31$$
$$5x + 4y + 2z = 35$$
$$-x + y - z = -4$$

Solution:

- **Step 1:** Decide which variable to first eliminate.

   The first step in solving a system of three equations with three unknowns is to examine the coefficients of each of the variables to see if the variables for any coefficient are matching numbers or have numbers that are multiples of the other coefficients. Any variable can be eliminated, but it is easier to eliminate a variable when the coefficients contain multiples of other coefficients.

   In this example, the z variables have the coefficients -2, $+2$, and -1. The z will be the variable we try to eliminate.

- **Step 2:** Combine the first and second equation to eliminate a variable.

   Since -2 is the opposite of $+2$, this is like the first example from the previous section. The variable z can be eliminated by adding the two equations. This step may require multiplying both sides of one or both equations by a constant before combining them.

$$
\begin{array}{r}
-3x - 7y - 2z = -31 \\
+\ 5x + 4y + 2z = 35 \\
\hline
2x - 3y = 4
\end{array}
$$

This equation $2x - 3y = 4$ will be used in step 4.

102   **Some Key Algebra II Facts and Skills**

- **Step 3:** Combine the second and third equations (or the first and third if it is easier) to eliminate a variable.

Since $+2z + -z$ will not eliminate the z, both sides of the second equation must first be multiplied by $+2$ so the z coefficient will become -2, the opposite of the coefficient of the other z, $+2$.

$$5x + 4y + 2z = 35$$
$$2(-x + y - z) = 2(-4)$$

Now the equations can be combined to eliminate the z.

$$
\begin{array}{r}
5x + 4y + 2z = 35 \\
+ \ -2x + 2y - 2z = -8 \\
\hline
3x + 6y = 27
\end{array}
$$

This equation, together with the one from step 2, is used in step 4.

- **Step 4:** Solve the resulting system of two equations with two unknowns

$$2x - 3y = 4$$
$$3x + 6y = 27$$

Since the $+6$ is a multiple of the -3, multiply both sides of the first equation by $+2$.

$$2(2x - 3y) = 2(4)$$
$$3x + 6y = 27$$

$$
\begin{array}{r}
4x - 6y = 8 \\
+ \ 3x + 6y = 27 \\
\hline
7x = 35
\end{array}
$$

$$\frac{7x}{7} = \frac{35}{7}$$

$$x = 5$$

Systems of Equations 103

Substitute 5 for x into either of the two variable equations.

$$2(5) - 3y = 4$$
$$10 - 3y = 4$$
$$\underline{-10 = -10}$$
$$-3y = -6$$
$$\frac{-3y}{-3} = \frac{-6}{-3}$$
$$y = 2$$

- **Step 5:** Substitute the solution of the two variable system of equations into one of the original equations to find the solution for the third variable.

Any of the original three equations can be used. Since the third equation has the smallest coefficients, it will be simplest.

$$-(5) + 2 - z = -4$$
$$-5 + 2 - z = -4$$
$$-3 - z = -4$$
$$\underline{+3 = +3}$$
$$-z = -1$$
$$\frac{-z}{-1} = \frac{-1}{-1}$$
$$z = 1$$

The solution is the ordered triple (5, 2, 1).

This can be checked by substituting $x = 5$, $y = 2$, and $z = 1$ into each of the three original equations.

$$-3 \cdot 5 - 7 \cdot 2 - 2 \cdot 1 \overset{?}{=} -31$$
$$-15 - 14 - 2 \overset{?}{=} -31$$
$$-31 \overset{\checkmark}{=} -31$$

$$5 \cdot 5 + 4 \cdot 2 + 2 \cdot 1 \overset{?}{=} 35$$
$$25 + 8 + 3 \overset{?}{=} 35$$
$$35 \overset{\checkmark}{=} 35$$

$$-5 + 2 - 1 \overset{?}{=} -4$$
$$-4 \overset{\checkmark}{=} -4$$

104   **Some Key Algebra II Facts and Skills**

Example

Find the solution to this system of equations

$$-x + 5y + z = -18$$
$$4x - 5y + 3z = 34$$
$$x + 5y - z = -12$$

Solution:

$(4, -3, 1)$

Since the y-variables have coefficients of $+5$, -5, and $+5$, the y should be eliminated. Combining equations 1 and 2 gets $3x + 4z = 16$. Combining equations 2 and 3 gets $5x + 2z = 22$.

Now solve the system

$$3x + 4z = 16$$
$$5x + 2z = 22$$

Multiply both sides of the second equation by -2 so the z coefficient becomes -4.

$$
\begin{aligned}
3x + 4z &= 16 \\
-10x - 4z &= -44 \\
\hline
-7x &= -28 \\
x &= 4
\end{aligned}
$$

$$
\begin{aligned}
3 \cdot 4 + 4z &= 16 \\
12 + 4z &= 16 \\
-12 &= -12 \\
\hline
4z &= 4 \\
z &= 1
\end{aligned}
$$

$$
\begin{aligned}
-4 + 5y + 1 &= -18 \\
-3 + 5y &= -18 \\
+3 &= +3 \\
\hline
5y &= -15 \\
y &= -3
\end{aligned}
$$

The solution is $(4, -3, 1)$.

105

Practice Exercises

1. Solve this system of equations.
   $y = 2x + 1$
   $3x + 4y = 15$
   (1) (3, 1) (3) (2, 4)
   (2) (4, 2) (4) (1, 3)

2. Solve this system of equations.
   $y = 3x - 2$
   $7x - 2y = 7$
   (1) (3, 7) (3) (4, 6)
   (2) (7, 3) (4) (6, 4)

3. Solve this system of equations.
   $3x + 5y = 37$
   $2x - 5y = 8$
   (1) (9, 2) (3) (8, 3)
   (2) (2, 9) (4) (3, 8)

Some Key Algebra II Facts and Skills

4. Solve this system of equations.
$$4x + 3y = -11$$
$$-4x + 5y = 35$$

(1) $(5, -3)$ (3) $(-5, 3)$

(2) $(-3, 5)$ (4) $(3, -5)$

5. Solve this system of equations.
$$2x + 3y = 11$$
$$3x - 6y = 6$$

(1) $(4, 1)$ (3) $(2, 3)$

(2) $(1, 4)$ (4) $(3, 2)$

6. Solve this system of equations.
$$5x - 3y = 27$$
$$6x - 7y = 46$$

(1) $(3, -4)$ (3) $(-3, 4)$

(2) $(-4, 3)$ (4) $(4, -3)$

7. How many solutions does this system of equations have?
$$5x + 3y = 10$$
$$10x + 6y = 20$$

(1) No solution   (3) Two solutions

(2) One solution   (4) Infinite solutions

8. If a system of two equations in two unknowns has no solution, what do the graphs of the two lines representing the two equations look like?

(1) They intersect at one point.

(2) They are parallel.

(3) They are the same line.

(4) They intersect at two points.

Systems of Equations   107

9. If $x = 2$ and $y = 1$ and $3x + 2y + 4z = 24$, what is the value of z?

10. Solve this system of equations.

$$3x + 2y + z = 15$$
$$2y + z = 6$$
$$z = 4$$

Solutions

1. Since the y is isolated in the first equation, use the substitution method by substituting $2x + 1$ for y in the second equation.

$$3x + 4(2x + 1) = 15$$
$$3x + 8x + 4 = 15$$
$$11x + 4 = 15$$
$$11x = 11$$
$$x = 1$$

To solve for y, replace the x with 1 in the first equation.

$$y = 2(1) + 1 = 2 + 1 = 3$$

The solution is $(1, 3)$.

The correct choice is **(4)**.

2. Since the y is isolated in the first equation, use the substitution method by substituting $3x - 2$ for y in the second equation.

$$6x - 3(3x - 2) = 3$$
$$6x - 9x + 6 = -3$$
$$-3x + 6 = -3$$
$$-3x = -9$$
$$x = 3$$

To solve for y, replace the x with 3 in the first equation.

$$y = 3(3) - 2 = 9 - 2 = 7$$

The solution is $(3, 7)$

The correct choice is **(1)**.

Some Key Algebra II Facts and Skills

3. Since the coefficient of the y in the first equation is the same number, but opposite sign, as the coefficient of the y in the second equation, the y-terms can be eliminated by adding the two equations.

$$3x + 5y = 37$$
$$\underline{2x - 5y = 8}$$
$$5x = 45$$
$$x = 9$$

Substitute 9 for x into either of the equations and solve for y.

$$3(9) + 5y = 37$$
$$27 + 5y = 37$$
$$5y = 10$$
$$y = 2$$

The solution is $(9, 2)$.

The correct choice is **(1)**.

4. Since the coefficient of the x in the first equation is the same number, but opposite sign, as the coefficient of the x in the second equation, the x-terms can be eliminated by adding the two equations.

$$4x + 3y = -11$$
$$\underline{-4x + 5y = 35}$$
$$8y = 24$$
$$y = 3$$

Substitute 3 for y into either of the equations, and solve for x.

$$4x + 3(3) = -11$$
$$4x + 9 = -11$$
$$4x = -20$$
$$x = -5$$

The solution is $(-5, 3)$.

The correct choice is **(3)**.

Systems of Equations 109

5. Since the 6 is a multiple of the 3, multiply both sides of the first equation by 2 so that the coefficient of the modified first equation will be +6. Then the y can be eliminated by adding the modified first equation to the second equation.

$$2(2x + 3y) = 2(11)$$
$$3x - 6y = 6$$

$$
\begin{array}{r}
4x + 6y = 22 \\
3x - 6y = 6 \\
\hline
7x = 28 \\
x = 4
\end{array}
$$

Substitute 4 for x into either of the original equations to solve for y.

$$2(4) + 3y = 11$$
$$8 + 3y = 11$$
$$3y = 3$$
$$y = 1$$

The solution is (4, 1).

The correct choice is (1).

6. Multiply the top equation by –6 and the bottom by +5 so that the coefficients of the x terms become –30 and +30. Then the modified equations can be added to eliminate the x.

$$-6(5x - 3y) = -6(27)$$
$$5(6x - 7y) = 5(46)$$

$$
\begin{array}{r}
-30x + 18y = -162 \\
30x - 35y = 230 \\
\hline
-17y = 68 \\
y = -4
\end{array}
$$

110   **Some Key Algebra II Facts and Skills**

Substitute −4 for y into either of the original equations to solve for x.

$$5x - 3(-4) = 27$$
$$5x + 12 = 27$$
$$5x = 15$$
$$x = 3$$

The solution is $(3, -4)$

The correct choice is **(1)**.

7. Multiply the first equation by −2 so that the coefficient on the x becomes −10.

$$-2(5x + 3y) = -2(10)$$
$$10x + 6y = 20$$

$$-10x - 6y = -20$$
$$\underline{10x + 6y = 20}$$
$$0 = 0$$

In trying to just eliminate the x, the y got eliminated, too. Since the equation that remains, $0 = 0$, is true, there are an infinite number of solutions. Had it been something like $0 = 5$, there would be no solutions.

The correct choice is **(4)**.

8. The intersection points of the two lines representing the two equations correspond to the solutions to the system of equations. If there are no solutions, then there are no intersection points. Lines can only have no intersection points if they are parallel.

The correct choice is **(2)**.

Systems of Equations 111

9. Substitute 2 for x and 1 for y and solve for z.

$$3(2) + 2(1) + 4z = 24$$
$$6 + 2 + 4z = 24$$
$$8 + 4z = 24$$
$$4z = 16$$
$$z = 4$$

10. Since the third equation says that $z = 4$, substitute 4 for z in the second equation to solve for y.

$$2y + 4 = 6$$
$$2y = 2$$
$$y = 1$$

Substitute 4 for z and 1 for y into the first equation and solve for x.

$$3x + 2(1) + 4 = 15$$
$$3x + 2 + 4 = 15$$
$$3x + 6 = 15$$
$$3x = 9$$
$$x = 3$$

The solution is $(3, 1, 4)$.

112   **Some Key Algebra II Facts and Skills**

7. FUNCTIONS

7.1 COMPOSITE FUNCTIONS

A **function** is a mathematical rule that generally takes a number as an input and then outputs a number. If the function is called f, the notation $f(4) = 11$ means that when the number 4 is put into the function, the number 11 is output from the function.

- Functions are often defined by a formula. For example $f(x) = 2x + 3$ means that when a number is put into the function, a number that is three more than twice that number will come out of the function. So $f(4) = 2 \cdot 4 + 3 = 8 + 3 = 11$.

- Not only can numbers be put into a function, but so can variables, or even other functions. Using the function f defined above, $f(a) = 2a + 3$ and $f(x^2 + 1) = 2(x^2 + 1) + 3$.

- If another function g is defined as $g(x)=3x - 2$, it is possible to create a function $f(g(x))$. By putting the $g(x)$ into the f function, it becomes $f(3x - 2) = 2(3x - 2) + 3$, which can be simplified to $f(3x - 2) = 6x - 4 + 3 = 6x - 1$.

When a function is put into another function, the result is called a **composite function.**

Example 1
If $f(x) = 5x + 2$ and $g(x) = 3x - 1$, what is the value of $f(g(4))$?

Solution:
Since $g(4) = 3 \cdot 4 - 1 = 11, f(g(4))=f(11) = 5 \cdot 11 + 2 = 57$.

Example 2
If $f(x) = 2x + 3$ and $g(x) = 3x - 2$, what is $g(f(x))$ and $f(g(x))$?

Solution:
$g(f(x)) = g(2x + 3) = 3(2x + 3) - 2 = 6x + 9 - 2 = 6x + 7$
$f(g(x)) = f(3x - 2) = 2(3x - 2) + 3 = 6x - 4 + 3 = 6x - 1$

Notice that, in this example, $g(f(x))$ is not equivalent to $f(g(x))$.

Functions   **113**

7.2 INVERSE FUNCTIONS

If a function has an **inverse function**, the inverse function can take what was output from the original function and turn it back into the original input. An inverse function is a function that "undoes" what some other function has done to a number. The notation for the inverse of a function called f is f^{-1}.

- If the function f is defined by $f(x) = x + 3$, then $f(7) = 7 + 3 = 10$. The inverse of function f is $f^{-1}(x) = x - 3$. When the number 10 is put into the inverse function $f^{-1}(10) = 10 - 3 = 7$, which is what was put into the original function, f, to get the number 10. Whatever the function f does to a number, the function f^{-1} undoes to that number.

Example

If $g(6) = 19$, what is the value of $g^{-1}(19)$?

(1) 6

(2) $\dfrac{1}{6}$

(3) 19

(4) −19

Solution:

Since the function g turned the number 6 into the number 19, then the inverse will turn the 19 back into the 6. With the given information, this is the only value for g^{-1} that can be determined.

The correct choice is (1).

Determining the Inverse Function of a Linear Function

A linear function of the form $f(x) = ax + b$, such as $f(x) = 2x + 3$, has an inverse function that is also a linear function. The process for finding the inverse function is to first rewrite the function, but with the $f(x)$ replaced with an x and the x replaced with an $f^{-1}(x)$.

$$x = 2f^{-1}(x) + 3$$

Then "solve" for $f^{-1}(x)$ by, in this case, subtracting 3 from both sides of the equation and dividing both sides of the equation by 2.

$$x = 2f^{-1}(x) + 3$$
$$\underline{-3 = -3}$$
$$x - 3 = 2f^{-1}(x)$$
$$\frac{x-3}{2} = \frac{2f^{-1}(x)}{2}$$
$$\frac{x-3}{2} = f^{-1}(x)$$

If you pick a number for x like 5, see that $f(5) = 2 \cdot 5 + 3 = 10 + 13$ and that $f^{-1}(13) = \frac{13-3}{2} = \frac{10}{2} = 5$. So this inverse function undoes what the function did to the number 5.

Graphs of Inverse Functions

If for some function f, $f(3) = 9$, then the point $(3, 9)$ will be on the graph for that function. If $f(3) = 9$, then $f^{-1}(9) = 3$, and the point $(9, 3)$ will be on the graph for the inverse function. In general, if (x, y) is a point on the graph of the original function, then (y, x) will be a graph of the inverse function. In terms of transformations, the point $(9, 3)$ is the reflection over the line $y = x$ of the point $(3, 9)$. When this transformation is done to every point in the graph of the function, the graph of the inverse will be a reflection of the entire graph of the original function over the line $y = x$.

115

Practice Exercises

1. If $f(x) = 2x + 3$ and $g(x) = 3x - 4$, what is $f(g(5))$?
   (1) 35   (3) 25
   (2) 30   (4) 20

2. Which of the following could be $f(x)$ and $g(x)$ if
   $f(g(x)) = (3x - 2)^2 + 1$?
   (1) $f(x) = x^2 + 1, g(x) = 3x - 2$
   (2) $f(x) = 3x - 2, g(x) = x^2 + 1$
   (3) $f(x) = x^2 - 1, g(x) = 3x - 2$
   (4) $f(x) = x^2 + 1, g(x) = 3x + 2$

3. If $g(x) = 3x^2 - 2$, what is $g(f(x))$?
   (1) $2[f(x)]^2 + 3$
   (2) $2[f(x)]^2 - 3$
   (3) $3[f(x)]^2 - 2$
   (4) $3[f(x)]^2 + 2$

Some Key Algebra II Facts and Skills

4. If $g(x)$ is the inverse of $f(x)$, what is the value of $f(g(9))$?
  (1) 3   (3) –9

  (2) $\dfrac{1}{9}$ (4) 9

5. What is the inverse of $f(x) = x - 3$?
  (1) $f^{-1}(x) = x + 3$ (3) $f^{-1}(x) = x - 3$

  (2) $f^{-1}(x) = \dfrac{1}{x+3}$ (4) $f^{-1}(3) = -1$

6. What is the inverse of $f(x) = 2x - 7$?
  (1) $f^{-1}(x) = 2x + 7$ (3) $f^{-1}(x) = \dfrac{x+7}{2}$

  (2) $f^{-1}(x) = \dfrac{1}{2x-7}$ (4) $f^{-1}(x) = \dfrac{x}{2} + 7$

7. If $f(3) = 11$, what must also be true?
  (1) $f^{-1}(11) = 3$ (3) $f^{-1}(3) = \dfrac{1}{11}$

  (2) $f^{-1}(3) = 11$ (4) $f^{-1}(3) = -11$

Functions 117

8. If the graph of $y = f(x)$ is shown below, what is the graph of $y = f^{-1}(x)$?

(1)

(3)

(2)

(4)

9. If $f(x) = 5x + 4$, what is $f^{-1}(34)$?

10. If $f(x) = 7x + 3$, what is $f^{-1}(x)$?

Some Key Algebra II Facts and Skills

118

Solutions

1. $g(5) = 3(5) - 4 = 15 - 4 = 11$ so

$$f(g(5)) = f(11) = 2(11) + 3 = 22 + 3 = 25$$

The correct choice is **(3)**.

2. The safest approach to this question is to test each of the four answer choices. For choice (4),

$$f(g(x)) = [g(x)]^2 + 1 = (3x - 2)^2 + 1$$

The correct choice is **(1)**.

3. If you replace the x in the $g(x)$ formula with $f(x)$, it becomes $g(f(x)) = 3[f(x)]^2 - 2$.

The correct choice is **(3)**.

4. If $g(x) = f^{-1}(x)$, then $f(g(9)) = f(f^{-1}(9))$, which must equal 9 since the inverse undoes whatever the function does and also the function undoes whatever the inverse does.

The correct choice is **(4)**.

5. To find the inverse of a function, create a new equation like the given equation but with the $f(x)$ replaced with an x and the x replaced with an $f^{-1}(x)$.

$$x = f^{-1}(x) - 3$$

Then solve for $f^{-1}(x)$:

$$
\begin{aligned}
x &= f^{-1}(x) - 3 \\
+3 &= +3 \\
\hline
x + 3 &= f^{-1}(x)
\end{aligned}
$$

The correct choice is **(1)**.

Functions 119

6. To find the inverse of a function, create a new equation like the given equation but with the $f(x)$ replaced with an x and the x replaced with an $f^{-1}(x)$.

$$x = 2f^{-1}(x) - 7$$

Then solve for $f^{-1}(x)$:

$$x = 2f^{-1}(x) - 7$$
$$\underline{+7 = +7}$$
$$x + 7 = 2f^{-1}(x)$$
$$\frac{x+7}{2} = \frac{2f^{-1}(x)}{2}$$
$$\frac{x+7}{2} = f^{-1}(x)$$

The correct choice is **(3)**.

7. If $f(3) = 11$, it means that when 3 is put into the function, 11 comes out. So if the inverse function undoes whatever the function does to a number then $f^{-1}(11)$ must be 3.

The correct choice is **(1)**.

8. The graph of the function $f(x)$ has $(-2, -5)$, $(1, 4)$, and $(7, 6)$ as three of the points. The inverse will have three points like these but with the x- and y-coordinates interchanged or $(-5, -2)$, $(4, 1)$ and $(6, 7)$. Of the choices, choice (3) has these three points.

Also the graph of the inverse function is the graph of the original after it has been reflected over the line $y = x$, which is true about choice (3).

The correct choice is **(3)**.

120   **Some Key Algebra II Facts and Skills**

9. The long way is to calculate $f^{-1}(x)$ and then use it to find $f^{-1}(34)$. The short cut for this question is to realize that $f^{-1}(34)$ is the number that would have to go into the original function to make 34 come out or what x value makes $f(x) = 34$. This can be solved with the equation.

$$f(x) = 5x + 4$$
$$34 = 5x + 4$$
$$\underline{-4 = -4}$$
$$30 = 5x$$
$$x = 6$$

10. To find the inverse of a function, create a new equation like the given equation but with the $f(x)$ replaced with an x and the x replaced with an $f^{-1}(x)$.

$$x = 7f^{-1}(x) + 3$$

Then solve for $f^{-1}(x)$:

$$x = 7f^{-1}(x) + 3$$
$$\underline{-3 = -3}$$
$$x - 3 = 7f^{-1}(x)$$
$$\frac{x-3}{7} = \frac{7f^{-1}(x)}{7}$$
$$\frac{x-3}{7} = f^{-1}(x)$$

Sequences   121

8. SEQUENCES

8.1 SEQUENCES AND SERIES

A **sequence** is a list of numbers that follow some pattern. An example of a sequence is a = 1, 4, 9, 16, ... The individual terms of the sequence are called a_1, a_2, a_3, etc., where the subscript is the term's position on the list.

Explicit Formula for a Sequence

An explicit formula is a type of function into which only positive integers can be input. Sometimes the explicit formula notation resembles function notation like $a(n)$ = $2n$ + 3. An alternative notation is with a subscript, like a_n = $2n$ + 3. By substituting values 1, 2, 3, and 4 for n, the first four terms of the sequence can be determined.

$$a_1 = 2 \cdot 1 + 3 = 5$$
$$a_2 = 2 \cdot 2 + 3 = 7$$
$$a_3 = 2 \cdot 3 + 3 = 9$$
$$a_4 = 2 \cdot 4 + 3 = 11$$

The sequence is 5, 7, 9, 11, ...

Example
What are the first three terms of the sequence defined by equation a_n = 5 + 3(n −1)?

Solution:
Substitute 1, 2, and 3 for n in the formula to get

$$a_1 = 5 + 3(1 - 1) = 5 + 3(0) = 5, \quad a_2 = 5 + 3(2 - 1) = 5 + 3(1) = 8$$

and

$$a_3 = 5 + 3(3 - 1) = 5 + 3(2) = 11$$

122   **Some Key Algebra II Facts and Skills**

Recursive Formula for a Sequence

A recursive formula has two parts. There is the *base case* where the first value (or the first few values) of the sequence are given, and there is the *recursive definition* where a rule for calculating more terms is described in terms of the previous terms.

For example,

$$a_1 = 5$$
$$a_n = 2 + a_{n-1}, \text{ for } n \geq 2$$

To get a_1, there is no calculation since a_1 is given as 5 in the base case part of the definition. For a_2, substitute 2 for n into the recursive part of the definition.

$$a_2 = 2 + a_{2-1} = 2 + a_1 = 2 + 5 = 7$$

Notice that the value of a_1 needed to be known to do this step. For $n = 3$, $a_3 = 2 + a_{3-1} = 2 + a_2 = 2 + 7 = 9$.
The first three terms are 5, 7, and 9.

Notice that this generates the same sequence as the explicit formula $a_n = 2n + 3$.

Example

What is the value of a_3 for the sequence defined by

$$a_1 = 3$$
$$a_n = n \cdot a_{n-1}, \text{ for } n \geq 2$$

Solution:
$a_1 = 3$ is given.

Substitute 2 for n in the recursive part of the definition to get

$$a_2 = 2 \cdot a_{2-1} = 2 \cdot a_1 = 2 \cdot 3 = 6$$

Substitute 3 for n in the recursive part of the definition to get

$$a_3 = 3 \cdot a_{3-1} = 3 \cdot a_2 = 3 \cdot 6 = 18$$

Sequences

Arithmetic and Geometric Sequences

A sequence like 5, 7, 9, 11, . . . is known as an *arithmetic sequence* since each term is equal to 2 more than the previous term. The number 2, in this case, is known as the *common difference*, usually denoted by the variable d.

A sequence like 5, 10, 20, 40, . . . is known as a *geometric sequence* since each term is equal to 2 times the previous term. The number 2, in this case, is known as the *common ratio*, usually denoted by the variable r.

The two formulas for the nth term of an arithmetic or geometric sequence are provided on the reference sheet given to you in the Regents booklet.

Arithmetic Sequence	Geometric Sequence
$a_n = a_1 + (n-1)d$	$a_n = a_1 r^{n-1}$

Example

What is the 50th term of the sequence 7, 11, 15, 19, . . . ?

Solution:

The explicit formula is $a_n = 7 + 4(n-1)$. Substitute 50 for n to get $a_{50} = 7 + 4(50 - 1) = 7 + 4(49) = 203$.

Sum of a Finite Geometric Series

A *finite series* is like a finite sequence except all the terms are added together. An example of a finite geometric series is $5 + 10 + 20 + 40 + 80 + 160$.

- The formula for calculating the sum of a finite geometric series is given in the reference sheet in the Regents booklet. The formula is:

$$S_n = \frac{a_1 - a_1 r^n}{1 - r}$$

124   **Some Key Algebra II Facts and Skills**

Example

Use the sum of a finite geometric series formula to calculate the sum of $5 + 10 + 20 + 40 + 80 + 160$.

Solution:

To use this formula to calculate $5 + 10 + 20 + 40 + 80 + 160$, substitute 6 for n (since there are 6 terms), 5 for a_1, and 2 for r.

$$S_6 = \frac{5 - 5 \cdot 2^6}{1 - 2} = \frac{5 - 5 \cdot 64}{-1} = \frac{-315}{-1} = 315$$

125

Practice Exercises

1. What are the first four terms of the sequence defined by $a_1 = 5$, $a_n = 3 + a_{n-1}$ for $n > 1$?
   (1) 5, 8, 11, 14
   (2) 5, 15, 45, 135
   (3) 5, 2, –1, –4
   (4) 5, 3, 1, –1

2. The sequence 3, 12, 48, 192, . . . can be defined by
   (1) $a_1 = 3$, $a_n = 8 + a_{n-1}$ for $n > 1$
   (2) $a_1 = 3$, $a_n = 4a_{n-1}$ for $n > 1$
   (3) $a_1 = 3$, $a_n = 12a_{n-1}$ for $n > 1$
   (4) $a_1 = 3$, $a_n = 12 + a_{n-1}$ for $n > 1$

3. What is the 20th term of the sequence 4, 11, 18, 25, . . . ?
   (1) 130   (3) 144
   (2) 137   (4) 151

4. What position is the number 247 in the sequence 4, 13, 22, 31, . . . ?
   (1) 26   (3) 28
   (2) 27   (4) 29

5. What is the 10th term of the sequence 4, 20, 100, 500, . . . ?
   (1) 7,812,500   (3) 195,312,500
   (2) 39,062,500   (4) 976,562,500

Some Key Algebra II Facts and Skills

6. What is the sum of the series

$$2 + 2 \cdot 3 + 2 \cdot 3^2 + 2 \cdot 3^3 + \cdots + 2 \cdot 3^{15}?$$

(1) $\dfrac{2(1-3^{15})}{1-3}$ (3) $\dfrac{2(1+3^{16})}{1+3}$

(2) $\dfrac{2(1+3^{15})}{1+3}$ (4) $\dfrac{2(1-3^{16})}{1-3}$

7. A series is defined as $d_1 = 0$, $d_2 = 1$, $d_n = (n-1)(d_{n-2} + d_{n-1})$ for $n > 2$. What is the value of d_5?
   (1) 2   (3) 44
   (2) 9   (4) 265

8. What is the sum of the first 20 terms of

$$2 - 6 + 18 - 54 + 162 - 486 + \cdots?$$

   (1) –1,743,392,000   (3) –1,743,392,200
   (2) –1,743,392,100   (4) –1,743,392,300

9. $\dfrac{1}{2} + \dfrac{1}{4} + \dfrac{1}{8} + \dfrac{1}{16} + \cdots + \dfrac{1}{2^{20}}$ is closest to

   (1) 0.99995   (3) 0.9999995
   (2) 0.999995   (4) 1

10. For 20 years Journey puts $1,000 into the bank every January 1. The bank pays 3% interest every December 31. If she starts on January 1, 2000, the amount of money she will have on January 2, 2020 is the sequence:

$$1,000 + 1,000(1.05) + 1,000(1.05^2) + \cdots + 1,000(1.05)^{20}$$

How much money is this?

Sequences

127

Solutions

1. The first term of the sequence, a_1, is given as 5.

For $n = 2$: $a_2 + 3 + a_{2-1} = 3 + a_1 = 3 + 5 = 8$

For $n = 3$: $a_3 + 3 + a_{3-1} = 3 + a_2 = 3 + 8 = 11$

For $n = 4$: $a_4 + 3 + a_{4-1} = 3 + a_3 = 3 + 11 = 14$

The first four terms of the sequence are 5, 8, 11, 14.

The correct choice is **(1)**.

2. In this sequence, the first term is 3 and each term is equal to 4 times the previous term. For the base case of the recursive definition, $a_1 = 3$. For the recursive part, the way you define each term as 4 times the previous term is $a_n = 4a_{n-1}$.

The correct choice is **(2)**.

3. This is an arithmetic series with a first term of 4 and a common difference of 7 since each term is 7 more than the previous term. The formula for the nth term of an arithmetic series with a first term of a_1 and a common difference of d is $a_n = a_1 + (n-1)d$. For the 20th term of this sequence, $n = 20$ and $a_{20} = 4 + (20-1)7 = 4 + 19 \cdot 7 = 137$.

The correct choice is **(2)**.

4. This is an arithmetic series with a constant difference of 9 since each term is 9 more than the previous term. To find the position of the number 247 in the sequence, use the formula $a_n = a_1 + (n-1)d$ to solve for n if $a_n = 247$, $a_1 = 4$, and $d = 9$.

$$247 = 4 + (n-1)9$$
$$\underline{-4 = -4}$$
$$243 = (n-1)9$$
$$\frac{243}{9} = \frac{(n-1)9}{9}$$
$$27 = n - 1$$
$$28 = n$$

The correct choice is **(3)**.

128   **Some Key Algebra II Facts and Skills**

5. Since each term is 5 times the previous term, this is a geometric sequence. The nth term of a geometric sequence can be found with the formula $a_n = a_1 r^{n-1}$. If $a_1 = 4$, $r = 5$, and $n = 10$, this becomes $a_{10} = 4 \cdot 5^{10-1} = 4 \cdot 5^9 = 7{,}812{,}500$. Notice that this is not the same as 20^9.

   The correct choice is (**1**).

6. Since each term is 3 times the previous term, this is a geometric series. The formula from the reference sheet given to you on the Regents for the sum of the first n terms of a geometric series is $S_n = \dfrac{a_1 - a_1 r^n}{1 - r}$. For this example, $n = 16$ and not 15 since the first term can be thought of as $2 \cdot 3^0$. Substituting $n = 16$, $a_1 = 2$, and $r = 3$, this becomes

   $$S_{16} = \frac{2 - 2 \cdot 3^{16}}{1 - 3} = \frac{2(1 - 3^{16})}{1 - 3}$$

   The correct choice is (**4**).

7. Using the recursive formula for $n = 3$, $n = 4$, and $n = 5$,

   $d_3 = (3 - 1)(d_{3-2} + d_{3-1}) = 2(d_1 + d_2) = 2(0 + 1) = 2$
   $d_4 = (4 - 1)(d_{4-2} + d_{4-1}) = 3(d_2 + d_3) = 3(1 + 2) = 9$
   $d_5 = (5 - 1)(d_{5-2} + d_{5-1}) = 4(d_3 + d_4) = 4(2 + 9) = 44$

   The correct choice is (**3**).

8. Since each term is -3 times the previous term, this is a geometric series with $a_1 = 2$, $r = -3$, and $n = 20$. Using the formula from the reference sheet given to you on the Regents, $S_n = \dfrac{a_1 - a_1 r^n}{1 - r}$, this becomes

   $$S_{20} = \frac{2 - 2 \cdot (-3)^{20}}{1 - (-3)} = -1{,}743{,}392{,}200$$

   The correct choice is (**3**).

Sequences   129

9. This is a geometric series with $a_1 = \dfrac{1}{2}$, $r = \dfrac{1}{2}$, and $n = 19$ (and not 20 since the first denominator is 2^1). According to the formula for the sum of the geometric series $S_n = \dfrac{a_1 - a_1 r^n}{1-r}$, for this example $S_{20} = \dfrac{\dfrac{1}{2} - \dfrac{1}{2} \cdot \dfrac{1}{2}^{20}}{1 - \dfrac{1}{2}} = 0.9999990463$. This is a bit closer to 0.9999995 than it is to 1.

   The correct choice is **(3)**.

10. This is a geometric series with $a_1 = 1{,}000$, $r = 1.05$, and $n = 21$ (and not 20 since the first term is like $1{,}000 \cdot (1.05)^0$. According to the formula for the sum of a geometric series given in the Regents reference sheet, $S_n = \dfrac{a_1 - a_1 r^n}{1-r}$.

   This becomes, for this example,

   $$S_{21} = \frac{1{,}000 - 1{,}000 \cdot 1.05^{21}}{1 - 1.05} = 35{,}719.25$$

9. PROBABILITY

9.1 CALCULATING PROBABILITY FROM KNOWN DATA

How likely it is that something happens is measured by a number between 0 and 1. Something certain to happen has a probability of 1, and something that is certain *not* to happen has a probability of 0. Most things have a probability between 0 and 1.

When things are randomly selected from a group, it is often possible to calculate the probability that the selected thing has a certain property.

The formula for calculating probability is

$$P = \frac{\text{number of favorable outcomes}}{\text{number of possible outcomes}}$$

If there is a room with 12 adults and 38 children, for example, and a random person is chosen, the probability that the selected person is a child is $\frac{12}{50} = 0.24$.

When information about a data set is collected on a Venn diagram, it is possible to answer probability questions about that set.

Example

Fifty people are surveyed and asked if they subscribe to Netflix, Amazon Prime, Both, or Neither. The results are collected on the Venn diagram below.

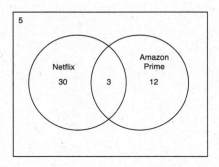

Probability 131

If a person is randomly selected, determine each probability:
1. $P(\text{Netflix})$
2. $P(\text{Amazon Prime})$
3. $P(\text{Both})$
4. $P(\text{Neither})$
5. $P(\text{Netflix or Amazon Prime})$
6. $P(\text{Netflix if it is known that the person subscribes to Amazon Prime})$
7. $P(\text{Amazon Prime if it is known that the person subscribes to Netflix})$

Solutions:
1. Since 30 + 3 = 33 people out of 50 subscribe to Netflix, the probability of a randomly chosen person subscribing to Netflix is $\dfrac{33}{50}$.

2. 12 + 3 = 15 so $\dfrac{15}{50}$

3. $\dfrac{3}{50}$

4. $\dfrac{5}{50}$

5. This is everyone except the 5 people who subscribe to neither so the answer is $\dfrac{45}{50}$.

6. There are 15 people who subscribe to Amazon Prime. Of those 15 people, 3 of them subscribe to Netflix (the people who are in the overlap) so the answer is $\dfrac{3}{15}$.

7. There are 33 people who subscribe to Netflix. Of those 33 people, 3 of them subscribe to Netflix so the answer is $\dfrac{3}{33}$.

132   **Some Key Algebra II Facts and Skills**

9.2 PROBABILITY OF INDEPENDENT EVENTS

Two events are said to be *independent* if one of the events happening (or not happening) has nothing to do with the other event happening (or not happening).

- An example of two independent events is "It rains in France on December 14, 1987" and "You will ace the Algebra II Regents." One thing has nothing to do with the other. It raining in France on December 14, 1987 does not, in any way, affect your score on the Algebra II Regents.

When events are not independent, they are known as *dependent* events. An example of dependent events is "You study from the Barron's Algebra II review book" and "You will ace the Algebra II Regents." If you study from this Barron's book, you will likely get a higher grade on the Regents than you would if you didn't study from this book.

MATH FACTS

When two events are independent, the probability of both events happening is equal to the product of the probability of each event on its own. In symbols, if A and B are independent events then $P(A \text{ and } B) = P(A) \cdot P(B)$.

For example, if a six-sided die is rolled, the probability of it landing with a 3 facing up is $\frac{1}{6}$ since there is one 3 out of six possible outcomes. If a fair coin is flipped, the probability of it landing with heads facing up is $\frac{1}{2}$ since there is one heads out of two possible outcomes. According to the formula, then, if the die is rolled and the coin is flipped, the probability that the die lands with the 3 facing up *AND* the coin lands with heads facing up is

$$P(3 \text{ and Heads}) = P(3) \cdot P(\text{Heads}) = \frac{1}{6} \cdot \frac{1}{2} = \frac{1}{12}$$

Probability **133**

MATH FACTS

For two events, A and B, the probability that A *or* B will happen is the sum of the probabilities of them happening individually *minus* the probability of them both happening.

$$P(A \text{ or } B) = P(A) + P(B) - P(A \text{ and } B)$$

If A and B are independent, this becomes

$$P(A \text{ or } B) = P(A) + P(B) - P(A) \cdot P(B)$$

Example
If the probability that it rains next Tuesday is 0.40 and the probability that the Knicks win the NBA championship is 0.03 and these events are independent, what is the probability that it rains next Tuesday *AND* the Knicks win the NBA championship?

Solution:
Since they are independent events, the probability of both happening is the product of the probabilities of each happening separately. $0.40 \cdot 0.03 = 0.012$.

9.3 PROBABILITY OF DEPENDENT EVENTS

Sometimes it is not obvious whether or not two events are independent or not. There is a way to use math to check. When two events, A and B, are *not* independent, it is no longer true that $P(A \text{ and } B) = P(A) \cdot P(B)$. Instead, it can be quite complicated to determine the probability that both happen. This idea can be used to actually check to see if two events are dependent or independent if we have sufficient information.

Some Key Algebra II Facts and Skills

> **MATH FACTS**
>
> When checking if two events are independent or not, first multiply $P(A)$ by $P(B)$.
> If $P(A \text{ and } B) = P(A) \cdot P(B)$, A and B are independent events.
> If $P(A \text{ and } B) \neq P(A) \cdot P(B)$, A and B are dependent events.

Example
If $P(A) = 0.30$, $P(B) = 0.40$, and $P(A \text{ and } B) = 0.24$, are events A and B independent or dependent?

Solution:
Since $P(A) \cdot P(B) = 0.30 \cdot 0.40 = 0.12 \neq 0.24$, $P(A) \cdot P(B) \neq P(A \text{ and } B)$ so they are dependent events.

Using the Netflix vs. Amazon Prime survey data from the diagram below, we can determine if event A = "A person subscribes to Netflix" and B = "A person subscribes to Amazon Prime" are independent.

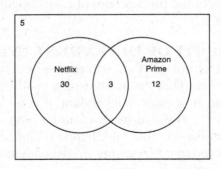

$P(A) = \dfrac{33}{50}$, $P(B) = \dfrac{15}{50}$, and $P(A \text{ and } B) = \dfrac{3}{50}$

Since $P(A) \cdot P(B) = \dfrac{33}{50} \cdot \dfrac{15}{50} = 0.198$ and

$P(A \text{ and } B) = 0.06$, $P(A \text{ and } B) \neq P(A) \cdot P(B)$

so they are not independent events.

135

Practice Exercises

1. Which two events are independent of one another?
   (1) A = It rains in Nevada on January 1, 1975.
       B = A New York resident goes to the movies on May 17, 2016.
   (2) A = It's Valentine's Day.
       B = Flower sales have increased.
   (3) A = A student eats a healthy breakfast.
       B = A student has energy at 10:00 A.M.
   (4) A = A teacher wins the lottery.
       B = The same teacher quits his job.

2. A and B are independent events. If $P(A) = 0.4$ and $P(B) = 0.5$, what is $P(A \text{ and } B)$?
   (1) 0.4
   (2) 0.5
   (3) 0.2
   (4) 0.9

136　　　　**Some Key Algebra II Facts and Skills**

Questions 3, 4, and 5 use the following information.

A fair coin is flipped, and a spinner with the letters A, B, C, D, and E, all equally likely, is spun.

3. What is the probability the coins will land heads and the spinner will land on D?

(1) $\dfrac{1}{2}$

(3) $\dfrac{1}{10}$

(2) $\dfrac{1}{5}$

(4) $\dfrac{7}{10}$

4. What is the probability that the coin will land on heads or the spinner will land on D?

(1) $\dfrac{4}{10}$

(3) $\dfrac{6}{10}$

(2) $\dfrac{5}{10}$

(4) $\dfrac{7}{10}$

5. What is the probability that the coin will land on heads given that the spinner landed on D?

(1) $\dfrac{1}{10}$

(3) $\dfrac{7}{10}$

(2) $\dfrac{3}{10}$

(4) $\dfrac{1}{2}$

6. If the probability that the Rangers win the Stanley Cup is 0.05 and the probability that the Yankees win the World Series is 0.15, what is the probability that both will happen?

(1) 0.20

(3) 0.05

(2) 0.0075

(4) 0.15

Probability 137

7. If $P(A) = 0.4$, $P(B) = 0.6$, and $P(A \text{ and } B) = 0.3$, what conclusion can you make about A and B?
   (1) $P(B|A) = 0.6$
   (2) A and B are independent events.
   (3) A and B are not independent events.
   (4) No conclusion is possible.

8. If $P(A) = 0.7$, $P(B) = 0.6$, and $P(A \text{ and } B) = 0.42$, what conclusion can you make?
   (1) $P(A|B) = 0.6$
   (2) A and B are independent events.
   (3) A and B are not independent events.
   (4) No conclusion can be made.

9. A group of 40 adults are asked if they like classical music, rock music, both, or neither. The results are collected in this Venn diagram.

   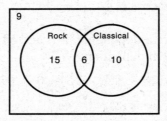

   Based on this data, what conclusion can be made about liking classical music and liking rock music?
   (1) $P(B) \neq P(B|A)$ (A = rock and B = classical)
   (2) They are not independent of each other.
   (3) They are independent of each other.
   (4) No conclusion can be made.

138   **Some Key Algebra II Facts and Skills**

10. 104 students were polled about whether they like soccer, basketball, both, or neither. Based on this data determine if liking basketball and liking soccer are independent events?

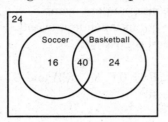

Solutions

1. In choices (2), (3), and (4) there is a cause-and-effect relationship. If the event A happens for those choices, then event B is more likely to happen than it would be if event A did not happen.

   The correct choice is **(1)**.

2. When events are independent, $P(A \text{ and } B) = P(A) \cdot P(B)$. Since these events are independent,

   $$P(A \text{ and } B) = P(A) \cdot P(B) = 0.4 \cdot 0.5 = 0.2$$

   The correct choice is **(3)**.

3. Since there are two sides to the coin and only one of them is heads, $P(\text{Heads}) = \frac{1}{2}$. Since there are five different letters on the spinner and only one of them is D, $P(\text{D}) = \frac{1}{5}$.

   The coin and spinner are independent of one another so the formula $P(A \text{ and } B) = P(A) \cdot P(B)$ can be used. For this example,

   $$P(\text{Heads and D}) = P(\text{Heads}) \cdot P(\text{D}) = \frac{1}{2} \cdot \frac{1}{5} = \frac{1}{10}$$

   The correct choice is **(3)**.

Probability
139

4. Since there are two sides to the coin and only one of them is heads, $P(\text{Heads}) = \frac{1}{2}$. Since there are five different letters on the spinner and only one of them is D, $P(\text{D}) = \frac{1}{5}$.

The coin and spinner are independent of one another so the formula $P(A \text{ and } B) = P(A) + P(B) - P(A) \cdot P(B)$ can be used. For this example

$$P(\text{Heads or D}) = P(\text{Heads}) + P(\text{D}) - P(\text{Heads}) \cdot P(\text{D})$$

$$= \frac{1}{2} + \frac{1}{5} - \frac{1}{2} \cdot \frac{1}{5} = \frac{5}{10} + \frac{2}{10} - \frac{1}{10} = \frac{6}{10}$$

The correct choice is **(3)**.

5. Since the events are independent, the probability of the coin landing on heads is not affected by whether or not the spinner lands on D. If it is given that the spinner landed on D, the probability of the coin landing on heads is still $\frac{1}{2}$.

The correct choice is **(4)**.

6. These are independent events since one happening has no affect on the other happening. For independent events, the formula $P(A \text{ and } B) = P(A) \cdot P(B)$ can be used.

For this example,

$$P(\text{Stanley Cup and World Series})$$
$$= P(\text{Stanley Cup}) \cdot P(\text{World Series})$$
$$= 0.05 \cdot 0.15 = 0.0075$$

The correct choice is **(2)**.

140　　　　**Some Key Algebra II Facts and Skills**

7. When events are independent, $P(A \text{ and } B) = P(A) \cdot P(B)$. In this example, $P(A) \cdot P(B) = 0.4 \cdot 0.6 = 0.24$, but it is given that $P(A \text{ and } B) = 0.3 \neq 0.24$. So these are not independent events.

   The correct choice is **(3)**.

8. When events are independent, $P(A \text{ and } B) = P(A) \cdot P(B)$. In this example, $P(A) \cdot P(B) = 0.7 \cdot 0.6 = 0.42$, and it is given that $P(A \text{ and } B) = 0.42$. So these are independent events.

   The correct choice is **(2)**.

9. The events are independent if

   $$P(\text{rock and classical}) = P(\text{rock}) \cdot P(\text{classical})$$

   Since there are 40 people and 6 of them like both,

   $$P(\text{rock and classical}) = \frac{6}{40} = 0.15$$

   From the Venn diagram, there are $15 + 6 = 21$ people who like rock and $10 + 6 = 16$ people who like classical so

   $$P(\text{rock}) = \frac{21}{40} \text{ and } P(\text{classical}) = \frac{16}{40}$$

   This means that $P(\text{rock}) \cdot P(\text{classical}) = \frac{21}{40} \cdot \frac{16}{40} = 0.21 \neq 0.15$ so these are not independent events.

   The correct choice is **(2)**.

10.

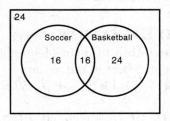

To see if liking soccer and liking basketball are independent, check to see if

$P(\text{soccer}) \cdot P(\text{basketball}) = P(\text{soccer and basketball})$

Because 16 + 16 = 32, 32 out of 40 people like soccer and $P(\text{soccer}) = \dfrac{32}{80}$.

Because 24 + 16 = 40, 40 out of 80 people like basketball and $P(\text{basketball}) = \dfrac{40}{80}$.

There are 16 out of 80 people who like both sports so

$$P(\text{soccer and basketball}) = \dfrac{16}{80} = 0.2$$

Multiplying $P(\text{soccer}) \cdot P(\text{basketball}) = \dfrac{32}{80} \cdot \dfrac{40}{80} = 0.2$.

Because $P(\text{soccer}) \cdot P(\text{basketball}) = P(\text{soccer and basketball})$, these are all independent events.

10. NORMAL DISTRIBUTION

10.1 STANDARD DEVIATION

The **standard deviation** of a set of numbers is a measure of how "spread out" the numbers in the set are.

The 14 numbers 1, 7, 14, 14, 20, 20, 20, 20, 20, 20, 26, 26, 33, 39 have a mean of 20.

The 14 numbers 18, 19, 19, 19, 20, 20, 20, 20, 20, 20, 21, 21, 21, 22 also have a mean of 20.

Though these two data sets have the same mean, the numbers in the second data set are less spread out. In statistics, we say that the second set of numbers has a lower standard deviation. The standard deviation for the second set is 1, whereas the standard deviation for the first set is 9.3. To have a standard deviation of zero, all the numbers in the data set would have to be the same number.

Calculating Standard Deviation with a Calculator

The standard deviation of a set of numbers can be quickly calculated with a graphing calculator.

- Here is how to calculate the standard deviation for the 14 numbers in the first data set 1, 7, 14, 14, 20, 20, 20, 20, 20, 20, 26, 26, 33, 39.

For the TI-84:

Press [STAT] [1] to get to the list editor. Go to the L1 cell and press [CLEAR] [ENTER]. Enter the 14 scores into L1. Press [STAT], move to CALC, press [1]. Make List: L1 and leave FreqList: blank. Go to Calculate, and press [ENTER]. The first value, 20, is the mean. The fifth value, next to the σx= is the standard deviation. The standard deviation for this data set is 9.3.

 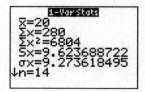

Normal Distribution 143

For the TI-Nspire:

From the home screen, select the Add Lists & Spreadsheet icon. Name column A x and fill in cells A1 through A14 with the 14 scores. Press [menu] [4] [1] [1] for One-Variable Statistics. Set the Number of Lists to 1. Select the [OK] button. Set the X1 List to x. In cell C2 will be the mean, 20. In cell C6 will be the standard deviation next to the σx=. The standard deviation for this data set is 9.3.

Standard Deviation for Data Sets with Repeated Values

Since there are repeated values in the data set, the values can be listed in a two-column chart. The first column is for the values, and the second column is for how many of that value is in the data set. For this set, it would look like this.

x_i	f_i
1	1
7	1
14	2
20	6
26	2
33	1
39	1

For the TI-84:

Press [STAT] [1] and enter the numbers 1, 7, 14, 20, 26, 33, and 39 into L1 and enter the numbers 1, 1, 2, 6, 2, 1, and 1 into L2. Press [STAT], move to CALC, and press [1]. In List: put "L1" by pressing [2ND] [1]. In FreqList: put "L2" by pressing [2ND] [2]. Move to Calculate and press [ENTER].

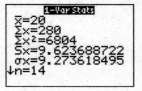

Normal Distribution

For the TI-Nspire:

From the home screen, select the Add Lists & Spreadsheet icon. Name column A x and fill in cells A1 through A7 with the 7 scores. Name column B y and fill in cells B1 through B7 with the 7 frequencies. Press [menu] [4] [1] [1] for One-Variable Statistics. Set the number of lists to 1. Select the [OK] button. Set the X1 List to x and the Frequency List to y. In cell C2 will be the mean, 20. In cell C6 will be the standard deviation next to the $\sigma x=$. The standard deviation for this data set is 9.3.

146   **Some Key Algebra II Facts and Skills**

- **z-scores**

  In the data set from before with a mean of 60 and a standard deviation of 4, the number 64 was one standard deviation above the mean since it is equal to the mean plus the standard deviation, $60 + 4 = 64$. The number 68 was two standard deviations above the mean since it is equal to the mean plus two times the standard deviation $60 + 2 \cdot 4 = 68$.

  In statistics, we say that a number that is one standard deviation above the mean has a *z-score* of 1, whereas a number that is two standard deviations above the mean has a *z-score* of 2.

MATH FACTS

The *z-score* of a number is how many standard deviations above the mean that number is. If the number is greater than the mean, the *z-score* will be positive. If it is less than the mean, the *z-score* will be negative. A formula for calculating *z-score* is

$$z = \frac{\text{number} - \text{mean}}{\text{standard deviation}}$$

Example

If the mean is 60 and the standard deviation is 4, what is the *z-score* of the number 52?

Solution:

Since 52 is less than 60, the *z-score* will be negative. For this example, it is possible to subtract the standard deviation, 4, from 60 to get to 56, which would have a *z-score* of −1. Then subtract it again to get to 52, which has a *z-score* of −2.

This can also be calculated with the formula

$$z = \frac{\text{number} - \text{mean}}{\text{standard deviation}} = \frac{52 - 60}{4} = \frac{-8}{4} = -2$$

10.2 NORMAL DISTRIBUTION

> **KEY IDEAS**
> The dot plot of real-world data often has the shape of a *bell curve* also known as a *normal distribution curve*. When this happens, it is possible to answer certain questions about the data that you would not be able to answer had the shape of the dot plot not been a bell curve.

The Normal Curve

Below is a dot plot with 100 data points representing the weights of 100 children. The mean value is approximately 60 and the standard deviation is approximately 4.

Since the mean is 60 and the standard deviation is 4, a score of 64 is said to be *one standard deviation above the mean*. A score of 68 is said to be *two standard deviations above the mean*. A score of 56 is said to be *one standard deviation below the mean*. A score of 52 is said to be *two standard deviations below the mean*.

Using the Calculator to Find the Percentage of Numbers Between Two Numbers If the Mean and Standard Deviation Are Known

- If the mean and the standard deviation of a data set are known, the graphing calculator can quickly determine what percent of the numbers in the data set will be between any two values.

The heights of 400 people are measured; the mean is approximately 60, and the standard deviation is approximately 4.

When the data has a normal distribution, the calculator can be used to determine the percent of numbers between two values. To find what percent of people in the above example are between 57.5 inches and 62 inches, do the following:

For the TI-84:

Press [2ND] [VARS] [2] for normalcdf. Fill in the fields with 57.5 for lower, 62 for upper, 60 for μ. Then select Paste.

Normal Distribution 149

For the TI-Nspire:

From the home screen, press [A]. Press [menu] [6] [5] [2] for Normal Cdf. Set the Lower Bound to 57.5, the Upper Bound to 62, the mean (µ) to 60, and the standard deviation (σ) to 4.

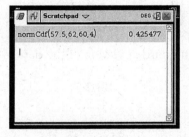

The percent is approximately 43%, which is approximately 0.43 · 400 = 172 people between 57.5 inches and 62 inches tall.

For examples when there is no lower bound, make the lower bound a large negative number like –9999. If there is no upper bound, make the upper bound a large positive number like 9999.

Practice Exercises

1. Each of these data sets has a mean of 5. Which has the smallest standard deviation?
   (1) 1, 2, 5, 7, 10
   (2) 2, 3, 5, 5, 10
   (3) 4, 5, 5, 5, 6
   (4) 3, 4, 5, 5, 8

2. What is the standard deviation of the data set 16, 19, 14, 12, 10, 15, 15, 16?
   (1) 2.5
   (2) 2.6
   (3) 2.7
   (4) 2.8

3. What is the standard deviation for this data set?

x_i	f_i
18	4
19	7
20	2
21	3
22	6

   (1) 1.41
   (2) 1.51
   (3) 1.54
   (4) 1.61

Normal Distribution 151

4. If the mean of a data set is 60 and the standard deviation is 8, what is the *z-score* for the number 72?
   (1) 0.67   (3) 1.25
   (2) 1.00   (4) 1.50

5. How many of these 8 numbers are more than two standard deviations from the mean?

   18, 40, 44, 45, 48, 50, 50, 68

   (1) 1   (3) 3
   (2) 2   (4) 0

6. If the mean of a set of normally distributed 200 numbers is 70 and the standard deviation is 6, approximately what percent of the numbers will be between 64 and 76?
   (1) 68%   (3) 76%
   (2) 70%   (4) 80%

7. The weights of 400 children are collected, and the data are normally distributed. What percent of the children were between 46 and 52 pounds if the mean was 50 pounds and the standard deviation was 3?
   (1) 57%   (3) 63%
   (2) 60%   (4) 66%

8. On the Algebra II Regents, the mean score was 75 and the standard deviation was 9. If the data are normally distributed, what percent of the scores were over 90?
   (1) 20%   (3) 10%
   (2) 15%   (4) 5%

152   **Some Key Algebra II Facts and Skills**

9. The amount of time a group of 300 adults exercised per week was collected. If the data were normally distributed, what percent of the people exercised less than 20 minutes a week if the mean was 40 minutes and the standard deviation was 15?
   (1) 9%   (3) 13%
   (2) 11%   (4) 15%

10. In a set of 500 samples, the mean is 90 and the standard deviation is 17. If the data are normally distributed, how may of the 500 are expected to have a value between 93 and 101?
    (1) 82   (3) 86
    (2) 84   (4) 88

Solutions
1. When the numbers are all very close together, the standard deviation is close to 0. The smallest possible standard deviation is 0 when all the numbers are equal to the mean, as for the data set 5, 5, 5, 5, 5. Of the four choices, choice (3) has three of the numbers equal to the mean, and the other two are just one away from the mean so it has the smallest standard deviation.

   The correct choice is **(3)**.

2. Enter the eight numbers into the graphing calculator and do the 1-Variable Stats. The calculator will output a bunch of values. Find the number next to $\sigma x=$, which, in this case, is 2.54666, which is closest to 2.5.

   The correct choice is **(1)**.

3. In the graphing calculator, enter the values 18, 19, 20, 21, 22 into the first column and the numbers 4, 7, 2, 3, 6 into the second column. Follow the instructions for lists with frequencies given in this section. Find the number next to $\sigma x=$, which, in this case, is 1.507556, which is closest to 1.51.

   The correct choice is **(2)**.

Normal Distribution   **153**

4. The z-*score* can be calculated with the formula

$$z = \frac{\text{number} - \text{mean}}{\text{standard deviation}}$$

For this example, $z = \frac{72-60}{8} = \frac{12}{8} = 1.5$. The number 72 has a z-*score* of 1.5 so it is 1.5 standard deviations above the mean. The correct choice is (**4**).

5. Using the graphing calculator, the mean for these 8 numbers is 45.375, and the standard deviation is 12.95. Numbers more than two standard deviations above the mean are over $45.375 + 2 \cdot 12.95 = 71.275$ or under $45.375 - 2 \cdot 12.95 = 19.475$. In the list of 8 numbers, there are no numbers above 71.275 and one number, 18, below 19.475. Only one number, 18, is more than two standard deviations from the mean.

The correct choice is (**1**).

6. Using the normal cdf function on the calculator, enter a mean of 70, a standard deviation of 6, a lower bound of 64, and an upper bound of 76. The output is 0.68, which means that in a normal distribution with a mean of 70 and a standard deviation of 6, about 68% of the values are expected to be between 64 and 76.

The correct choice is (**1**).

7. Using the normal cdf function on the calculator, enter a mean of 50, a standard deviation of 3, a lower bound of 46, and an upper bound of 52. The output is 0.6563, which means that in a normal distribution with a mean of 50 and a standard deviation of 3, about 66% of the values are expected to be between 46 and 52.

The correct choice is (**4**).

Some Key Algebra II Facts and Skills

8. Using the normal cdf function on the calculator, enter a mean of 75, a standard deviation of 9, a lower bound of 90, and, since there is no upper bound, an upper bound of 999. The output is 0.04779, which means that in a normal distribution with a mean of 75 and a standard deviation of 9, about 5% of the values are expected to be greater than 90.

    The correct choice is **(4)**.

9. Using the normal cdf function on the calculator, enter a mean of 40, a standard deviation of 15, an upper bound of 20, and, since there is no lower bound, a lower bound of −999. The output is 0.09121, which means that in a normal distribution with a mean of 40 and a standard deviation of 15, about 9% of the values are expected to be less than 20.

    The correct choice is **(1)**.

10. Using the normal cdf function on the calculator, enter a mean of 90, a standard deviation of 17, a lower bound of 93, and an upper bound of 101. The output is 0.1711, which means that in a normal distribution with a mean of 90 and a standard deviation of 17, about 17.12% of the values are expected to be between 93 and 101. Since there are 500 samples, find 17.12% of 500, which is $0.1712 \cdot 500 = 85.6 \approx 86$.

    The correct choice is **(3)**.

Inferential Statistics   155

11. INFERENTIAL STATISTICS

11.1 DIFFERENT TYPES OF STATISTICAL STUDIES

The three types of statistical studies you should be familiar with are surveys, observational studies, and experimental studies.

- A **survey** is a bunch of questions that are asked about the subjects you are studying.
- In an **observational study**, the subjects in the study get observed and data about them is recorded, but the observer does not try to intervene and control anything about the subjects.
- In an **experimental study**, the observer separates the subjects into groups and requires the different groups to do different things.

If Barron's wanted to know how well the Algebra II red review book is working, they could use any of these three methods.

If they wanted to use a survey they could send out letters to a random list of students and ask two questions: (1) Did you study with the Algebra II red review book? (2) What did you get on the Algebra II Regents? They would want to be careful to try to choose a random group of students and not introduce bias by sending to people more or less likely to have used the book.

If they wanted to do an observational study, they could monitor a group of people and observe whether or not they studied with the Algebra II red review book and then observe them taking the Regents and reacting to their eventual score.

If they wanted to do an experimental study, they could pick 100 random students and give 50 of them the Algebra II red review book and tell the other students not to use the Algebra II red review book. Data would then be collected to see how the students in the different groups performed. It is important that the groups are randomly selected so that there is no bias to distort the results.

156   **Some Key Algebra II Facts and Skills**

11.2 ANALYZING COMPUTER GENERATED HISTOGRAMS

There are three situations where you might see a computer-generated histogram that looks something like an approximate bell curve.

When asked to analyze this, the general guideline is that if the thing you are being asked about is either not represented on the histogram at all or represented by an extremely small bar, then that event is not very likely.

On the other hand, if the thing you are being asked about is represented by a bar that is not small, then it is at least somewhat likely that the thing will happen.

Sometimes in addition to the picture of the histogram, additional statistics are given like the mean and the standard deviation. With these numbers, it is possible to tell if the thing you are asked about is likely or unlikely, even without looking at the histogram. If the thing you are asked about is more than two standard deviations from the mean, it is very unlikely. If it is less than two standard deviations from the mean, then it is at least somewhat likely that the thing will happen.

Example

Thirty students in a focus group are asked to give their opinion on a new style of backpack. Only 9 of the 30 students polled, or 30%, liked the new backpack. The manufacturer wants to know if is likely that out of the entire population, 40% would like the backpack.

A computer is used to generate 100 random sets of 30 students from an imaginary group of students from which exactly 40% of them do like the backpack and the percent who like the backpack for each sample set is graphed on a dot plot or histogram and the mean and the standard deviation of those numbers is calculated.

Inferential Statistics

The mean value was 0.404 and the standard deviation was 0.089.

Is it likely that 40% of the entire population actually would like the backpack even though only 30% of the 30-person sample liked it?

Solution:
On the histogram it can be seen that the bars representing 30% or less were not particularly small. This means that even if the percent of the full sample who liked the backpack was 40%, it would be somewhat likely for samples of 30 to have 30% or less liking them.

This could also be done without looking at the histogram, but instead seeing if 30% was within two standard deviations of the mean.

For this example, 0.404 − 2 · 0.089 = 0.226 and 0.404 + 2 · 0.089 = 0.582. Since 30% is between 22.6% and 58.2%, it is likely for 30% of a random sample of 30 people to like the backpack even if really 40% of the entire population liked it.

Practice Exercises

1. You want to find out what type of movie is the favorite among the residents of New York City. If you want to do this by conducting a survey, what are some ways that you can reduce bias?

2. You want to study how the amount of sleep a student gets the night before the Regents relates to their Regents score. How can this be studied as an observational study? How can this be studied as an experiment?

3. There are 500 10th graders at Regentsville High School. You want to do an experiment where 20 students study for the Regents with just the Red Barron's book and 20 other students study for the Regents with the Blue and the Red Barron's books. How can the groups be chosen to reduce the chance of skewed data through bias?

4. A survey is conducted to learn what subjects students enjoy most at school. One question reads: "What is your most boring subject?" Could the wording of this question skew the results of the survey? Explain.

5. Using a population as of 10,000 fish in a lake, an ecologist takes 500 samples of size 50. The average for all these samplings is 0.24 with a standard deviation of 0.06.

This is a histogram of the sampling distribution of the sample proportion.

Using this data, with a 95% confidence interval, we can determine the percent of fish in the lake that are cod is which of the following?
(1) Between 0.12 and 0.36.
(2) Between 0.14 and 0.34.
(3) Between 0.20 and 0.28.
(4) Exactly 0.24.

6. A town has 300,000 households and a statistician at a phone company wants to know what percent of them have a landline. She takes 300 samplings, each with size 100 and calculates the percent of households with a landline for each of the samplings. The average of all the means was 0.73 and the standard deviation was 0.05.

This is a histogram of the sampling distribution of the sample proportion.

Using this data, with a 95% confidence interval, we can determine that the percent of homes in the town that have a land line is

(1) between 0.63 and 0.83.
(2) between 0.68 and 0.78.
(3) between 0.70 and 0.76.
(4) exactly 0.73.

7. A sports doctor conducts an observational study to learn the average amount of time that 3,000 swimmers in the town can hold their breath underwater. He creates 150 random samplings of 60 people per sampling where for each sampling the average amount of time that the swimmers can hold their breath underwater was calculated. The average of all the means of all the samplings was 72.7, and the standard deviation was 0.92.

This is a histogram of the sampling distribution of the sample mean.

Based on this data, with a 95% confidence interval, the researchers can determine that the actual average amount of time the entire population can hold its breath under water is

(1) exactly 72.7.
(2) between 72 and 73.4.
(3) between 71.28 and 73.12.
(4) between 70.86 and 74.54.

8. A data specialist at a social media website wants to estimate the average age of the 20,000 people attending a Taylor Swift concert. For 150 random samplings of 80 people per sample, the average age of the 80 person sample was calculated. The average of all the means of all the samplings was 25.5, and the standard deviation was 1.5.

This is a histogram of the sampling distribution of the sample mean.

Based on this data, with a 95% confidence interval, the researchers can determine that the average age of the entire 20,000 person population is

(1) exactly 25.5.
(2) between 22.5 and 28.5.
(3) between 23.5 and 27.5.
(4) between 24.5 and 26.5.

9. Amy Hogan rolls a six-sided die 12 times and gets a 6 on four of the rolls. She wants to examine if this might suggest that the die is defective since, on average, the number of expected sixes in 12 rolls is two. She uses a computer to simulate 500 times of rolling a fair die 12 times. A histogram of the results is shown below.

Based on this histogram, Amy concludes that,

(1) yes, the die is defective because two 6s is most common.
(2) yes, the die is defective because four 6s is not common.
(3) no, the die is not necessarily defective because two 6s is not very common.
(4) no, the die is not necessarily defective because four 6s is fairly common.

10. A new plant food is developed that the inventors claim helps plants grow faster. They cite an experiment they did where out of 40 randomly chosen plants, 20 of them were randomly chosen to get the new plant food while the other 20 plants got a different kind of plant food.

The 20 plants that got the new plant food grew an average of 2 inches more than the plants that got a different plant food. To test if this difference is statistically significant, the data for the 40 plants are entered into a computer which, then randomly separates them into two groups of 20. This is done 50 times, and the results are shown in the histogram below where the number is the difference between the first group and the second group of the randomly chosen groupings.

Based on this data, can you conclude that the new plant food causes plants to grow faster?

(1) No, the difference is not statistically significant. +2 is uncommon.
(2) No, the difference is not statistically significant. +2 is fairly common.
(3) Yes, the difference is statistically significant. +2 is uncommon.
(4) Yes, the difference is statistically significant. +2 is fairly common.

Inferential Statistics 165

Solutions

1. Bias in a survey happens when people more likely to answer one way are more frequently surveyed than people more likely to answer the other way. To make sure this doesn't happen in the movie survey in New York City, be sure to not find people for the survey who live in the same neighborhood or are about the same age.

   The best way to reduce bias is to take a list of all the possible subjects for the survey and have a computer randomly select people for the survey from the list.

2. For an observational survey, you would observe who gets what amount of sleep and what their Regents score is. You would not, in any way, control who gets what amount of sleep.

   For an experiment, you would get a random group of people and then randomly separate those people into groups. For one group, you would tell them to get more than 10 hours of sleep. For another group, you would tell them to get between 7 and 10 hours of sleep, and for a third group you would tell them to get less than 7 hours of sleep. Later, you would record what scores the people from the different groups got on the test and look at that data for patterns or trends.

3. The best way to reduce bias is to use a computer with a random number generator to pick who uses which book. You do not want to let the participants select which group they are in since that could distort the results.

4. The wording of a question could introduce bias. The word "boring" has connotations that could influence the person taking the survey to answer differently than they might with a more neutrally worded question.

166   **Some Key Algebra II Facts and Skills**

5. According to the histogram, the percent of cod in the lake is somewhere between approximately 10% and 40% since anything smaller than 10% or bigger than 40% does not have a bar on it. Since they give the mean and the standard deviation, you can approximate the true percent by calculating mean − 2 · standard deviation and mean + 2 · standard deviation. For this example, that is

$$0.24 - 2 \cdot 0.06 = 0.12 \text{ and } 0.24 + 2 \cdot 0.06 = 0.36$$

The correct choice is **(1)**.

6. According to the histogram, the percent of people having a land line is somewhere between approximately 55% and 85% since anything smaller than 55% or bigger than 85% does not have a bar on it. Since they give the mean and the standard deviation, you can approximate the true percent by calculating mean − 2 · standard deviation and mean + 2 · standard deviation. For this example, that is

$$0.73 - 2 \cdot 0.05 = 0.63 \text{ and } 0.73 + 2 \cdot 0.05 = 0.83$$

The correct choice is **(1)**.

7. According to the histogram, the amount of time people can hold their breath is somewhere between 70 and 76 seconds since anything less than 70 or greater than 76 does not have a bar on it. Since they give the mean and the standard deviation, you can approximate the true percent by calculating mean − 2 · standard deviation and mean + 2 · standard deviation. For this example, that is

$$72.7 - 2 \cdot 0.92 = 70.86 \text{ and } 72.7 + 2 \cdot 0.92 = 74.54$$

The correct choice is **(4)**.

Inferential Statistics   167

8. According to the histogram, the average age of the people at the concert is somewhere between 22 and 30 since anything less than 22 or greater than 30 does not have a bar on it. Since they give the mean and the standard deviation, you can approximate the true percent by calculating mean $-2 \cdot$ standard deviation and mean $+2 \cdot$ standard deviation. For this example, that is

$$25.5 - 2 \cdot 1.5 = 22.5 \text{ and } 22.5 + 2 \cdot 1.5 = 28.5$$

The correct choice is (2).

9. Based on the histogram, it is most likely to get one or two 6s when a fair die is rolled 12 times. However, notice that there is a bar for four 6s and that that bar is more on the medium to small side, but would not be considered "tiny." Because of this, it can't be concluded that the die is defective.

The correct choice is (4).

10. According to the histogram, randomly separating all the plants into two groups usually makes it so there is 0 difference between the two groups. However, notice that there is a visible bar over the 2 that is not relatively "tiny" compared to the others. This means that a difference of 2 is fairly common so we would say that the difference is not statistically significant.

The correct choice is (2).

Glossary of Terms

Algebraic identity Two expressions usually involving polynomials that can be shown to be equal to each another.

Amplitude The distance between the middle of sine or cosine curve to the top of that sine or cosine curve. When the equation is $y = a \sin(bx) + d$ or $y = a \cos(bx) + d$, the amplitude is a.

Arithmetic sequence A sequence of numbers like 3, 7, 11, 15, . . . , where each number is the same amount more (or less) than the previous number. That difference is called d. In this case, $d = 4$.

Arithmetic series The sum of the terms in an arithmetic sequence. $3 + 7 + 11 + 15$ is an arithmetic series.

Asymptote A line that a curve gets closer and closer to without ever reaching.

Base In an exponential expression, the base is the number being raised to a power. In the expression $5 \cdot 2^x$, 2 is the base. In a logarithmic expression, the base is the small subscript number after the word "log." In the expression $\log_3 81 = 4$, the 3 is the base.

Binomial A polynomial with two terms. The expression $2x + 5$ is a binomial.

170　　　　　　　　**Glossary of Terms**

Coefficient The number that gets multiplied by a variable in a mathematical expresssion. In the expression $5x^2$, the coefficient of the x^2 term is 5.

Complex number A number of the form $a + bi$ where a and b are real numbers. $5 + 3i$ is a complex number.

Composite function When the expression that defines one function is put into another function. If $f(x) = 3x$ and $g(x) = x^2$, then $g(f(x)) = (3x)^2$ is a composite function.

Compound interest When money in a bank gets interest on an initial deposit and then later gets more interest based on the original deposit and on the interest already earned.

Constant A number; the term in a polynomial that has no variable part. In the polynomial $2x + 3$, the constant is 3.

Cosine The ratio between the adjacent side and hypotenuse of a right triangle. Also the x-coordinate of a point on the unit circle.

Cubic equation An equation of the form $ax^3 + bx^2 + cx + d = 0$.

Cubic function A function of the form $f(x) = ax^3 + bx^2 + cx + d$.

Degree The highest exponent in a polynomial. In $x^3 + 5x^2 + 7x + 2$ the degree is 3.

Degree A unit of angle measurement that is $\dfrac{1}{360}$ of a circle.

Dependent events When the probability of one event is affected by whether or not some other event happened, they are called dependent events. The event "The baseball game will be canceled" and the event "It will rain" are dependent events.

Difference of perfect squares A way of factoring a polynomial like $x^2 - 3^2$ into $(x - 3)(x + 3)$. In general, an expression of the form $a^2 - b^2$ can be factored into $(a - b)(a + b)$.

Directrix Every point on a parabola is equidistant from a point and a line. That line is called the directrix.

e A mathematical constant that is involved in many questions related to modeling growth. It is approximately 2.72.

Elimination method Multiplying one or more equations in a system of equations by a constant and then adding the equations

Glossary of Terms

in a way that makes one of the variables "cancel out." The system $x + 2y = 5$ and $3x - 2y = 7$ becomes $4x = 12$ when the two equations are added together.

Even function A function whose graph is symmetric with respect to the y-axis.

Experimental study When a researcher has one group of subjects do one thing and another group do another thing and studies how the two groups react. An example is giving half of the subjects one type of medicine and the rest of them another type of medicine.

Explicit sequence formula A way of describing a sequence based on relating the position of a term to the number in that position. $a_n = 5n + 8$ is an explicit sequence formula.

Exponential equation An equation that involves an exponential expression. $3 \cdot 2^x = 96$ is an exponential equation.

Exponential expression An expression that has a variable as an exponent. $3 \cdot 2^x$ is an exponential expression.

Exponential function A function that has a variable as an exponent. $f(x) = 3 \cdot 2^x$ is an exponential function.

Factor by grouping A way of factoring that involves separating a polynomial into two or more groups, finding a common factor in each group, and, if possible, factoring a common factor out of what remains. The expression

$$x^3 + 3x^2 + 2x + 6 = x^2(x + 3) + 2(x + 3) = (x + 3)(x^2 + 2)$$

is an example of factoring by grouping.

Factoring Finding two things whose product is a given thing. $15 = 3 \cdot 5$ is a way of factoring 15. $x^2 - 5x + 6 = (x - 2)(x - 3)$ is a way of factoring $x^2 - 5x + 6$.

Factors When a number or polynomial is divided by a number or polynomial and there is no remainder, the divisor is a factor of the dividend. Since $\dfrac{15}{3} = 5$ remainder 0, 3 is a factor of 15.

Factor theorem A theorem that states that if the remainder when you divide $P(x)$ by $x - a$ is 0, then $x - a$ is a factor of $P(x)$.

172   **Glossary of Terms**

For example, if you divide $P(x) = x^2 + 5x + 6$ by $x + 2$, you get $x + 3$ remainder 0 so $x + 2$ is a factor of $P(x)$.

Finite arithmetic series formula A formula $S_n = \dfrac{(a_1 + a_L)n}{2}$ for summing the terms of a finite arithmetic series.

Finite geometric series formula A formula $S_n = \dfrac{a_1(1 - r^n)}{1 - r}$ for summing the terms of a finite geometric series.

Focus Every point on a parabola is equidistant from a point and a line. That point is called the focus of the parabola.

FOIL A way to remember the four multiplications when multiplying one binomial by another binomial. When multiplying $(2x + 3)(4x + 5)$, F (for firsts) represents $2x \cdot 4x$, O (for outers) represents $2x \cdot 5$, I (for inners) represents $3 \cdot 4x$, and L (for lasts) represents $3 \cdot 5$.

Frequency The number of sine or cosine curves that can fit into a 360-degree interval.

Function A rule that takes a number or expression as an input and converts it into a number or expression in a predicable way. The function $f(x) = x^2$ takes a number or expression and turns it into the number or expression squared, for example, $f(3) = 3^2 = 9$.

Geometric sequence A sequence of numbers like 3, 6, 12, 24, . . . , where each number is equal to the previous number in the sequence multiplied by the same number. That number, the common ratio, is called r. In this case, $r = 2$.

Geometric series The sum of the terms in a geometric sequence. $3 + 6 + 12 + 24$ is a geometric series.

Given When calculating the probability of B given A, the "given" means that the event A has already happened, which may or may not affect the probability that B will also happen.

Greatest common factor Of all the common factors between two numbers or polynomials, the one that is the largest. For example, the numbers 12 and 18 have common factors 1, 2, 3, and 6 so 6 is the greatest common factor between 12 and 18.

Glossary of Terms 173

Growth rate In an exponential expression $a(1 + r)^x$, the r is the growth rate. For example, in $100(1 + 0.07)^x$, the growth rate is 0.07.

Horizontal shift When all the points of a graph move right or left the same number of units.

Horizontal shrink When the x-coordinates of every point in a graph are divided by the same number.

i A shorthand way of writing $\sqrt{-1}$.

Imaginary number A number that is a multiple of the imaginary unit i. $5i$ is an imaginary number.

Independent events When the probability of one event is not affected by whether or not some other event happened. The event "I wake up at 7:00 A.M." and "It will rain today" are independent events.

Inverse function If a function $f(x)$ has an inverse, $f^{-1}(x)$, then $f^{-1}(f(x)) = x$. For example, if $f(x) = x + 1$, $f^{-1}(x) = x - 1$. An inverse function "undoes" whatever the original function did to it.

Like terms When two monomials with the same variable part are added or subtracted. In the sum $2x^2 + 5x + 3x^2 + 7$, the $2x^2$ and the $3x^2$ are like terms and can be combined to $5x^2$.

Linear function A function whose largest exponent is 1. $f(x) = 3x + 7$ is a linear function.

ln x A shorthand for writing $\log_e x$.

$\log_b x$ The number to which b would have to be raised to get a result of x. For example, $\log_5 125 = 3$ since $5^3 = 125$.

Logarithm An exponent that tells what a base needs to be raised to in order to become another number. The logarithm of 25 base 5 or $\log_5 25 = 2$ since $5^2 = 25$.

Monomial A polynomial that has just one term. The polynomial $3x^5$ is a monomial.

Glossary of Terms

Normal distribution When a set of data points resembles a bell curve with 68% being within one standard deviation and 95% being within two standard deviations.

Observational study When a statistical researcher collects data by studying the behavior of the subjects without interfering with their environment.

Odd function A function whose graph is symmetric with respect to the origin.

Perfect square trinomial A polynomial of the form $x^2 + 2ax + a^2$ or $x^2 - 2ax + a^2$ that can be factored into $(x + a)^2$ or $(x - a)^2$. For example, $x^2 + 6x + 9$ can be factored into $(x + 3)^2$.

Period The number of degrees (or radians) in one cycle of a sine or cosine curve.

Polynomial An expression like $2x^3 + 5x^2 - 3x + 7$ that consists of one or more terms, each term having a coefficient and a variable raised to a power.

Population mean The average of an entire population for some characteristic. For example, if out of 10,000 people studied, their average height is 64 inches, 64 is the population mean.

Population proportion The percent of an entire population that has some characteristic. For example, if out of 10,000 people studied, 52% are boys, 0.52 is the population proportion.

Probability The chance of something happening ranging from 0 (impossible) to 1 (certain).

Quadratic equation An equation of the form $ax^2 + bx + c = 0$.

Quadratic formula The formula $\dfrac{-b \pm \sqrt{b^2 - 4ac}}{2a}$ used to find the solutions to the equation $ax^2 + bx + c = 0$.

Quadratic function A function whose largest exponent is 2. $f(x) = x^2 - 3x + 7$ is a quadratic function.

Radian A unit of angle measurement that is equal to approximately 57 degrees.

Glossary of Terms 175

Radical equation An equation involving a radical expression. $\sqrt{x+2} = 5$ is a radical equation.

Radical expression An expression involving a radical symbol. $\sqrt{x+2}$ is a radical expression.

Randomization test When a computer is used to simulate an ideal experiment with the goal of seeing if something that already happened was likely to happen or not.

Rational equation An equation that involves one or more rational expressions. $\dfrac{3}{x-2} + \dfrac{4}{x-2} = 5$ is a rational equation.

Rational expression An expression that has a fraction with a polynomial of degree at least one in the denominator. $\dfrac{3}{x+1}$ is a rational expression.

Rational function A function that has a fraction with a polynomial of degree at least one in the denominator. $f(x) = \dfrac{3}{x+1}$ is a rational function.

Recursive sequence formula A way of describing a sequence based on how each term relates to the preceding term. $a_1 = 3$, $a_n = 5 + a_{n-1}$ is a recursive sequence formula.

Remainder theorem A theorem that states that when you divide a polynomial $P(x)$ by $x - a$, the remainder will always equal $P(a)$. For example, if you divide the polynomial $P(x) = x^2 + 5x + 6$ by $x - 3$, you will get a remainder of $P(3) = 30$.

Sample mean The average of a sample that has some characteristic. For example, if out of 10,000 people, 100 people are randomly chosen and, of those 100 people, their average height is 61 inches, then 61 is the sample mean for that sample.

Sample proportion The percent of a sample that has some characteristic. For example, if out of 10,000 people, 100 people are randomly chosen and, of those 100 people, 38% are boys, then 0.38 is the sample proportion.

Sample space A list of all the possible outcomes of a probability experiment.

Glossary of Terms

Sampling distribution of the sample mean When multiple groups of samples are taken from a larger population and the average of some attribute of each sample group is calculated and graphed in a histogram.

Sampling distribution of the sample proportion When multiple groups of samples are taken from a larger population and the percent of some attribute of each sample group is calculated and graphed in a histogram.

Simplified radical form When the number inside a square root (or cube root) sign has no factors that are perfect squares (or perfect cubes). The expression $\sqrt{18}$ is not in simplified radical form since $18 = 2 \cdot 9$. $3\sqrt{2}$, which is equal to $\sqrt{18}$, is in simplest radical form.

Sine The ratio between the opposite side and hypotenuse of a right triangle. Also the y-coordinate of a point on the unit circle.

Standard deviation A way of measuring how "spread out" a data set is. The smallest standard deviation is 0 when all the elements of a set are equal.

Standard form of a quadratic expression An expression of the form $y = ax^2 + bx + c$.

Statistical significance When data from an experiment are randomly mixed together multiple times with the goal of seeing if a difference between two groups was likely to happen.

Substitution method In a system of equations, when one of the variables is solved in terms of the others and then substituted into the other equation. For example, the system of equations $y = 2x + 1$, $3x + 2y = 10$, the y in the second equation can be replaced by $2x + 1$ to become $3x + 2(2x + 1) = 10$.

Survey A question or set of questions answered by a subject in a statistical study.

System of equations Two or more equations that have two or more variables. A system of equations generally has one solution, which is an ordered pair (or triple) with numbers that satisfy each of the equations.

Glossary of Terms 177

Tangent The ratio between the opposite side and the adjacent side of a right triangle.

Trigonometric function A function that involves a trigonometry ratio, like sine, cosine, or tangent. $f(x) = 3 \sin(x) + 1$ is a trigonometric function.

Trigonometric identity Two expressions involving trigonometric ratios that can be shown to be equal to one another.

Trinomial A polynomial with three terms. The expression $x^2 + 5x + 6$ is a trinomial.

Two-way frequency table A chart used for listing the different ways that things can have or not have certain attributes.

Unit circle A circle with a radius of 1 and center at $(0, 0)$.

Variable A letter that is used to represent a number. Often this letter represents the thing to be solved for in an equation. In the expression $2x + 5$, the variable is x.

Venn diagram A way of organizing data that have two or more intersecting circles surrounded by a rectangle.

Vertex form of a quadratic expression An expression of the form $y = a(x - h)^2 + k$. When graphed, it is a parabola with vertex (h, k).

Vertical shift When all the points of a graph move up or down the same number of units.

Vertical stretch When the y-coordinates of every point in a graph are multiplied by the same number.

z-score A measure of how many standard deviations above (positive *z-score*) or below (negative *z-score*) a number is from the mean.

Regents Examinations, Answers, and Self-Analysis Charts

181

Examination
June 2016
Algebra II

HIGH SCHOOL MATH REFERENCE SHEET

Conversions

1 inch = 2.54 centimeters
1 meter = 39.37 inches
1 mile = 5280 feet
1 mile = 1760 yards
1 mile = 1.609 kilometers

1 kilometer = 0.62 mile
1 pound = 16 ounces
1 pound = 0.454 kilogram
1 kilogram = 2.2 pounds
1 ton = 2000 pounds

1 cup = 8 fluid ounces
1 pint = 2 cups
1 quart = 2 pints
1 gallon = 4 quarts
1 gallon = 3.785 liters
1 liter = 0.264 gallon
1 liter = 1000 cubic centimeters

Formulas

Triangle $A = \dfrac{1}{2}bh$

Parallelogram $A = bh$

Circle $A = \pi r^2$

Circle $C = \pi d$ or $C = 2\pi r$

EXAMINATION June 2016

Formulas (continued)

General Prisms	$V = Bh$
Cylinder	$V = \pi r^2 h$
Sphere	$V = \dfrac{4}{3}\pi r^3$
Cone	$V = \dfrac{1}{3}\pi r^2 h$
Pyramid	$V = \dfrac{1}{3}Bh$
Pythagorean Theorem	$a^2 + b^2 = c^2$
Quadratic Formula	$x = \dfrac{-b \pm \sqrt{b^2 - 4ac}}{2a}$
Arithmetic Sequence	$a_n = a_1 + (n-1)d$
Geometric Sequence	$a_n = a_1 r^{n-1}$
Geometric Series	$S_n = \dfrac{a_1 - a_1 r^n}{1 - r}$ where $r \neq 1$
Radians	$1 \text{ radian} = \dfrac{180}{\pi} \text{ degrees}$
Degrees	$1 \text{ degree} = \dfrac{\pi}{180} \text{ radians}$
Exponential Growth/Decay	$A = A_0 e^{k(t - t_0)} + B_0$

EXAMINATION June 2016   183

PART I

Answer all 24 questions in this part. Each correct answer will receive 2 credits. No partial credit will be allowed. For each statement or question, write in the space provided the numeral preceding the word or expression that best completes the statement or answers the question.   [48 credits]

1 When $b > 0$ and d is a positive integer, the expression $(3b)^{\frac{2}{d}}$ is equivalent to

(1) $\dfrac{1}{\left(\sqrt[d]{3b}\right)^2}$

(2) $\left(\sqrt{3b}\right)^d$

(3) $\dfrac{1}{\sqrt{3b^d}}$

(4) $\left(\sqrt[d]{3b}\right)^2$

1 _____

2 Julie averaged 85 on the first three tests of the semester in her mathematics class. If she scores 93 on each of the remaining tests, her average will be 90. Which equation could be used to determine how many tests, T, are left in the semester?

(1) $\dfrac{255+93T}{3T} = 90$

(2) $\dfrac{255+90T}{3T} = 93$

(3) $\dfrac{255+93T}{T+3} = 90$

(4) $\dfrac{255+90T}{T+3} = 93$

2 _____

3 Given i is the imaginary unit, $(2 - yi)^2$ in simplest form is

(1) $y^2 - 4yi + 4$

(2) $-y^2 - 4yi + 4$

(3) $-y^2 + 4$

(4) $y^2 + 4$

3 _____

4 Which graph has the following characteristics?
- three real zeros
- as $x \to -\infty$, $f(x) \to -\infty$
- as $x \to \infty$, $f(x) \to \infty$

(1)

(3)

(2)

(4)

4 _____

5 The solution set for the equation is $\sqrt{56-x} = x$ is
(1) {−8, 7}   (3) {7}
(2) {−7, 8}   (4) { }

5 _____

6 The zeros for $f(x) = x^4 - 4x^3 - 9x^2 + 36x$ are
(1) {0, ±3, 4}   (3) {0, ±3, −4}
(2) {0, 3, 4}   (4) {0, 3, −4}

6 _____

7 Anne has a coin. She does not know if it is a fair coin. She flipped the coin 100 times and obtained 73 heads and 27 tails. She ran a computer simulation of 200 samples of 100 fair coin flips. The output of the proportion of heads is shown below.

Given the results of her coin flips and of her computer simulation, which statement is most accurate?

(1) 73 of the computer's next 100 coin flips will be heads.
(2) 50 of her next 100 coin flips will be heads.
(3) Her coin is not fair.
(4) Her coin is fair.

7 _____

8 If $g(c) = 1 - c^2$ and $m(c) = c + 1$, then which statement is *not* true?

(1) $g(c) \cdot m(c) = 1 + c - c^2 - c^3$
(2) $g(c) + m(c) = 2 + c - c^2$
(3) $m(c) - g(c) = c + c^2$
(4) $\dfrac{m(c)}{g(c)} = \dfrac{-1}{1-c}$

8 _____

186   **EXAMINATION June 2016**

9 The heights of women in the United States are normally distributed with a mean of 64 inches and a standard deviation of 2.75 inches. The percent of women whose heights are between 64 and 69.5 inches, to the *nearest whole percent*, is

(1) 6   (3) 68
(2) 48   (4) 95   9 _____

10 The formula below can be used to model which scenario?

$$a_1 = 3000$$
$$a_n = 0.80a_{n-1}$$

(1) The first row of a stadium has 3000 seats, and each row thereafter has 80 more seats than the row in front of it.
(2) The last row of a stadium has 3000 seats, and each row before it has 80 fewer seats than the row behind it.
(3) A bank account starts with a deposit of $3000, and each year it grows by 80%.
(4) The initial value of a specialty toy is $3000, and its value each of the following years is 20% less.   10 _____

11 Sean's team has a baseball game tomorrow. He pitches 50% of the games. There is a 40% chance of rain during the game tomorrow. If the probability that it rains given that Sean pitches is 40%, it can be concluded that these two events are

(1) independent   (3) mutually exclusive
(2) dependent   (4) complements   11 _____

EXAMINATION June 2016 — 187

12 A solution of the equation $2x^2 + 3x + 2 = 0$ is

(1) $-\dfrac{3}{4} + \dfrac{1}{4} i\sqrt{7}$

(3) $-\dfrac{3}{4} + \dfrac{1}{4}\sqrt{7}$

(2) $-\dfrac{3}{4} + \dfrac{7}{4} i$

(4) $\dfrac{1}{2}$

12 _____

13 The Ferris wheel at the landmark Navy Pier in Chicago takes 7 minutes to make one full rotation. The height, H, in feet, above the ground of one of the six-person cars can be modeled by

$H(t) = 70 \sin\left(\dfrac{2\pi}{7}(t - 1.75)\right) + 80$, where t is time, in minutes. Using $H(t)$ for one full rotation, this car's minimum height, in feet, is

(1) 150

(3) 10

(2) 70

(4) 0

13 _____

14 The expression $\dfrac{4x^3 + 5x + 10}{2x + 3}$ is equivalent to

(1) $2x^2 + 3x - 7 + \dfrac{31}{2x + 3}$

(3) $2x^2 + 2.5x + 5 + \dfrac{15}{2x + 3}$

(2) $2x^2 - 3x + 7 - \dfrac{11}{2x + 3}$

(4) $2x^2 - 2.5x - 5 - \dfrac{20}{2x + 3}$

14 _____

15 Which function represents exponential decay?

(1) $y = 2^{0.3t}$

(3) $y = \left(\dfrac{1}{2}\right)^{-t}$

(2) $y = 1.2^{3t}$

(4) $y = 5^{-t}$

15 _____

188   **EXAMINATION June 2016**

16 Given $f^{-1}(x) = -\dfrac{3}{4}x + 2$, which equation represents $f(x)$?

(1) $f(x) = \dfrac{4}{3}x - \dfrac{8}{3}$ (3) $f(x) = \dfrac{3}{4}x - 2$

(2) $f(x) = -\dfrac{4}{3}x + \dfrac{8}{3}$ (4) $f(x) = -\dfrac{3}{4}x + 2$ 16 _____

17 A circle centered at the origin has a radius of 10 units. The terminal side of an angle, θ, intercepts the circle in quadrant II at point C. The y-coordinate of point C is 8. What is the value of cos θ?

(1) $-\dfrac{3}{5}$ (3) $\dfrac{3}{5}$

(2) $-\dfrac{3}{4}$ (4) $\dfrac{4}{5}$ 17 _____

18 Which statement about the graph of $c(x) = \log_6 x$ is *false*?

(1) The asymptote has equation $y = 0$.
(2) The graph has no y-intercept.
(3) The domain is the set of positive reals.
(2) The range is the set of all real numbers.   18 _____

19 The equation $4x^2 - 24x + 4y^2 + 72y = 76$ is equivalent to

(1) $4(x - 3)^2 + 4(y + 9)^2 = 76$
(2) $4(x - 3)^2 + 4(y + 9)^2 = 121$
(3) $4(x - 3)^2 + 4(y + 9)^2 = 166$
(2) $4(x - 3)^2 + 4(y + 9)^2 = 436$ 19 _____

20 There was a study done on oxygen consumption of snails as a function of pH, and the result was a degree 4 polynomial function whose graph is shown below.

Which statement about this function is *incorrect*?

(1) The degree of the polynomial is even.
(2) There is a positive leading coefficient.
(3) At two pH values, there is a relative maximum value.
(4) There are two intervals where the function is decreasing.

20 _____

21 Last year, the total revenue for Home Style, a national restaurant chain, increased 5.25% over the previous year. If this trend were to continue, which expression could the company's chief financial officer use to approximate their monthly percent increase in revenue? [Let m represent months.]

(1) $(1.0525)^m$

(3) $(1.00427)^m$

(2) $(1.0525)^{\frac{12}{m}}$

(4) $(1.00427)^{\frac{m}{12}}$

21 _____

190   **EXAMINATION June 2016**

22 Which value, to the *nearest tenth*, is *not* a solution of
$p(x) = q(x)$ if $p(x) = x^3 + 3x^2 - 3x - 1$ and $q(x) = 3x + 8$?

(1) -3.9 (3) 2.1

(2) -1.1 (4) 4.7 22 _____

23 The population of Jamesburg for the years
2010–2013, respectively, was reported as follows:

250,000   250,937   251,878   252,822

How can this sequence be recursively modeled?

(1) $j_n = 250{,}000(1.00375)^{n-1}$

(2) $j_n = 250{,}000 + 937^{(n-1)}$

(3) $j_1 = 250{,}000$
    $j_n = 1.00375j_{n-1}$

(4) $j_1 = 250{,}000$
    $j_n = j_{n-1} + 937$ 23 _____

24 The voltage used by most households can be modeled
by a sine function. The maximum voltage is 120 volts,
and there are 60 cycles *every second*. Which equation best represents the value of the voltage as it flows
through the electric wires, where t is time in seconds?

(1) $V = 120 \sin (t)$ (3) $V = 120 \sin (60\pi t)$

(2) $V = 120 \sin (60t)$ (4) $V = 120 \sin (120\pi t)$ 24 _____

EXAMINATION June 2016   191

PART II

Answer all 8 questions in this part. Each correct answer will receive 2 credits. Clearly indicate the necessary steps, including appropriate formula substitutions, diagrams, graphs, charts, etc. For all questions in this part, a correct numerical answer with no work shown will receive only 1 credit.   [16 credits]

25 Solve for x: $\dfrac{1}{x} - \dfrac{1}{3} = -\dfrac{1}{3x}$

EXAMINATION June 2016

26 Describe how a controlled experiment can be created to examine the effect of ingredient X in a toothpaste.

27 Determine if $x - 5$ is a factor of $2x^3 - 4x^2 - 7x - 10$. Explain your answer.

28 On the axes below, graph *one* cycle of a cosine function with amplitude 3, period $\frac{\pi}{2}$, midline $y = -1$, and passing through the point (0, 2).

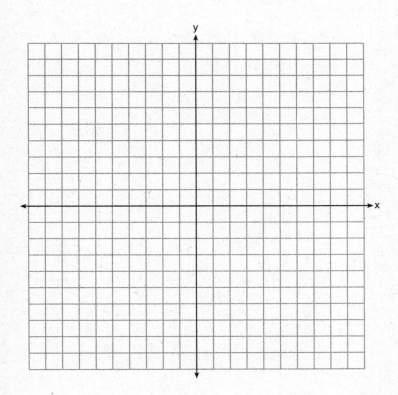

194   **EXAMINATION June 2016**

29 A suburban high school has a population of 1376 students. The number of students who participate in sports is 649. The number of students who participate in music is 433. If the probability that a student participates in either sports or music is $\frac{974}{1376}$, what is the probability that a student participates in both sports and music?

EXAMINATION June 2016   195

30 The directrix of the parabola $12(y + 3) = (x - 4)^2$ has the equation $y = -6$. Find the coordinates of the focus of the parabola.

31 Algebraically prove that $\dfrac{x^3 + 9}{x^3 + 8} = 1 + \dfrac{1}{x^3 + 8}$, where $x \neq -2$.

196   **EXAMINATION June 2016**

32 A house purchased 5 years ago for $100,000 was just sold for $135,000. Assuming exponential growth, approximate the annual growth rate, to the *nearest percent*.

EXAMINATION June 2016   **197**

PART III

Answer all 4 questions in this part. Each correct answer will receive 4 credits. Clearly indicate the necessary steps, including appropriate formula substitutions, diagrams, graphs, charts, etc. For all questions in this part, a correct numerical answer with no work shown will receive only 1 credit.   [16 credits]

33 Solve the system of equations shown below algebraically.

$$(x - 3)^2 + (y + 2)^2 = 16$$
$$2x + 2y = 10$$

198   **EXAMINATION June 2016**

34 Alexa earns $33,000 in her first year of teaching and earns a 4% increase in each successive year.

Write a geometric series formula, S_n, for Alexa's total earnings over n years.

Use this formula to find Alexa's total earnings for her first 15 years of teaching, to the *nearest cent*.

35 Fifty-five students attending the prom were randomly selected to participate in a survey about the music choice at the prom. Sixty percent responded that a DJ would be preferred over a band. Members of the prom committee thought that the vote would have 50% for the DJ and 50% for the band.

A simulation was run 200 times, each of sample size 55, based on the premise that 60% of the students would prefer a DJ. The approximate normal simulation results are shown below.

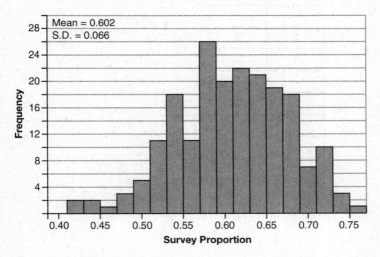

Using the results of the simulation, determine a plausible interval containing the middle 95% of the data. Round all values to the *nearest hundredth*.

Members of the prom committee are concerned that a vote of all students attending the prom may produce a 50%–50% split. Explain what statistical evidence supports this concern.

EXAMINATION June 2016

36 Which function shown below has a greater average rate of change on the interval [−2, 4]? Justify your answer.

x	f(x)
−4	0.3125
−3	0.625
−2	1.25
−1	2.5
0	5
1	10
2	20
3	40
4	80
5	160
6	320

$g(x) = 4x^3 - 5x^2 + 3$

EXAMINATION June 2016

201

PART IV

Answer the question in this part. A correct answer will receive 6 credits. Clearly indicate the necessary steps, including appropriate formula substitutions, diagrams, graphs, charts, etc. A correct numerical answer with no work shown will receive only 1 credit. [6 credits]

37 Drugs break down in the human body at different rates and therefore must be prescribed by doctors carefully to prevent complications, such as overdosing. The breakdown of a drug is represented by the function $N(t) = N_0(e)^{-rt}$, where $N(t)$ is the amount left in the body, N_0 is the initial dosage, r is the decay rate, and t is time in hours. Patient A, $A(t)$, is given 800 milligrams of a drug with a decay rate of 0.347. Patient B, $B(t)$, is given 400 milligrams of another drug with a decay rate of 0.231.

Write two functions, $A(t)$ and $B(t)$, to represent the breakdown of the respective drug given to each patient.

Question 37 is continued on the next page.

Question 37 continued.

Graph each function on the set of axes below.

To the *nearest hour*, t, when does the amount of the given drug remaining in patient B begin to exceed the amount of the given drug remaining in patient A?

The doctor will allow patient A to take another 800 milligram dose of the drug once only 15% of the original dose is left in the body. Determine, to the *nearest tenth of an hour*, how long patient A will have to wait to take another 800 milligram dose of the drug.

Answers
June 2016
Algebra II

Answer Key

PART I

1. (4)
2. (3)
3. (2)
4. (3)
5. (3)
6. (1)
7. (3)
8. (4)
9. (2)
10. (4)
11. (1)
12. (1)
13. (3)
14. (2)
15. (4)
16. (2)
17. (1)
18. (1)
19. (4)
20. (2)
21. (3)
22. (4)
23. (3)
24. (4)

PART II

25. 4
26. The experimenter must randomly split subjects into a control group and an experimental group.
27. No
28.

29. $\dfrac{108}{1376}$

30. (4, 0)
31. Yes
32. 6%

PART III

33. (3, 2) and (7, −2)
34. $S_n = \dfrac{33{,}000 - 33{,}000 \cdot 1.04^n}{1 - 1.04}$, 660,778.39
35. Between 0.47 and 0.73
36. g

PART IV

37. $A(t) = 800(e)^{-0.347t}$,
    $B(t) = 400(e)^{-0.231t}$, 6, 5.5

In **Parts II–IV**, you are required to show how you arrived at your answers. For sample methods of solutions, see the *Answers Explained* section.

204   **ANSWERS June 2016**

Answers Explained

PART I

1. An expression of the form $x^{\frac{a}{b}}$ involving a fractional exponent is equivalent to the expression $\left(\sqrt[b]{x}\right)^a$. So $(3b)^{\frac{2}{d}} = \left(\sqrt[d]{3b}\right)^2$.

   The correct choice is **(4)**.

2. Use the formula that relates the average, the total points, and the number of tests:

   $$\frac{\text{total points}}{\text{number of tests}} = \text{average}$$

   Use this formula to determine how many total points Julie had after taking the first three tests:

   $$\frac{\text{total points}}{3} = 85$$
   $$3 \times \frac{\text{total points}}{3} = 3 \times 85$$
   $$= 255$$

   If Julie gets a 93 on T more tests, her total points will increase by $93T$ and the number of tests will increase by T. If the new average is known to be 90, the equation relating the new average, the new total points, and the new number of tests can be determined:

   $$\frac{255 + 93T}{3 + T} = 90$$

   The correct choice is **(3)**.

3. Expand $(2 - yi)^2$ as $(2 - yi)(2 - yi)$ and multiply with FOIL:

   $$(2 - yi)(2 - yi) = 4 - 2yi - 2yi + y^2 i^2$$

   Since $i^2 = -1$, this is equivalent to:

   $$4 - 2yi - 2yi - y^2 = 4 - 4yi - y^2$$
   $$= -y^2 - 4yi + 4$$

   The correct choice is **(2)**.

ANSWERS June 2016

4. Real zeros of a function correspond to x-intercepts of the graph of the function. Each of the four choices have three x-intercepts, so they each have three real zeros. For the other two characteristics, the graph must decrease as x gets more and more negative and must increase as x gets more and more positive.

The correct choice is **(3)**.

5. Since the radical is already isolated, square both sides and solve the resulting quadratic equation:

$$\sqrt{56-x} = x^2$$
$$\left(\sqrt{56-x}\right)^2 = x^2$$
$$56 - x = x^2$$
$$0 = x^2 + x - 56$$
$$0 = (x + 8)(x - 7)$$
$$x + 8 = 0 \quad \text{or} \quad x - 7 = 0$$
$$x = -8 \text{ or} \qquad x = 7$$

When squaring both sides of a radical equation, there is a chance that an extraneous solution is introduced. So you must check both answer choices:

Check $x = 7$:

$$\sqrt{56-7} = 7$$
$$\sqrt{49} = 7$$
$$7 = 7 \quad ✔$$

Check $x = -8$:

$$\sqrt{56-(-8)} = -8$$
$$\sqrt{64} = -8$$
$$8 = -8 \quad \text{NO}$$

The -8 must be rejected.

The correct choice is **(3)**.

6. The zeros of a function f are values of x that make $f(x) = 0$. The zeros of this function are the solutions to the equation $0 = x^4 - 4x^3 - 9x^2 + 36x$. There are three ways to find the zeros of this function.

Method 1: Factoring

$0 = x^4 - 4x^3 - 9x^2 + 36x$

$0 = x(x^3 - 4x^2 - 9x + 36)$ Factor out the greatest common factor

$0 = x(x^2(x - 4) - 9(x - 4))$ Factor by grouping
$0 = (x - 4)(x^2 - 9)$

$0 = x(x - 4)(x - 3)(x + 3)$ Difference of perfect square factoring

$x = 0$ or $x - 4 = 0$ or $x - 3 = 0$ or $x + 3 = 0$
$x = 0$ or $x = 4$ or $x = 3$ or $x = -3$

Method 2: Graphing Calculator

Find the x-intercepts of the graph of $y = x^4 - 4x^3 - 9x^2 + 36x$. If the x-intercepts are integers, as they are for this graph, they can be easily seen on the graph.

For the TI-84:

For the TI-Nspire:

ANSWERS June 2016

Method 3: Test the Answer Choices

Find the values of $f(0)$, $f(-3)$, $f(3)$, $f(-4)$, and $f(4)$ to see which of them equal 0.

$$f(0) = (0)^4 - 4(0)^3 - 9(0)^2 + 36(0)$$
$$= 0 \quad ✔$$

$$f(-3) = (-3)^4 - 4(-3)^3 - 9(-3)^2 + 36(-3)$$
$$= 0 \quad ✔$$

$$f(3) = (3)^4 - 4(3)^3 - 9(3)^2 + 36(3)$$
$$= 0 \quad ✔$$

$$f(-4) = (-4)^4 - 4(-4)^3 - 9(-4)^2 + 36(-4)$$
$$= 224 \quad \textbf{NO}$$

$$f(4) = (4)^4 - 4(4)^3 - 9(4)^2 + 36(4)$$
$$= 0 \quad ✔$$

The correct choice is **(1)**.

7. In 200 simulations there were never any cases where more than 65 heads happened. This suggests that it is very unlikely for a fair coin to land on heads 73 out of 100 flips.

The numbers from the simulations indicate that it is very unlikely for an event to happen that is more than 2 standard deviations (SD) from the mean or less than 2 standard deviations from the mean. In this case, mean + 2SD = 0.597 and mean − 2SD = 0.397. This means that it is very unlikely for there to be more than 60 heads or less than 40 heads if a fair coin is tossed 100 times. Since 73 is more than 60 heads, it is very likely that the coin Anne has is not fair.

The correct choice is **(3)**.

ANSWERS June 2016

8. Test the four answer choices:

$$g(c) \cdot m(c) = (1 - c^2) \cdot (c + 1)$$
$$= c + 1 - c^3 - c^2$$
$$= 1 + c - c^2 - c^3 \quad \checkmark$$

$$g(c) + m(c) = (1 - c^2) + (c + 1)$$
$$= 1 - c^2 + c + 1$$
$$= 2 + c - c^2 \quad \checkmark$$

$$m(c) - g(c) = (c + 1) - (1 - c^2)$$
$$= c + 1 - 1 + c^2$$
$$= c + c^2 \quad \checkmark$$

$$\frac{m(c)}{g(c)} = \frac{c + 1}{1 - c^2}$$
$$= \frac{c + 1}{(1 - c)(1 + c)}$$
$$= \frac{(1 + c)}{(1 - c)(1 + c)}$$
$$= \frac{1}{1 - c}$$
$$\neq \frac{-1}{1 - c} \qquad \text{NO}$$

The correct choice is **(4)**.

9. The normal cdf function of the graphing calculator is used for this type of question.

For the TI-84:

Press [2nd] [VARS] [2] to get the normalcdf menu. In the "lower" field, put 64. In the "upper" field, put 69.5. In the "μ" field put the mean, which is 64. In the "σ" field, put the standard deviation, which is 2.75. Select "Paste." Press [ENTER].

 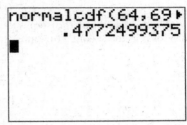

For the TI-Nspire:

Press [A] to open the calculator scratchpad. Press [menu] [6] [5] [2] for the normal cdf menu. In the "Lower Bound" field, put 64. In the "Upper Bound" field put 69.5. In the "μ" field, put the mean, which is 64. In the "σ" field, put the standard deviation, which is 2.75. Select the OK button, and press [enter].

 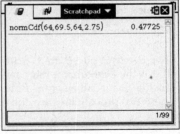

Since 0.47725 rounds to 0.48, 48% of the women have heights between 64 and 69.5 inches.

The correct choice is **(2)**.

210   **ANSWERS June 2016**

10. The first number in the sequence is given as $a_1 = 3,000$. The second term of the sequence, a_2, can be calculated by substituting 2 for n in the second equation:

$$a_2 = 0.80a_{2-1}$$
$$= 0.80a_1$$
$$= 0.80 \cdot 3,000$$
$$= 2,400$$

The third term of the sequence, a_3, can be calculated by substituting 3 for n in the second equation:

$$a_3 = 0.80a_{3-1}$$
$$= 0.80a_2$$
$$= 0.80 \cdot 2,400$$
$$= 1,920$$

The first five terms of this sequence are 3,000; 2,400; 1,920; 1,536; and 1,228.8. Compare this sequence to the four choices:

Choice (1): 3,000; 3,080; 3,160; 3,240; 3,320

Choice (2): 3,000; 2,920; 2,840; 2,760; 2,680

Choice (3): 3,000; 5,400; 9,720; 17,496; 31,492.80

Choice (4): 3,000; 2,400; 1,920; 1,536; 1,228.80

The correct choice is **(4)**.

11. Two events A and B are considered independent if $P(A$ given $B) = P(A)$. If A is the event that it rains and if B is the event that Sean pitches, based on the given information $P(A) = 0.40$ and $P(A$ given $B) = 0.40$. Since these two probabilities are equal, the events are independent.

The correct choice is **(1)**.

ANSWERS June 2016

12. Use the quadratic formula with $a = 2$, $b = 3$, and $c = 2$:

$$x = \frac{-b \pm \sqrt{b^2 - 4ac}}{2a}$$

$$= \frac{-3 \pm \sqrt{3^2 - 4 \cdot 2 \cdot 2}}{2 \cdot 2}$$

$$= \frac{-3 \pm \sqrt{-7}}{4}$$

$$= \frac{-3 \pm i\sqrt{7}}{4}$$

$$= -\frac{3}{4} \pm \frac{1}{4} i\sqrt{7}$$

The correct choice is **(1)**.

13. Since the minimum value of the sine function is -1, the lowest value of the H function is $H(t) = 70(-1) + 80 = 10$.

Another way to solve this is to graph the function with the graphing calculator and locate the lowest point of the graph. Make sure the calculator is in radian mode.

For the TI-84:

For the TI-Nspire:

The correct choice is **(3)**.

14. Use polynomial long division:

$$\begin{array}{r}
2x^2 - 3x + 7 \\
2x+3 \overline{\smash{)}4x^3 + 0x^2 + 5x + 10} \\
\underline{-(4x^3 + 6x^2)} \\
-6x^2 + 5x \\
\underline{-(-6x^2 - 9x)} \\
14x + 10 \\
\underline{-(14x + 21)} \\
-11
\end{array}$$

This remainder, −11, becomes the numerator of the last term. The divisor, $2x + 3$, becomes the denominator of the last term.

The correct choice is **(2)**.

15. Graph each of the four choices on the graphing calculator to see which graph is always decreasing. Of the four choices, only $y = 5^{-t}$ is always decreasing.

For the TI-84:

For the TI-Nspire:

Another way to do this problem is to know that a function with exponential decay is often in the form $y = a^x$ where $0 < a < 1$. Using the properties of exponents, $y = 5^{-t}$ can be expressed as:

$$y = 5^{-1 \cdot t} = \left(5^{-1}\right)^t = \left(\frac{1}{5}\right)^t$$

This has the form of exponential decay.

The correct choice is **(4)**.

16. The inverse of the function $f^{-1}(x)$ is $f(x)$. To find the inverse of a function, follow these three steps:

Step 1: Write the original equation as $y = -\dfrac{3}{4}x + 2$.

Step 2: Swap the x- and the y-variables.

$$x = -\dfrac{3}{4}y + 2$$

Step 3: Isolate the y in the new equation to find the inverse.

$$x = -\dfrac{3}{4}y + 2$$
$$\underline{-2 = \qquad\qquad -2}$$
$$x - 2 = -\dfrac{3}{4}y$$
$$-\dfrac{4}{3}(x - 2) = -\dfrac{4}{3}\left(-\dfrac{3}{4}y\right)$$
$$-\dfrac{4}{3}x + \dfrac{8}{3} = y$$

This question can also be answered with the graphing calculator. A graph of an inverse function is the graph of the original function reflected over the line $y = x$.

Graph the given inverse function on the same set of axes as each of the choices to see which answer is the reflection of the original equation over the $y = x$-line. The default window for the TI-84 will give a distorted graph, so do [ZOOM] [6] [ZOOM] [5] when graphing.

For the TI-84:   For the TI-Nspire:

The correct choice is **(2)**.

17.

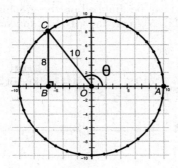

In the diagram, $\angle\theta$ is $\angle AOC$. Triangle OBC is a right triangle with right angle OBC, with $BC = 8$, and with $OC = 10$. Use the sine ratio:

$$\sin \angle BOC = \frac{\text{opposite}}{\text{hypotenuse}} = \frac{8}{10}$$

$$m\angle BOC = \sin^{-1}\left(\frac{8}{10}\right) \approx 53.13°$$

$\angle\theta$ is supplementary to $\angle BOC$, so $m\angle\theta \approx 180° - 53.13° = 126.87°$.
The question asks for the value of $\cos\theta$, and $\cos 126.87° \approx -0.6 = -\frac{3}{5}$.
The correct choice is **(1)**.

18. The graph of $c(x)$ can be graphed on the graphing calculator. For the TI-84, use [MATH] [A] to get the logBASE function. For the TI-Nspire, press [ctrl] [10^x].

For the TI-84:

For the TI-Nspire:

Based on the graph, there is a vertical asymptote at $x = 0$. The y-intercept is $(0, 1)$. The domain is $x > 0$, which is the set of positive reals. The range is the set of all real numbers. There is no horizontal asymptote.

The correct choice is **(1)**.

19. This equation can be converted to an equivalent equation that looks like the answer choices with a technique called "completing the square."

   Step 1: Group the x-terms together, and factor out the common factor, 4.
   $$4(x^2 - 6x) + 4y^2 + 72y = 76$$

   Step 2: Group the y-terms together, and factor out the common factor, 4.
   $$4(x^2 - 6x) + 4(y^2 + 18y) = 76$$

   Step 3: A trinomial of the form $x^2 + bx + c$ is a perfect square trinomial if $c = \left(\dfrac{b}{2}\right)^2$. In order to make each of the expressions inside the parentheses

ANSWERS June 2016   217

into perfect square trinomials, the proper value of c needs to be added inside each set of parentheses. For the first set of parentheses, +9 must be added because $\left(\dfrac{-6}{2}\right)^2 = 9$. For the second set of parentheses, +81 must be added because $\left(\dfrac{18}{2}\right)^2 = 81$. On the right side of the equal sign, you must add $4 \times 9 = 36$ and $4 \times 81 = 324$ to keep both sides equal because of the 4s outside the parentheses.

$$4(x^2 - 6x + 9) + 4(y^2 + 18y + 81) = 76 + 36 + 324$$

Step 4: Factor the expressions inside the parentheses, and simplify.

$$4(x - 3)^2 + 4(y + 9)^2 = 436$$

A way to do this question without using completing the square is to simplify each of the answer choices to see which of them becomes the given equation.

The correct choice is **(4)**.

20. Check each of the answer choices.

Choice (1): When a polynomial has an even degree, like 2 or 4, the end behavior at both ends of the graph will be the same. In this graph, the branches at the right and at the left are both decreasing, which suggests that the degree is even. When a polynomial has an odd degree, the end behavior on the two ends will be opposite. One will be increasing, and the other will be decreasing.

Choice (2): When a polynomial has a positive leading coefficient, the end behavior on the right will be increasing. In this graph, the end behavior on the right is decreasing. So the function does not have a positive leading coefficient.

Choice (3): A relative maximum value happens when there is a hump with a highest point on it. In this graph, there is a relative maximum around pH = 7 and another around pH = 9. So there are two pH values where there is a relative maximum value.

Choice (4): Going from left to right, the graph of a decreasing function will be going down. Like a roller coaster that starts as the leftmost point, it would be climbing until it hits pH = 7 and then go down, decreasing, until it reaches pH = 8. Then it would climb again until it reaches pH = 9 and then be decreasing again from pH = 9 and greater. There are two intervals where the function is decreasing.

The correct choice is **(2)**.

21. Since the revenue increases by 5.25% each year, the yearly growth factor is $1 + 0.0525 = 1.0525$. An equation that could be used to calculate the growth factor for y years is $(1.0525)^y$. This could be used to find the percent increase for any number of years. For example, if $y = 10$, the growth factor would be $(1.0525)^{10} = 1.668 = 1 + 0.668$, which would be a percent increase of 66.8%.

To find an equation that involves m for months, replace y with $\frac{m}{12}$, $(1.0525)^{\frac{m}{12}}$. This can be simplified using the rules of exponents.

$$(1.0525)^{\frac{m}{12}} = (1.0525)^{\frac{1}{12} \cdot m}$$
$$= \left((1.0525)^{\frac{1}{12}}\right)^m$$
$$= (1.00427)^m$$

The correct choice is **(3)**.

22. The x-coordinates of the intersection points of the graphs of $p(x)$ and $q(x)$ are the solutions to the equation $p(x) = q(x)$. Using the graphing calculator, graph both functions on the same set of axes. Then find the intersection points. There are three intersection points: -3.9, -1.1, and 2.1.

For the TI-84:

For the TI-Nspire:

The correct choice is **(4)**.

23. Recursive equations have a base case and a recursive part that uses the results of a previous calculation. Choices **(1)** and **(2)** are not recursive equations.

    Test choices **(3)** and **(4)** to see which will generate the given numbers.

    Choice (3):

    $$j_1 = 250{,}000$$
    $$j_2 = 1.00375 j_{2-1}$$
    $$= 1.00375 j_1$$
    $$= 1.00375 \cdot 250{,}000$$
    $$= 250{,}937.5$$
    $$j_3 = 1.00375 j_{3-1}$$
    $$= 1.00375 j_2$$
    $$= 1.00375 \cdot 250{,}937.5$$
    $$= 251{,}878.5$$
    $$j_4 = 1.00375 j_{4-1}$$
    $$= 1.00375 j_3$$
    $$= 1.00375 \cdot 251{,}878.5$$
    $$= 252{,}823.0$$

220 ANSWERS June 2016

Choice (4):

$$j_1 = 250,000$$
$$j_2 = j_{2-1} + 937$$
$$= j_1 + 937$$
$$= 250,000 + 937$$
$$= 250,937$$
$$j_3 = j_{3-1} + 937$$
$$= j_2 + 937$$
$$= 250,937 + 937$$
$$= 251,874$$
$$j_4 = j_{4-1} + 937$$
$$= j_3 + 937$$
$$= 251,874 + 937$$
$$= 252,811$$

The numbers generated by choice (3) are closer to the given numbers than the numbers generated by choice (4).

The correct choice is (3).

24. In the equation $V = A \sin(B_t)$, A represents the amplitude. The amplitude is 120 in this scenario since the maximum value is 120. To calculate the B-value, use the equation period $= \dfrac{2\pi}{B}$ when using radians and period $= \dfrac{360°}{B}$ when using degrees. As there is no mention of degrees, it is implied that this equation is intended to be in radians. The period is $\dfrac{1}{60}$ because there are 60 cycles in one second.

Use the formula, $\dfrac{1}{60} = \dfrac{2\pi}{B}$. Solve for B by cross multiplying:

$$B = 60 \cdot 2\pi = 120\pi$$

The correct choice is (4).

ANSWERS June 2016   221

PART II

25. Multiply both sides of the equation by the lowest common multiple (LCM) of x, 3, and $3x$, which is $3x$. Then solve for x.

$$3x \cdot \frac{1}{x} - 3x \cdot \frac{1}{3} = 3x \cdot -\frac{1}{3x}$$
$$3 - x = -1$$
$$-3 = -3$$
$$\overline{-x = -4}$$
$$x = 4$$

In a rational equation, you should check your solution(s) by substituting the solution(s) into the original equation to see if any don't make the original equation true.

$$\frac{1}{4} - \frac{1}{3} = -\frac{1}{3 \cdot 4}$$
$$\frac{3}{12} - \frac{4}{12} = -\frac{1}{12}$$
$$-\frac{1}{12} = -\frac{1}{12}$$

The correct answer is $x = 4$.

26. A controlled experiment is when the experimenter makes two groups of subjects. One group is the control group, and the other is the experimental group. Members of the experimental group in this scenario would brush their teeth only with a toothpaste that contains ingredient X. Members of the control group would brush their teeth only with a toothpaste that is the same as the other toothpaste except that it does not contain ingredient X. After some time, data from both groups would be collected and analyzed.

A controlled experiment differs from an observational experiment where the researcher does not control who is using which toothpaste.

ANSWERS June 2016

27. There are two ways to determine if $x - 5$ is a factor. The long way is to divide $2x^3 - 4x^2 - 7x - 10$ by $x - 5$ and see if there is no remainder.

$$
\begin{array}{r}
2x^2 + 6x + 23 \\
x - 5 \overline{)\, 2x^3 - 4x^2 - 7x - 10} \\
-(2x^3 - 10x^2) \\
\hline
6x^2 - 7x \\
-(6x^2 - 30x) \\
\hline
23x - 10 \\
-(23x - 115) \\
\hline
105
\end{array}
$$

Since the remainder is 105 and not zero, $x - 5$ is not a factor.

The shortcut for this question is to use the factor theorem. It says that if $x - a$ is a factor of $P(x)$, then $P(a) = 0$:

$$P(5) = 2 \cdot 5^3 - 4 \cdot 5^2 - 7 \cdot 5 - 10$$

$$= 250 - 100 - 35 - 10$$

$$= 105$$

Since $105 \neq 0$, $x - 5$ is not a factor.

28. The graph of a cosine function with midline $y = -1$ and amplitude 3 stays between the lines $y = -1 + 3 = 2$ and $y = -1 - 3 = -4$.

The curve passes through (0, 2), which is the highest point on the graph. So (0, 2) can be the starting point on the graph.

Since the frequency is $\frac{\pi}{2}$, the ending point of the cycle is $\frac{\pi}{2}$ units to the right of the starting point (0, 2), which is $\left(\frac{\pi}{2}, 2\right)$. Since $\frac{\pi}{2}$ is only about 1.5, mark the horizontal scale so that $\frac{\pi}{2}$ is equal to about 5 boxes.

The low point is on the line $y = -4$, which is halfway between the two high points. The two points on the midline are halfway between the high and low points. Connect the 5 points to form a cosine curve.

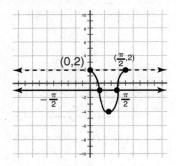

29. The number of students who participate in sports is 649, and the number who participate in music is 433. Together, this is $649 + 433 = 1082$ students. According to the given information, the probability that a student participates in either sports or music is $\frac{974}{1376}$. This means that 974 students participate in sports or in music (or in both sports and music). This means that in the 1082 students calculated by adding the number of students who participate in sports to the number of students who participate in music, some students have been counted twice. These are the students who participate in both sports and music. Subtract 974 from 1082 to get the number of students who have been counted twice: $1082 - 974 = 108$. Since 108 students out of 1376 participate in both sports and music, the probability that a student participates in both is $\frac{108}{1376}$.

This information can be organized on a Venn diagram. The two circles divide the rectangle into four regions. A is the region of students who participate in sports but not in music. C is the region of students who participate in music but not in sports. B is the region of students who participate in both sports and music. D is the region of students who participate in neither sports nor music.

- Regions A and B combined equal 649 students participating in sports.
- Regions B and C combined equal 433 students participating in music.
- Regions A, B, and C combined equal 974 students.

Adding 649 and 433 is the sum of regions A, B, another region B, and region C. This is the sum of 649 and 433, which is 1082. Region B, then, is 1082 (which has two region Bs in it) minus 974. This is given as the sum of regions A, B, and C.

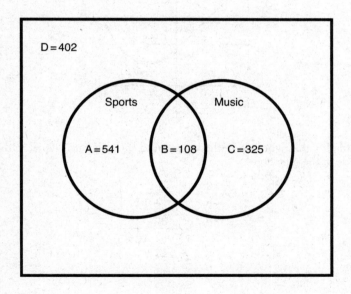

30. When the equation of a parabola is of the form $4p(y - k) = (x - h)^2$, the vertex is (h, k). The graph of this equation is a parabola with a vertex at $(4, -3)$.

The focus of the parabola is p units above the vertex (or below if p is negative). Since $12 = 4p$, $p = 3$. So the vertex is 3 units above $(4, -3)$ at $(4, 0)$.

Even without knowing this formula, this question can be answered another way. Since the directrix of this parabola is 3 units away from the vertex of the parabola, the focus of the parabola will also be 3 units away from the vertex of the parabola but in the opposite direction. The directrix is 3 units below the vertex, so the focus will be 3 units above the vertex, which is located at $(4, 0)$.

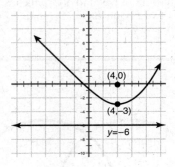

31. Simplify the right side of the equation to show that it is equal to the left side of the equation:

$$\frac{x^3 + 9}{x^3 + 8} = 1 + \frac{1}{x^3 + 8}$$

$$= \frac{x^3 + 8}{x^3 + 8} + \frac{1}{x^3 + 8}$$

$$= \frac{x^3 + 8 + 1}{x^3 + 8}$$

$$= \frac{x^3 + 9}{x^3 + 8}$$

ANSWERS June 2016

32. The formula new value = old value$(1 + r)^t$ can be used where t is the time in years and r is the annual growth rate:

$$135,000 = 100,000(1+r)^5$$

$$\frac{135,000}{100,000} = \frac{100,000(1+r)^5}{100,000}$$

$$1.35 = (1+r)^5$$

$$\sqrt[5]{1.35} = \sqrt[5]{(1+r)^5}$$

$$1.06 = 1+r$$

$$\frac{-1 = -1}{0.06 = r}$$

The annual growth rate is 6%.

PART III

33. To use the substitution method of solving a system of equations, one of the equations must be converted into an equivalent form in which one of the variables is isolated. The second equation can be quickly solved for x in terms of y:

$$2x + 2y = 10$$
$$\underline{-2y = -2y}$$
$$2x = -2y + 10$$
$$\frac{2x}{2} = \frac{-2y + 10}{2}$$
$$x = -y + 5$$

Replace the x in the first equation with $-y + 10$, and solve for y:

$$(-y + 5 - 3)^2 + (y + 2)^2 = 16$$
$$(-y + 2)^2 + (y + 2)^2 = 16$$
$$y^2 - 4y + 4 + y^2 + 4y + 4 = 16$$
$$2y^2 + 8 = 16$$
$$\underline{-8 = -8}$$
$$2y^2 = 8$$
$$\frac{2y^2}{2} = \frac{8}{2}$$
$$y^2 = 4$$
$$y = \pm 2$$

For each y-value, substitute into either original equation and solve for the corresponding x-value.

For $y = 2$:

$$2x + 2(2) = 10$$
$$2x + 4 = 10$$
$$\underline{-4 = -4}$$
$$2x = 6$$
$$\frac{2x}{2} = \frac{6}{2}$$
$$x = 3$$

ANSWERS June 2016

For $y = -2$:

$$2x + 2(-2) = 10$$
$$2x - 4 = 10$$
$$\underline{+4 = +4}$$
$$2x = 14$$
$$\frac{2x}{2} = \frac{14}{2}$$
$$x = 7$$

The solutions are $(3, 2)$ and $(7, -2)$.

34. If Alexa earns \$33,000 the first year, in the second year her salary will increase by $0.04 \cdot \$33,000 = \$1320$. So her salary in the second year will be $\$33,000 + \$1320 = \$34,320$. A quicker way to get this value would be to multiply \$33,000 by 1.04:

$$\$33,000 \cdot 1.04 = \$34,320$$

To get to the salary for the third year, multiply again by 1.04:

$$\$34,320 \cdot 1.04 = \$35,692.80$$

To get this number without first calculating the salary for the second year, you could calculate $\$33,000 \cdot 1.04^2$.

In general, Alexa's salary in year n will be $33,000 \cdot 1.04^{n-1}$. To find her total earnings after n years, you have to use the formula to calculate the sum of the geometric series:

$$S_n = 33,000 + 33,000 \cdot 1.04^1 + 33,000 \cdot 1.04^2 + 33,000 \cdot 1.04^3 + \cdots + 33,000 \cdot 1.04^{n-1}$$

In the reference sheet, the formula for the sum of a geometric series

$S_n = \dfrac{a_1 - a_1 r^n}{1 - r}$ is given. For this example, $a_1 = 33,000$ and $r = 1.04$.

So the formula becomes:

$$S_n = \frac{33,000 - 33,000 \cdot 1.04^n}{1 - 1.04}$$

For $n = 15$:

$$S_{15} = \frac{33,000 - 33,000 \cdot 1.04^{15}}{1 - 1.04} = 660,778.39$$

Alexa's total earnings for her first 15 years of teaching is \$660,778.39.

230   ANSWERS June 2016

35. The plausible interval containing the middle 95% of the data is between mean − 2 · SD and mean + 2 · SD. For this example, that is between $0.602 - 2 \cdot 0.066 = 0.47$ and $0.602 + 2 \cdot 0.066 = 0.73$.

The members of the prom committee are correct to be concerned since 0.50 is between 0.47 and 0.73. Alternately, it can be seen on the histogram that 0.50 is something that will happen sometimes, even when 60% of the total population prefers a DJ. If the histogram showed no values equal to or less than 0.50, it would suggest that a 50%–50% was very unlikely.

36. The average rate of change of a function f on an interval $[a, b]$ can be calculated with the formula average rate of change $= \dfrac{f(b) - f(a)}{b - a}$.

The average rate of change for the given function f on the interval $[-2, 4]$ is:

$$\frac{f(4) - f(-2)}{4 - (-2)} = \frac{80 - 1.25}{6}$$
$$= \frac{78.75}{6}$$
$$= 13.125$$

The average rate of change for the given function g on the interval $[-2, 4]$ is:

$$\frac{g(4) - g(-2)}{4 - (-2)} = \frac{[4(4)^3 - 5(4)^2 + 3] - [4(-2)^3 - 5(-2)^2 + 3]}{6}$$
$$= \frac{179 - (-49)}{6}$$
$$= \frac{228}{6}$$
$$= 38$$

Since $38 > 13.125$, the average rate of change on the interval $[-2, 4]$ is greater for g than it is for f.

PART IV

37. For patient A, $N_0 = 800$ and $r = 0.347$. So the equation is $A(t) = 800(e)^{-0.347t}$.

For patient B, $N_0 = 400$ and $r = 0.231$. So the equation is $B(t) = 400(e)^{-0.231t}$.

When the graph of both functions is graphed on a graphing calculator and the intersect feature is used, the intersection point is approximately (6, 100). Before hour 6, patient A has more of the drug in his or her body. After hour 6, patient B has more of the drug in his or her body.

For the TI-84:

For the TI-Nspire:

Make your scale so the graph includes the two y-intercepts and also the intersection point. On the axes below, each vertical unit is 40 milligrams and each horizontal unit is $\frac{1}{2}$ hour.

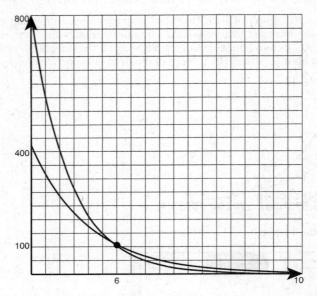

Since 15% of 800 is 120, the next dose will be given when $A(t)=120$. This becomes an exponential equation:

$$120 = 800(e)^{-0.347t}$$

$$\frac{120}{800} = \frac{800(e)^{-0.347t}}{800}$$

$$0.15 = e^{-0.347t}$$

$$\ln 0.15 = \ln e^{-0.347t}$$

$$\ln 0.15 = -0.347t$$

$$\frac{\ln 0.15}{-0.347} = \frac{-0.347t}{-0.347}$$

$$5.5 \approx t$$

The next dose can be administered after 5.5 hours.

This can also be solved with the graphing calculator by finding the x-coordinate of the intersection of $A(x)$ and $y = 120$.

For the TI-84:

For the TI-Nspire:

SELF-ANALYSIS CHART June 2016

Topic	Question Numbers	Number of Points	Your Points	Your Percentage
1. Polynomial Expressions and Equations	6, 12, 22, 27, 30, 36	2 + 2 + 2 + 2 + 2 + 4 = 14		
2. Complex Numbers	3	2		
3. Exponential Expressions and Equations	1, 15, 21, 32, 34, 37	2 + 2 + 2 + 4 + 4 + 4 = 18		
4. Rational Expressions and Equations	14, 25, 31	2 + 2 + 2 = 6		
5. Radical Expressions and Equations	5	2		
6. Trigonometric Expressions and Equations	13, 17, 24, 28	2 + 2 + 2 + 2 = 8		
7. Graphing	4, 18, 20	2 + 2 + 2 = 6		
8. Functions	8, 16, 19	2 + 2 + 2 = 6		
9. Systems of Equations	33	4		
10. Sequences and Series	10, 23	2 + 2 = 4		
11. Probability	11, 19	2 + 2 = 4		
12. Statistics	2, 7, 9, 26, 35	2 + 2 + 2 + 2 + 4 = 12		

SELF-ANALYSIS CHART June 2016   **235**

HOW TO CONVERT YOUR RAW SCORE TO YOUR ALGEBRA II REGENTS EXAMINATION SCORE

The accompanying conversion chart must be used to determine your final score on the June 2016 Regents Examination in Algebra II. To find your final exam score, locate in the column labeled "Raw Score" the total number of points you scored out of a possible 86 points. Since partial credit is allowed in Parts II, III, and IV of the test, you may need to approximate the credit you would receive for a solution that is not completely correct. Then locate in the adjacent column to the right the scale score that corresponds to your raw score. The scale score is your final Algebra II Regents Examination score.

236   SELF-ANALYSIS CHART June 2016

Regents Examination in Algebra II—June 2016
Chart for Converting Total Test Raw Scores to Final
Examination Scores (Scaled Scores)

Raw Score	Scale Score	Performance Level	Raw Score	Scale Score	Performance Level	Raw Score	Scale Score	Performance Level
86	100	5	57	82	4	28	68	3
85	99	5	56	82	4	27	67	3
84	98	5	55	81	4	26	66	3
83	97	5	54	81	4	25	65	3
82	97	5	53	81	4	24	64	2
81	96	5	52	80	4	23	63	2
80	95	5	51	80	4	22	61	2
79	94	5	50	80	4	21	60	2
78	94	5	49	79	4	20	58	2
77	93	5	48	79	4	19	55	1
76	92	5	47	79	4	18	54	1
75	91	5	46	78	4	17	53	1
74	91	5	45	78	4	16	51	1
73	90	5	44	77	3	15	49	1
72	89	5	43	77	3	14	47	1
71	89	5	42	77	3	13	44	1
70	88	5	41	76	3	12	42	1
69	88	5	40	76	3	11	39	1
68	87	5	39	76	3	10	37	1
67	87	5	38	75	3	9	34	1
66	86	5	37	75	3	8	31	1
65	86	5	36	74	3	7	27	1
64	85	5	35	73	3	6	24	1
63	84	4	34	73	3	5	20	1
62	84	4	33	72	3	4	17	1
61	84	4	32	72	3	3	13	1
60	83	4	31	71	3	2	9	1
59	83	4	30	70	3	1	4	1
58	82	4	29	69	3	0	0	1

237

Examination
August 2016
Algebra II

HIGH SCHOOL MATH REFERENCE SHEET

Conversions

1 inch = 2.54 centimeters	1 cup = 8 fluid ounces
1 meter = 39.37 inches	1 pint = 2 cups
1 mile = 5280 feet	1 quart = 2 pints
1 mile = 1760 yards	1 gallon = 4 quarts
1 mile = 1.609 kilometers	1 gallon = 3.785 liters
	1 liter = 0.264 gallon
1 kilometer = 0.62 mile	1 liter = 1000 cubic centimeters
1 pound = 16 ounces	
1 pound = 0.454 kilogram	
1 kilogram = 2.2 pounds	
1 ton = 2000 pounds	

Formulas

Triangle	$A = \frac{1}{2}bh$
Parallelogram	$A = bh$
Circle	$A = \pi r^2$
Circle	$C = \pi d$ or $C = 2\pi r$

Formulas (continued)

General Prisms	$V = Bh$
Cylinder	$V = \pi r^2 h$
Sphere	$V = \dfrac{4}{3}\pi r^3$
Cone	$V = \dfrac{1}{3}\pi r^2 h$
Pyramid	$V = \dfrac{1}{3}Bh$
Pythagorean Theorem	$a^2 + b^2 = c^2$
Quadratic Formula	$x = \dfrac{-b \pm \sqrt{b^2 - 4ac}}{2a}$
Arithmetic Sequence	$a_n = a_1 + (n-1)d$
Geometric Sequence	$a_n = a_1 r^{n-1}$
Geometric Series	$S_n = \dfrac{a_1 - a_1 r^n}{1-r}$ where $r \neq 1$
Radians	1 radian = $\dfrac{180}{\pi}$ degrees
Degrees	1 degree = $\dfrac{\pi}{180}$ radians
Exponential Growth/Decay	$A = A_0 e^{k(t - t_0)} + B_0$

EXAMINATION August 2016   239

PART I

Answer all 24 questions in this part. Each correct answer will receive 2 credits. No partial credit will be allowed. For each statement or question, write in the space provided the numeral preceding the word or expression that best completes the statement or answers the question.   [48 credits]

1 Which equation has $1 - i$ as a solution?

(1) $x^2 + 2x - 2 = 0$ (3) $x^2 - 2x - 2 = 0$

(2) $x^2 + 2x + 2 = 0$ (4) $x^2 - 2x + 2 = 0$ 1 _____

2 Which statement(s) about statistical studies is true?

   I. A survey of all English classes in a high school would be a good sample to determine the number of hours students throughout the school spend studying.

   II. A survey of all ninth graders in a high school would be a good sample to determine the number of student parking spaces needed at that high school.

   III. A survey of all students in one lunch period in a high school would be a good sample to determine the number of hours adults spend on social media websites.

   IV. A survey of all Calculus students in a high school would be a good sample to determine the number of students throughout the school who don't like math.

(1) I, only   (3) I and III

(2) II, only   (4) III and IV   2 _____

240　　　　　**EXAMINATION August 2016**

3 To the *nearest tenth*, the value of x that satisfies $2^x = -2x + 11$ is

(1) 2.5　　　　　　　(3) 5.8
(2) 2.6　　　　　　　(4) 5.9　　　　　　　3 _____

4 The lifespan of a 60-watt lightbulb produced by a company is normally distributed with a mean of 1450 hours and a standard deviation of 8.5 hours. If a 60-watt lightbulb produced by this company is selected at random, what is the probability that its lifespan will be between 1440 and 1465 hours?

(1) 0.3803　　　　　　(3) 0.8415
(2) 0.4612　　　　　　(4) 0.9612　　　　　　4 _____

5 Which factorization is *incorrect*?

(1) $4k^2 - 49 = (2k + 7)(2k - 7)$
(2) $a^3 - 8b^3 = (a - 2b)(a^2 + 2ab + 4b^2)$
(3) $m^3 + 3m^2 - 4m + 12 = (m - 2)^2(m + 3)$
(4) $t^3 + 5t^2 + 6t + t^2 + 5t + 6 = (t + 1)(t + 2)(t + 3)$　　　5 _____

6 Sally's high school is planning their spring musical. The revenue, R, generated can be determined by the function $R(t) = -33t^2 + 360t$, where t represents the price of a ticket. The production cost, C, of the musical is represented by the function $C(t) = 700 + 5t$. What is the highest ticket price, to *the nearest dollar*, they can charge in order to *not* lose money on the event?

(1) $t = 3$　　　　　　(3) $t = 8$
(2) $t = 5$　　　　　　(4) $t = 11$　　　　　　6 _____

EXAMINATION August 2016   **241**

7 The set of data in the table below shows the results of a survey on the number of messages that people of different ages text on their cell phones each month.

Age Group	Text Messages per Month		
	0–10	11–50	Over 50
15–18	4	37	68
19–22	6	25	87
23–60	25	47	157

If a person from this survey is selected at random, what is the probability that the person texts over 50 messages per month given that the person is between the ages of 23 and 60?

(1) $\dfrac{157}{229}$ (3) $\dfrac{157}{384}$

(2) $\dfrac{157}{312}$ (4) $\dfrac{157}{456}$ 7 _____

8 A recursive formula for the sequence 18, 9, 4.5, . . . is

(1) $g_1 = 18$ (3) $g_1 = 18$

   $g_n = \dfrac{1}{2} g_{n-1}$ $g_n = 2g_{n-1}$

(2) $g_n = 18\left(\dfrac{1}{2}\right)^{n-1}$ (4) $g_n = 18(2)^{n-1}$ 8 _____

EXAMINATION August 2016

9 Kristin wants to increase her running endurance. According to experts, a gradual mileage increase of 10% per week can reduce the risk of injury. If Kristin runs 8 miles in week one, which expression can help her find the total number of miles she will have run over the course of her 6-week training program?

(1) $\sum_{n=1}^{6} 8(1.10)^{n-1}$ (3) $\dfrac{8-8(1.10)^6}{0.90}$

(2) $\sum_{n=1}^{6} 8(1.10)^{n}$ (4) $\dfrac{8-8(0.10)^n}{1.10}$ 9 _____

10 A sine function increasing through the origin can be used to model light waves. Violet light has a wavelength of 400 nanometers. Over which interval is the height of the wave *decreasing*, only?

(1) (0, 200)   (3) (200, 400)
(2) (100, 300)   (4) (300, 400)   10 _____

11 The expression $\dfrac{x^3+2x^2+x+6}{x+2}$ is equivalent to

(1) x^2+3 (3) $2x^2+x+6$
(2) $x^2+1+\dfrac{4}{x+2}$ (4) $2x^2+1+\dfrac{4}{x+2}$ 11 _____

12 A candidate for political office commissioned a poll. His staff received responses from 900 likely voters and 55% of them said they would vote for the candidate. The staff then conducted a simulation of 1000 more polls of 900 voters, assuming that 55% of voters would vote for their candidate. The output of the simulation is shown in the diagram below.

Given this output, and assuming a 95% confidence level, the margin of error for the poll is closest to

(1) 0.01 (3) 0.06
(2) 0.03 (4) 0.12

12 _____

13 An equation to represent the value of a car after t months of ownership is $v = 32{,}000(0.81)^{\frac{t}{12}}$. Which statement is *not* correct?

(1) The car lost approximately 19% of its value each month.
(2) The car maintained approximately 98% of its value each month.
(3) The value of the car when it was purchased was $32,000.
(4) The value of the car 1 year after it was purchased was $25,920.

13 _____

14 Which equation represents an odd function?
   (1) $y = \sin x$ (3) $y = (x + 1)^3$
   (2) $y = \cos x$ (4) $y = e^{5x}$ 14 _____

15 The completely factored form of
   $2d^4 + 6d^3 - 18d^2 - 54d$ is
   (1) $2d(d^2 - 9)(d + 3)$ (3) $2d(d + 3)^2(d - 3)$
   (2) $2d(d^2 + 9)(d + 3)$ (4) $2d(d - 3)^2(d + 3)$ 15 _____

16 Which diagram shows an angle rotation of 1 radian on the unit circle?

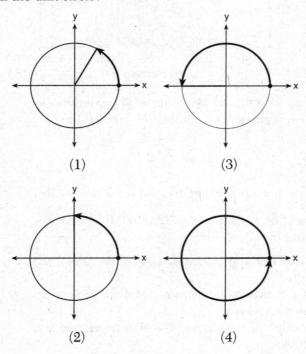

(1)   (3)

(2)   (4)   16 _____

EXAMINATION August 2016   245

17 The focal length, F, of a camera's lens is related to the distance of the object from the lens, J, and the distance to the image area in the camera, W, by the formula below.

$$\frac{1}{J} + \frac{1}{W} = \frac{1}{F}$$

When this equation is solved for J in terms of F and W, J equals

(1) $F - W$ (3) $\dfrac{FW}{W-F}$

(2) $\dfrac{FW}{F-W}$ (4) $\dfrac{1}{F} - \dfrac{1}{W}$ 17 ____

18 The sequence $a_1 = 6$, $a_n = 3a_{n-1}$ can also be written as

(1) $a_n = 6 \cdot 3^n$ (3) $a_n = 2 \cdot 3^n$

(2) $a_n = 6 \cdot 3^{n+1}$ (4) $a_n = 2 \cdot 3^{n+1}$ 18 ____

19 Which equation represents the set of points equidistant from line l and point R shown on the graph below?

(1) $y = -\frac{1}{8}(x+2)^2 + 1$ (3) $y = -\frac{1}{8}(x-2)^2 + 1$

(2) $y = -\frac{1}{8}(x+2)^2 - 1$ (4) $y = -\frac{1}{8}(x-2)^2 - 1$ 19 _____

20 Mr. Farison gave his class the three mathematical rules shown below to either prove or disprove. Which rules can be proved for all real numbers?

I $(m+p)^2 = m^2 + 2mp + p^2$
II $(x+y)^3 = x^3 + 3xy + y^3$
III $(a^2 + b^2)^2 = (a^2 - b^2)^2 + (2ab)^2$

(1) I, only
(2) I and II
(3) II and III
(4) I and III   20 _____

21 The graph of $p(x)$ is shown below.

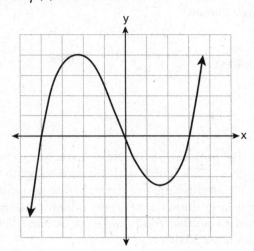

What is the remainder when $p(x)$ is divided by $x + 4$?
(1) $x - 4$ (3) 0
(2) -4 (4) 4   21 _____

22 A payday loan company makes loans between $100 and $1000 available to customers. Every 14 days, customers are charged 30% interest with compounding. In 2013, Remi took out a $300 payday loan. Which expression can be used to calculate the amount she would owe, in dollars, after one year if she did not make payments?

(1) $300(.30)^{\frac{14}{365}}$ (3) $300(.30)^{\frac{365}{14}}$

(2) $300(1.30)^{\frac{14}{365}}$ (4) $300(1.30)^{\frac{365}{14}}$ 22 _____

248   **EXAMINATION August 2016**

23 Which value is *not* contained in the solution of the system shown below?

$$a + 5b - c = -20$$
$$4a - 5b + 4c = 19$$
$$-a - 5b - 5c = 2$$

(1) –2   (3) 3

(2) 2   (4) –3   23 _____

24 In 2010, the population of New York State was approximately 19,378,000 with an annual growth rate of 1.5%. Assuming the growth rate is maintained for a large number of years, which equation can be used to predict the population of New York State t years after 2010?

(1) $P_t = 19,378,000(1.5)t$

(2) $P_0 = 19,378,000$
    $P_t = 19,378,000 + 1.015P_{t-1}$

(3) $P_t = 19,378,000 (1.015)^{t-1}$

(4) $P_0 = 19,378,000$
    $P_t = 1.015P_{t-1}$ 24 _____

EXAMINATION August 2016   249

PART II

Answer all 8 questions in this part. Each correct answer
will receive 2 credits. Clearly indicate the necessary steps,
including appropriate formula substitutions, diagrams,
graphs, charts, etc. For all questions in this part, a correct
numerical answer with no work shown will receive only 1
credit.   [16 credits]

25 The volume of air in a person's lungs, as the person breathes in
and out, can be modeled by a sine graph. A scientist is studying
the differences in this volume for people at rest compared to
people told to take a deep breath. When examining the graphs,
should the scientist focus on the amplitude, period, or midline?
Explain your choice.

250   **EXAMINATION August 2016**

26 Explain how $\left(3^{\frac{1}{5}}\right)^2$ can be written as the equivalent radical expression $\sqrt[5]{9}$.

EXAMINATION August 2016   251

27 Simplify $xi(i - 7i)^2$, where i is the imaginary unit.

28 Using the identity $\sin^2 \theta + \cos^2 \theta = 1$, find the value of $\tan \theta$, to the *nearest hundredth*, if $\cos \theta$ is -0.7 and θ is in Quadrant II.

252   **EXAMINATION August 2016**

29 Elizabeth waited for 6 minutes at the drive thru at her favorite fast-food restaurant the last time she visited. She was upset about having to wait that long and notified the manager. The manager assured her that her experience was very unusual and that it would not happen again.

A study of customers commissioned by this restaurant found an approximately normal distribution of results. The mean wait time was 226 seconds and the standard deviation was 38 seconds. Given these data, and using a 95% level of confidence, was Elizabeth's wait time unusual? Justify your answer.

EXAMINATION August 2016   253

30 The x-value of which function's x-intercept is larger, f or h?
Justify your answer.

$f(x) = \log(x - 4)$

x	h(x)
−1	6
0	4
1	2
2	0
3	−2

254   **EXAMINATION August 2016**

31 The distance needed to stop a car after applying the brakes varies directly with the square of the car's speed. The table below shows stopping distances for various speeds.

Speed (mph)	10	20	30	40	50	60	70
Distance (ft)	6.25	25	56.25	100	156.25	225	306.25

Determine the average rate of change in braking distance, in ft/mph, between one car traveling at 50 mph and one traveling at 70 mph.

Explain what this rate of change means as it relates to braking distance.

EXAMINATION August 2016   255

32 Given events A and B, such that $P(A) = 0.6$, $P(B) = 0.5$, and $P(A \cup B) = 0.8$, determine whether A and B are independent or dependent.

PART III

Answer all 4 questions in this part. Each correct answer will receive 4 credits. Clearly indicate the necessary steps, including appropriate formula substitutions, diagrams, graphs, charts, etc. For all questions in this part, a correct numerical answer with no work shown will receive only 1 credit. [16 credits]

33 Find algebraically the zeros for $p(x) = x^3 + x^2 - 4x - 4$.

On the set of axes below, graph $y = p(x)$.

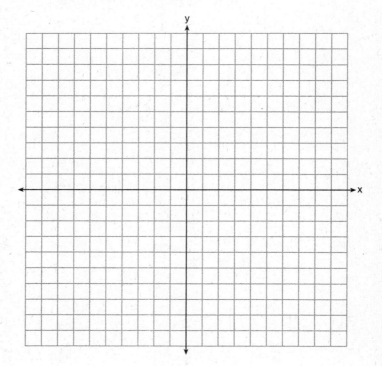

EXAMINATION August 2016   257

34 One of the medical uses of Iodine–131 (I–131), a radioactive isotope of iodine, is to enhance x-ray images. The half-life of I–131 is approximately 8.02 days. A patient is injected with 20 milligrams of I–131. Determine, to the *nearest day*, the amount of time needed before the amount of I–131 in the patient's body is approximately 7 milligrams.

258   **EXAMINATION August 2016**

35 Solve the equation $\sqrt{2x-7} + x = 5$ algebraically, and justify the solution set.

EXAMINATION August 2016   **259**

36 Ayva designed an experiment to determine the effect of a new energy drink on a group of 20 volunteer students. Ten students were randomly selected to form group 1 while the remaining 10 made up group 2. Each student in group 1 drank one energy drink, and each student in group 2 drank one cola drink. Ten minutes later, their times were recorded for reading the same paragraph of a novel. The results of the experiment are shown below.

Group 1 (seconds)	Group 2 (seconds)
17.4	23.3
18.1	18.8
18.2	22.1
19.6	12.7
18.6	16.9
16.2	24.4
16.1	21.2
15.3	21.2
17.8	16.3
19.7	14.5
Mean = 17.7	Mean = 19.1

a) Ayva thinks drinking energy drinks makes students read faster. Using information from the experimental design or the results, explain why Ayva's hypothesis may be *incorrect*.

Question 36 continued on next page.

Question 36 continued.

Using the given results, Ayva randomly mixes the 20 reading times, splits them into two groups of 10, and simulates the difference of the means 232 times.

b) Ayva has decided that the difference in mean reading times is *not* an unusual occurence. Support her decision using the results of the simulation. Explain your reasoning.

EXAMINATION August 2016

261

PART IV

Answer the question in this part. A correct answer will receive 6 credits. Clearly indicate the necessary steps, including appropriate formula substitutions, diagrams, graphs, charts, etc. A correct numerical answer with no work shown will receive only 1 credit. [6 credits]

37 Seth's parents gave him $5000 to invest for his 16th birthday. He is considering two investment options. Option A will pay him 4.5% interest compounded annually. Option B will pay him 4.6% compounded quarterly.

Write a function of option A and option B that calculates the value of each account after n years.

Seth plans to use the money after he graduates from college in 6 years. Determine how much more money option B will earn than option A to the *nearest cent*.

Algebraically determine, to the *nearest tenth of a year*, how long it would take for option B to double Seth's initial investment.

Answers
August 2016
Algebra II

Answer Key

PART I

1. (4)	**5.** (3)	**9.** (1)	**13.** (1)	**17.** (3)	**21.** (3)
2. (1)	**6.** (3)	**10.** (2)	**14.** (1)	**18.** (3)	**22.** (4)
3. (2)	**7.** (1)	**11.** (2)	**15.** (3)	**19.** (4)	**23.** (2)
4. (3)	**8.** (1)	**12.** (2)	**16.** (1)	**20.** (4)	**24.** (4)

PART II

25. Amplitude

26. $\sqrt[5]{9} = 9^{\frac{1}{5}} = \left(3^2\right)^{\frac{1}{5}} = 3^{\frac{2}{5}} = \left(3^{\frac{1}{5}}\right)^2$

27. $-36xi$

28. -1.02

29. Yes, Elizabeth's wait time was unusual.

30. f

31. 7.5

32. Independent

PART III

33. $2, -2, -1$

34. 12 days

35. 4

36. (a) The hypothesis might be incorrect.

(b) The difference is not an unusual occurrence.

PART IV

37. $A(x) = 5000\left(1 + \dfrac{0.045}{1}\right)^{1x}$;

$B(x) = 5000\left(1 + \dfrac{0.046}{4}\right)^{4x}$;

$67.57; 15.2 years

In **Parts II–IV**, you are required to show how you arrived at your answers. For sample methods of solutions, see the *Answers Explained* section.

ANSWERS August 2016

Answers Explained

PART I

1. Use the quadratic formula, $x = \dfrac{-b \pm \sqrt{b^2 - 4ac}}{2a}$, with each choice.

Choice (1):

$$x = \frac{-2 \pm \sqrt{2^2 - 4 \cdot 1 \cdot (-2)}}{2 \cdot 1}$$

$$= \frac{-2 \pm \sqrt{4 + 8}}{2}$$

$$= \frac{-2 \pm \sqrt{12}}{2}$$

$$= \frac{-2 \pm 2\sqrt{3}}{2}$$

$$= -1 \pm \sqrt{3}$$

Choice (2):

$$x = \frac{-2 \pm \sqrt{2^2 - 4 \cdot 1 \cdot 2}}{2 \cdot 1}$$

$$= \frac{-2 \pm \sqrt{4 - 8}}{2}$$

$$= \frac{-2 \pm \sqrt{-4}}{2}$$

$$= \frac{-2 \pm 2i}{2}$$

$$= -1 \pm i$$

Choice (3):

$$x = \frac{-(-2) \pm \sqrt{(-2)^2 - 4 \cdot 1 \cdot (-2)}}{2 \cdot 1}$$

$$= \frac{2 \pm \sqrt{4 + 8}}{2}$$

$$= \frac{2 \pm \sqrt{12}}{2}$$

$$= \frac{2 \pm 2\sqrt{3}}{2}$$

$$= 1 \pm \sqrt{3}$$

Choice (4):

$$x = \frac{-(-2) \pm \sqrt{(-2)^2 - 4 \cdot 1 \cdot 2}}{2 \cdot 1}$$

$$= \frac{2 \pm \sqrt{4 - 8}}{2}$$

$$= \frac{2 \pm \sqrt{-4}}{2}$$

$$= \frac{2 \pm 2i}{2}$$

$$= 1 \pm i$$

The correct choice is (4).

2. Check each of the four statements.

Check I: The most accurate sample you can use for a survey is the entire population. Since this will be a survey of all English classes, it is a good sample.

Check II: Since ninth graders do not drive cars, they are not a good sample to survey about parking spaces.

Check III: High school students won't know how many hours adults spend on social media. A better option would be to use a random sample that includes adults.

Check IV: Using a random sample of all students rather than just choosing students who are taking Calculus would be better. The students in Calculus probably like math more than students who are not in Calculus, so the results of a survey using only Calculus will be skewed.

The correct choice is (1).

3. There is not a simple way to solve this equation using just Algebra. One way to solve this would be to use the graphing calculator to graph $y = 2^x$ and $y = -2x + 11$ on the same set of axes. The x-coordinate of the intersection point is the solution to the equation. For this example, that x-coordinate is approximately 2.557, which rounds to 2.6.

For the TI-84:

For the TI-Nspire:

Another way would be to test the four choices to see which makes 2^x closest to $-2x + 11$. For $x = 2.6$, $2^{2.6} \approx 6.1$ while $-2 \cdot 2.6 + 11 = 5.8$.

The correct choice is **(2)**.

4. This question requires you use the normal cdf function of the graphing calculator. Four numbers need to be entered into the function. These are the mean (1450), the standard deviation (8.5), the lower bound (1440), and the upper bound (1465). The calculator will use these numbers to calculate the probability that a randomly chosen bulb will have a lifespan between 1440 and 1465 hours.

For the TI-84:

Press [2ND] [VARS] [2] for normalcdf. Fill in the fields with 1440 for lower, 1465 for upper, 1450 for μ, and 8.5 for σ. Then select "Paste."

For the TI-Nspire:

From the home screen, press [A]. Press [menu] [6] [5] [2] for Normal Cdf. Set the Lower Bound to 1440, the Upper Bound to 1465, the mean (μ) to 1450, and the standard deviation (σ) to 8.5.

The probability is approximately 0.8415.

The correct choice is **(3)**.

ANSWERS August 2016   267

5. Factor each of the answer choices to find out which one is incorrect.

Choice (1): This choice uses the difference of two squares factoring pattern.

$$4k^2 - 49 = (2k)^2 - 7^2 = (2k + 7)(2k - 7)$$

Choice (2): This choice can be checked by multiplying the expression on the right side of the equal sign to see if it simplifies to the expression on the left side of the equal sign.

$$(a - 2b)(a^2 + 2ab + 4b^2)$$
$$= a \cdot a^2 + a \cdot 2ab + a \cdot 4b^2 + (-2b)a^2 + (-2b)2ab + (-2b)4b^2$$
$$= a^3 + 2a^2b + 4ab^2 - 2a^2b - 4ab^2 - 8b^3 = a^3 - 8b^3$$

Choice (3): This choice can be checked by multiplying the expression on the right side of the equal to sign to see if it simplifies to the expression on the left side of the equal sign.

$$(m - 2)^2(m + 3) = (m - 2)(m - 2)(m + 3) = (m^2 - 4m + 4)(m + 3)$$
$$= m^2 \cdot m + m^2 \cdot 3 + (-4m)m + (-4m)3 + 4 \cdot m + 4 \cdot 3$$
$$= m^3 + 3m^2 - 4m^2 - 12m + 4m + 12 = m^3 - m^2 - 8m + 12$$

This does not equal the expression on the left side of the equal sign.

Choice (4): The expression on the left side of the equal sign can be factored by grouping.

$$t^3 + 5t^2 + 6t + t^2 + 5t + 6 = t(t^2 + 5t + 6) + 1(t^2 + 5t + 6)$$
$$= (t + 1)(t^2 + 5t + 6) = (t + 1)(t + 2)(t + 3)$$

The correct choice is **(3)**.

6. The high school will not lose money when the revenue minus the cost is greater than 0. Use an algebraic inequality to model this situation:

$$R(t)^3 - C(t) = (-33t^2 + 360t) - (700 + 5t)$$
$$= -33t + 360t - 700 - 5t$$
$$= -33t^2 + 355t - 700 \geq 0$$

Algebraically solving a quadratic inequality like this can be very difficult. Fortunately, this is a multiple-choice question. Since you want to know which answer choice is the highest price that can be charged without losing money on the event, start by seeing if choice (4) makes the inequality true. It is the highest number.

Choice (4): $-33 \cdot 11^2 + 355 \cdot 11 - 700 = -788$. If the ticket price is $11, the high school will lose $788.

268   **ANSWERS August 2016**

Choice (3): $-33 \cdot 8^2 + 355 \cdot 8 - 700 = +28$. If the ticket price is \$8, the high school will lose \$28.

Choices (1) and (2) also result in a positive value. Since those choices are lower than choice (3) and the question asks for the highest ticket price that will not lose money, $t = 8$ is the answer.

The correct choice is **(3)**.

7. It is "given" that the person is between the ages of 23 and 60. So add together the numbers in the last row, $25 + 47 + 157 = 229$, to get the denominator of the fraction. Exactly 157 people who are between the ages of 23 and 60 send over 50 text messages per month. So the probability that a randomly selected person texts over 50 messages per month given that the person is between the ages of 23 and 60 is $\frac{157}{229}$.

The correct choice is **(1)**.

8. A recursive formula first defines one or more terms of the sequence and then describes a formula for obtaining other terms by substituting values from terms earlier in the sequence into the recursive part of the formula. Since choices (2) and (4) are direct formulas, only choices (1) and (3) are recursive formulas. Check choices (1) and (3) to see if they result in the given sequence.

Choice (1): $g_1 = 18$, $g_2 = \frac{1}{2} g_{2-1} = \frac{1}{2} g_1 = \frac{1}{2} \cdot 18 = 9$

This agrees with the first two terms of the sequence.

Choice (3): $g_1 = 18$, $g_2 = 2g_{2-1} = 2g_1 = 2 \cdot 18 = 36$

This agrees with the first term of the sequence but not with the second term.

Choice (2) is a direct formula for the given sequence. However, the question specifically asks for the recursive formula.

The correct choice is **(1)**.

ANSWERS August 2016   269

9. In week 1, Kristin runs 8 miles. In week 2, she runs 10% more. Since 10% of 8 is 0.8, Kristin runs 8 + 0.8 = 8.8 miles in week 2. Another way to calculate the number of miles Kristin runs in the second week is to multiply the number of miles she ran the first week by 1.1 to get 8 · 1.1 = 8.8 miles. To find the number of miles she runs in week 3, multiply the number of miles she ran in the second week by 1.1, which gives 8.8 · 1.1 = 9.68. Continuing this pattern and multiplying each number by 1.1 to get the next number generates the following series:

$$8 + 8.8 + 9.68 + 10.648 + 11.7128 + 12.88408$$

The total number of miles Kristin runs in 6 weeks is the sum of these numbers.

Choice (1) can be expanded by beginning with $n = 1$ and ending with $n = 6$ to become the following:

For $n = 1$: $8(1.10)^{1-1} - 8(1.10)^0 - 8(1) = 8$

For $n = 2$: $8(1.10)^{2-1} - 8(1.10)^1 - 8(1.10) = 8.8$

For $n = 3$: $8(1.10)^{3-1} - 8(1.10)^2 - 8(1.21) = 9.68$

Inserting $n = 4, 5$, and 6 will match the numbers in the series calculated above.

Choice (2) would create a similar series but starting with 8.8 and ending with 14.172488 instead of 12.88408. So this choice does not match the series calculated above.

Choices (3) and (4) are not correct. However, there is an expression that resembles those choices that would be correct if it were an option. This is because the series can be written as the geometric series $8 + 8(1.1) + 8(1.1)^2 + 8(1.1)^3 + 8(1.1)^4 + 8(1.1)^5$. The sum of a geometric series is given in the reference sheet in the back of the Regents booklet:

$$S_n = \frac{a_1 - a_1 r^n}{1 - r}$$

For this example, $a_1 = 8$, $r = 1.1$, and $n = 6$, so

$$S_n = \frac{8 - 8(1.10)^6}{1 - 1.1} = \frac{8 - 8(1.10)^6}{-0.10}$$

This looks similar to choice (3) but not exactly.

The correct choice is (1).

10. A sine curve with a period of 400 is graphed below. As a point moves on the curve from left to right, the value is increasing if the graph is going up and is decreasing if the graph is going down. This curve is increasing from 0 to 100, decreasing from 100 to 300, and then increasing again from 300 to 400.

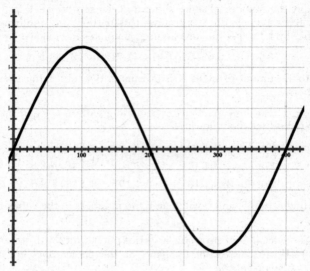

The correct choice is (**2**).

11. Use polynomial long division:

$$x+2\overline{\smash{\big)}\,x^3+2x^2+x+6} \quad \begin{array}{r} x^2+0x+1+\dfrac{4}{x+2} \\ \end{array}$$

$$\begin{array}{r}
x^2 + 0x + 1 + \dfrac{4}{x+2} \\
x+2\overline{\smash{\big)}\,x^3 + 2x^2 + x + 6} \\
-\left(x^3 + 2x^2\right) \\
\hline
0x^2 + x \\
-\left(0x^2 + 0x\right) \\
\hline
x + 6 \\
-(x+2) \\
\hline
4
\end{array}$$

The correct choice is (**2**).

ANSWERS August 2016   271

12. For a 95% confidence interval, nearly all of the values will be between the mean plus 2 times the margin of error and the mean minus 2 times the margin of error. The mean in this case is 0.55. Since the maximum value is about 0.06 above the mean and the minimum value is about 0.06 below the mean, the margin of error is $\frac{0.06}{2} = 0.03$.

The correct choice is **(2)**.

13. Test the four answer choices.

Choice (1): When an equation is in the form $y = a(1 - r)^x$, the variable r is the percent decrease. This equation can be rewritten:

$$v = 32{,}000\left((0.81)^{\frac{1}{12}}\right)^x = 32{,}000(0.98)^x = 32{,}000(1 - 0.02)^x$$

So the car loses approximately 2% of its value each month, not 19%.

Choice (2): Rewrite the equation:

$$v = 32{,}000\left((0.81)^{\frac{1}{12}}\right)^x = 32{,}000(0.98)^x$$

So the car does retain 98% of its value each month.

Choice (3): The value of the car when it was purchased can be calculated by substituting $t = 0$:

$$v = 32{,}000(0.81)^{\frac{0}{12}} = 32{,}000(0.81)^0 = 32{,}000(1) = 32{,}000$$

Choice (4): To find the value of the car 1 year after purchasing it, substitute $t = 12$ since t is measured in months:

$$v = 32{,}000\,(0.81)^{\frac{12}{12}} = 32{,}000(0.81)^1 = 32{,}000(0.81) = 25{,}920$$

The correct choice is **(1)**.

14. The graph of an odd function is symmetric with respect to the origin. Here are the graphs for the four functions in the answer choices:

Choice (1):

Choice (2):

Choice (3):

Choice (4):

Only choice (1) is symmetric with respect to the origin. Choice (2) is an even function because it is symmetric with respect to the y-axis. Choice (3) is symmetric with respect to the point $(-1, 0)$, which is not the origin. Choice (4) does not have any obvious symmetry.

The correct choice is **(1)**.

15. This expression can be factored by grouping by using the and difference of perfect squares:

$$2d^4 + 6d^3 - 18d^2 - 54d$$
$$2d^3(d+3) - 18d(d+3)$$
$$(2d^3 - 18d)(d+3)$$
$$2d(d^2 - 9)(d+3)$$
$$2d(d-3)(d+3)(d+3)$$
$$2d(d-3)(d+3)^2$$

Instead of factoring, you can answer this question by multiplying out each of the answer choices to see which one is equivalent to the given equation.

The correct choice is **(3)**.

16. The formula for converting radians to degrees is degrees = radians $\cdot \dfrac{180}{\pi}$.

Using this formula, the number of degrees in 1 radian is equal to approximately 57.3 degrees. Only in choice (1) is the angle of rotation less than 90 degrees.

The correct choice is **(1)**.

274   **ANSWERS August 2016**

17. To solve for J in terms of W and F, start by isolating the term that contains J:

$$\frac{1}{J} + \frac{1}{W} = \frac{1}{F}$$

$$-\frac{1}{W} = -\frac{1}{W}$$

$$\frac{1}{J} = \frac{1}{F} - \frac{1}{W}$$

Now simplify the right side of the equation by using the common denominator FW:

$$\frac{1}{J} = \frac{1}{F} - \frac{1}{W}$$

$$\frac{1}{J} = \frac{W}{FW} - \frac{F}{FW}$$

$$\frac{1}{J} = \frac{W - F}{FW}$$

Finally, take the reciprocal of both sides so the left side becomes J:

$$\frac{1}{J} = \frac{W - F}{FW}$$

$$\frac{J}{1} = \frac{FW}{W - F}$$

The correct choice is (**3**).

18. The terms of the sequence can first be determined with the recursive formula. Because $a_1 = 6$, the first term is 6. When $n = 2$, use the recursive part of the definition:

$$a_2 = 3a_{2-1} = 3a_1 = 3 \cdot 6 = 18$$

For $n = 3$, again use the recursive part of the definition:

$$a_3 = 3a_{3-1} = 3a_2 = 3 \cdot 18 = 54$$

Each term becomes 3 times the previous term to create the sequence 6, 18, 54, 162, 486, . . .

The answer choices are all direct formulas for calculating the terms of a sequence. Substitute $n = 1$, 2, and 3 into each formula to see which one leads to a sequence starting as 6, 18, and 54.

Choice (1): For $n = 1$:
$$a_1 = 6 \cdot 3^1 = 6 \cdot 3 = 18 \neq 6$$
Choice (2): For $n = 1$:
$$a_1 = 6 \cdot 3^{1+1} = 6 \cdot 3^2 = 6 \cdot 9 = 54 \neq 6$$
Choice (3): For $n = 1$, $n = 2$, and $n = 3$:
$$a_1 = 2 \cdot 3^1 = 2 \cdot 3 = 6$$
$$a_2 = 2 \cdot 3^2 = 2 \cdot 9 = 18$$
$$a_3 = 2 \cdot 3^3 = 2 \cdot 27 = 54$$
Choice (4): For $n = 1$:
$$a_1 = 2 \cdot 3^{1+1} = 2 \cdot 3^2 = 2 \cdot 9 = 18 \neq 6$$

The correct choice is **(3)**.

19. The set of points equidistant from a line and a point is a parabola that has the line as its directrix and the point as its focus. The vertex of the parabola is halfway between the focus and the directrix. Since the focus in this question is 4 units below the directrix, the vertex of the parabola will be 2 units above point R. So the vertex has the coordinates (2, –1).

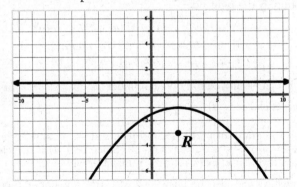

All four answer choices are in vertex form $y = a(x - h)^2 + k$, where the coordinates of the vertex are (h, k). Since the a-value for each of the answer choices is the same, the only difference is the vertex.

Since the vertex is (2, –1), $h = 2$ and $k = -1$. So the equation must be in the form
$$y = a(x - 2)^2 + (-1) = a(x - 2)^2 - 1$$

The correct choice is **(4)**.

276   **ANSWERS August 2016**

20. The left and right sides of each equation should be simplified to see if they are equivalent.

Check I:

$$(m + p)^2 = (m + p)(m + p) = m^2 + mp + mp + p^2 = m^2 + 2mp + p^2$$

This rule can be proved for all real numbers.

Check II:

$$(x + y)^3 = (x + y)(x + y)(x + y) = (x^2 + 2xy + y^2)(x + y)$$
$$= x^2 \cdot x + x^2 \cdot y + 2xy \cdot x + 2xy \cdot y + y^2 \cdot x + y^2 \cdot y$$
$$= x^3 + x^2y + 2x^2y + 2xy^2 + xy^2 + y^3$$
$$= x^3 + 3x^2y + 3xy^2 + y^3 \neq x^3 + 3xy + y^3$$

This rule cannot be proved for all real numbers.

Check III:

First simplify the left side:

$$(a^2 + b^2)^2 = (a^2 + b^2)(a^2 + b^2) = a^2 \cdot a^2 + a^2 \cdot b^2 + a^2 \cdot b^2 + b^2 \cdot b^2$$
$$= a^4 + 2a^2b^2 + b^4$$

Then simplify the right side:

$$(a^2 - b^2)^2 + (2ab)^2 = (a^2 - b^2)(a^2 - b^2) + (2ab)^2$$
$$= a^2 \cdot a^2 - a^2 \cdot b^2 - a^2 \cdot b^2 + b^2 \cdot b^2 + 4a^2b^2$$
$$= a^4 - 2a^2b^2 + b^4 + 4a^2b^2 = a^4 + 2a^2b^2 + b^4$$

This rule can be proved for all real numbers.

The correct choice is (**4**).

21. Because the graph of $p(x)$ has x-intercepts at -4, 0, and 3, the equation has the form $p(x) = a(x + 4)(x)(x - 3)$. Since $(x + 4)$ is a factor of $p(x)$, when you divide $p(x)$ by $(x + 4)$, the remainder will be 0.

This question can also be answered with the remainder theorem. The remainder theorem states that the remainder when $p(x)$ is divided by $(x - a)$ will be equal to $p(a)$. For this example, the remainder when $p(x)$ is divided by $(x + 4) = (x - (-4))$ will be equal to $p(-4)$. According to the graph, $p(-4) = 0$. So the remainder will also be 0.

The correct choice is (**3**).

ANSWERS August 2016   277

22. The easiest way to solve this question is to calculate the numerical value of each of the choices to see which is the most reasonable answer. With interest of 30% compounding every 14 days, the amount owed after a year should be a lot more than the original $300.

Choice (1):

$$300(.30)^{\frac{14}{365}} = 286.46$$

Choice (2):

$$300(1.30)^{\frac{14}{365}} = 303.03$$

Choice (3):

$$300(.30)^{\frac{365}{14}} \approx 0$$

Choice (4):

$$300(1.30)^{\frac{365}{14}} = 280,405.95$$

Even though choice (4) seems unbelievably high, all the other choices are way too low. Choice (3), which is nearly 0, is especially low. Choice (1) is actually lower than the original loan of $300.

The correct choice is **(4)**.

23. This system of three equations in three unknowns can be solved using the elimination method. Notice that the b-term in the first equation is $+5$ while the b-term in the second and third equations is -5.

Add the first and second equations to get a new equation with no b-term:

$$\begin{array}{r} a+5b-c=-20 \\ +\quad 4a-5b+4c=19 \\ \hline 5a+3c=-1 \end{array}$$

Add the first and third equations to make another new equation with no b-term:

$$\begin{array}{r} a+5b-c=-20 \\ +\quad -a-5b-5c=2 \\ \hline -6c=-18 \end{array}$$

ANSWERS August 2016

In the second new equation, the a-term also got eliminated. Solve for c by dividing both sides of the equation by -6:

$$\frac{-6c}{-6} = \frac{-18}{-6}$$
$$c = 3$$

Substitute 3 for c into the first new equation, $5a + 3c = -1$, to solve for a:

$$5a + 3 \cdot 3 = -1$$
$$5a + 9 = -1$$
$$\underline{-9 = -9}$$
$$5a = -10$$
$$\frac{5a}{5} = \frac{-10}{5}$$
$$a = -2$$

Substitute 3 for c and -2 for a into any of the original equations to solve for b:

$$a + 5b - c = -20$$
$$-2 + 5b - 3 = -20$$
$$5b - 5 = -20$$
$$\underline{+5 = +5}$$
$$5b = -15$$
$$\frac{5b}{5} = \frac{-15}{5}$$
$$b = -3$$

The correct choice is (2).

24. You are given that the population in 2010 was 19,378,000. So $P_0 = 19,378,000$. After 1 year, the population increased by 1.5%. To calculate the increase in population, multiply 19,378,000 by 0.015. Then add that amount to the original population to find the population after 1 year, P_1:

$$19{,}378{,}000 \cdot 0.015 = 290{,}670$$
$$19{,}378{,}000 + 290{,}670 = 19{,}668{,}670$$
$$P_1 = 19{,}668{,}670$$

Test each answer choice to see which one has $P_0 = 19,378,000$ and $P_1 = 19,668,670$.

ANSWERS August 2016

279

Choice (1):

$$P_0 = 19,378,000(1.5)^0 = 19,378,000 \cdot 1 = 19,378,000$$
$$P_1 = 19,378,000(1.5)^1 = 19,378,000 \cdot 1.5 = 29,067,000$$

Since the value of P_1 is wrong, this equation is not the correct answer.

Choice (2):

$$P_0 = 19,378,000$$
$$P_1 = 19,378,000 + 1.015P_{1-1} = 19,378,000 + 1.015P_0$$
$$= 19,378,000 + 1.015 \cdot 19,378,000 = 19,378,000 + 19,668,670$$
$$= 39,046,670$$

Since the value of P_1 is wrong, this equation is not the correct answer.

Choice (3):

$$P_0 = 19,378,000(1.015)^{0-1} = 19,378,000(1.015)^{-1}$$
$$= 19,378,000 \cdot 0.9852216749 \approx 19,091,625$$

Since the value of P_0 is wrong, this equation is not the correct answer.

Choice (4):

$$P_0 = 19,378,000$$
$$P_1 = 1.015P_{1-1} = 1.015P_0 = 1.015 \cdot 19,378,000$$
$$= 19,668,670$$

Since the values of P_0 and P_1 are correct, this equation is the correct answer.

A shorter way to solve this question is to know that when something grows by 1.5%, it gets multiplied by $1 + 0.015 = 1.015$. Choice (4) is a recursive sequence in which each term is equal to 1.015 multiplied by the previous term. This is the same as what happens in the model.

The correct choice is **(4)**.

PART II

25. When a person inhales, the volume of air in his or her lungs will increase until it reaches a maximum. When that person exhales, the volume of air in his or her lungs will decrease until it reaches a minimum.

Since breathing alternates between inhaling and exhaling, a graph of this scenario would look like this.

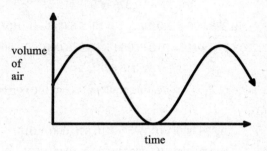

The amplitude of the sine curve is the difference between the maximum value and a middle value. Depending on the person, this amplitude can be large if the person's lungs can hold a lot of air or small if the person's lungs cannot hold a lot of air.

The scientist should focus on the amplitude.

26. $\left(3^{\frac{1}{5}}\right)^2$ can be simplified to $3^{\frac{1}{5} \cdot 2} = 3^{\frac{2}{5}}$.

$\sqrt[5]{9}$ is equivalent to $9^{\frac{1}{5}}$. Since $9 = 3^2$, $9^{\frac{1}{5}} = \left(3^2\right)^{\frac{1}{5}} = 3^{\frac{2}{5}} = \left(3^{\frac{1}{5}}\right)^2$.

27. $xi(i - 7i)^2 = xi(-6i)^2 = xi \cdot 36i^2 = xi \cdot 36(-1) = -36xi$.

The key step is replacing i^2 with -1 since that is how i is defined.

So $xi(i - 7i)^2$ simplifies to $-36xi$.

ANSWERS August 2016   281

28. If cos θ = –0.7, start by substituting it into the given formula to solve for sin θ.

$$\sin^2\theta + (-0.7)^2 = 1$$

$$\sin^2\theta + 0.49 = 1$$

$$\underline{-0.49 = -0.49}$$

$$\sin^2\theta = 0.51$$

$$\sqrt{\sin^2\theta} = \pm\sqrt{0.51}$$

$$\sin\theta \approx \pm 0.7141428$$

Sine is positive only in quadrants I and II, while cosine is positive only in quadrants I and IV. Since it is given that θ is in quadrant II and since sine is positive in quadrant II, use the positive solution sin θ ≈ +0.7141428. Now solve for tan θ to the *nearest hundreth*:

$$\tan\theta = \frac{\sin\theta}{\cos\theta}$$

$$\tan\theta \approx \frac{0.7141428}{-0.7} \approx -1.02$$

29. This question is asking whether or not 6 minutes, or 360 seconds, is more than 2 standard deviations above the mean wait time.

You are given that the mean time is 226 seconds and the standard deviation is 38 seconds. A wait time more than 2 standard deviations, which is 76 seconds, above the mean or less than 2 standard deviations below the mean should be unusual.

Expected wait times are between 226 – 76 = 150 seconds and 226 + 76 = 302 seconds. Six minutes, which is 360 seconds, is greater than 302 seconds. So a wait time of 6 minutes is more than 2 standard deviations above the mean.

Elizabeth's wait time was unusual.

282   **ANSWERS August 2016**

30. An x-intercept is the x-coordinate of the point on a graph that has a y-coordinate of 0. For the graph of $f(x)$, the x-intercept can be found by solving the equation:

$$0 = \log_{10}(x-4)$$
$$10^0 = x - 4$$
$$1 = x - 4$$
$$\underline{+4 = +4}$$
$$5 = x$$

So the x-intercept for $f(x)$ is 5.

According to the chart for $h(x)$, $h(2) = 0$. So the x-intercept for $h(x)$ is 2.

The x-intercept for function f is greater than the x-intercept for function h.

Function f has an x-intercept of 5. Function h has an x-intercept of 2.

31. The average rate of change can be calculated by dividing the change in braking distance divided by the change in speed. For this question, the calculation becomes:

$$\text{average rate of change} = \frac{306.25 - 156.25}{70 - 50} = \frac{150}{20} = 7.5$$

This rate of change means that for every 1 mile per hour above 50 miles per hour, the braking distance increases, on average, by about 7.5 feet.

32. Events are independent if $P(A \cap B) = P(A) \cdot P(B)$.

To check if events are independent in this question, organize the information on a Venn diagram. In the diagram below, $P(A)$ is the sum of the regions labeled II and III; $P(B)$ is the sum of the two regions labeled III and IV; $P(A \cup B)$ is the sum of the three regions labeled II, III, and IV; and $P(A \cap B)$ is the overlapping region labeled III.

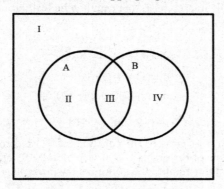

To find the probability for region III, $P(A \cup B)$, notice that

$$P(A) + P(B) = 1.1 = II + III + III + IV$$

Also it is given that $P(A \cup B) = 0.8 = II + III + IV$.

The difference between these two numbers will be the probability for region III, which is $P(A \cap B) = 1.1 - 0.8 = 0.3$.

Since $P(A) \cdot P(B) = 0.6 \cdot 0.5 = 0.3$ and this is the same as $P(A \cap B)$, these are independent events.

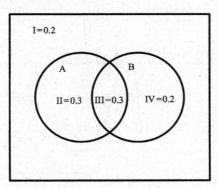

PART III

33. The zeros of a function p are the values of x for which $p(x) = 0$. To find the zeros of this function, solve the equation $0 = x^3 + x^2 - 4x - 4$. A polynomial equation with four terms like this can sometimes be factored with a technique called "factor by grouping." Factor an x^2 out of the first two terms, and factor -4 out of the last two terms. Luckily, there ends up being an $(x + 1)$ in each term that can then be further factored out:

$$0 = x^3 + x^2 - 4x - 4$$
$$0 = x^2(x+1) - 4(x+1)$$
$$0 = (x^2 - 4)(x+1)$$
$$0 = (x-2)(x+2)(x+1)$$

The solutions are $x = 2$, $x = -2$, and $x = -1$.

For the graph, the zeros of the function are related to the x-intercepts. The x-intercepts of the graph of p are $(2, 0)$, $(-2, 0)$, and $(-1, 0)$. Another useful point on the graph is the y-intercept, which you get by finding $p(0)$:

$$p(0) = 0^3 + 0^2 - 4(0) - 4 = -4$$

So the y-intercept is $(0, -4)$.

Plot the x-intercepts and the y-intercept on the graph. Connect them with a sideways "S" shape. You can use the graphing calculator to get an idea of what the graph should look like before you draw it.

For the TI-84:

For the TI-Nspire:

34. The formula for half-life is shown:

$$\text{amount remaining} = \text{initial amount} \cdot \left(\frac{1}{2}\right)^{\frac{t}{\text{half-life}}}$$

Substitute the given information into the formula:

$$7 = 20 \cdot \left(\frac{1}{2}\right)^{\frac{t}{8.02}}$$

Solving this equation for t will require you to use logarithms:

$$7 = 20 \cdot \left(\frac{1}{2}\right)^{\frac{t}{8.02}}$$

$$\frac{7}{20} = \frac{20 \cdot \left(\frac{1}{2}\right)^{\frac{t}{8.02}}}{20}$$

$$\frac{7}{20} = \left(\frac{1}{2}\right)^{\frac{t}{8.02}}$$

$$\log_{\frac{1}{2}}\left(\frac{7}{20}\right) = \frac{t}{8.02}$$

$$8.02 \cdot \log_{\frac{1}{2}}\left(\frac{7}{20}\right) = t$$

Enter this into either graphing calculator. The solution is $t \approx 12.14$, which rounds to 12 days.

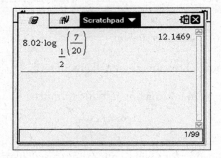

35. Isolate the radical expression by subtracting x from both sides of the equation and then squaring both sides of the equation:

$$\sqrt{2x-7} + x = 5$$
$$\underline{-x = -x}$$
$$\sqrt{2x-7} = 5 - x$$
$$\left(\sqrt{2x-7}\right)^2 = (5-x)^2$$
$$2x - 7 = 25 - 10x + x^2$$

Solve the quadratic equation by first bringing every term over to one side. Then try to factor the resulting expression:

$$2x - 7 = 25 - 10x + x^2$$
$$\underline{-2x + 7 = -2x + 7}$$
$$0 = x^2 - 12x + 32$$
$$0 = (x-4)(x-8)$$

$$x - 4 = 0 \text{ or } x - 8 = 0$$
$$x = 4 \text{ or } x = 8$$

ANSWERS August 2016

When solving a radical equation, there is a chance that an extraneous solution was formed in the step where both sides of the equation was squared. Both solutions must be substituted back into the original equation to see if this has happened.

Check $x = 4$:

$$\sqrt{2 \cdot 4 - 7} + 4 = 5$$
$$\sqrt{1} + 4 = 5$$
$$1 + 4 = 5$$
$$5 = 5$$

This is a solution to the equation.

Check $x = 8$:

$$\sqrt{2 \cdot 8 - 7} + 8 = 5$$
$$\sqrt{9} + 4 = 5$$
$$3 + 4 = 5$$
$$7 = 5$$

This is not a solution to the equation.

The $x = 8$ solution must be rejected, leaving $x = 4$ as the only solution.

36. a) There are several reasons why Ayva's hypothesis might be incorrect. One is that the sample size is very small, just 10 students per group. Another is that she cannot be sure that 10 minutes is enough time for the different drinks to impact reading speed.

b) In Ayva's initial experiment, the difference in the means, group 1 – group 2, was –1.4. The computer-generated simulations using the same 20 people split up into two groups randomly show that a difference of –1.4 is not so rare. A difference of 4 would be incredibly rare.

Since randomly splitting the 20 people into equal groups is not unlikely to produce a difference of means of –1.4, the difference is not statistically significant. So the difference is not an unusual occurrence.

PART IV

37. The formula for the amount of money at the end of x years is $f(x) = P\left(1+\dfrac{r}{n}\right)^{nx}$, where P is the initial deposit, r is the interest rate, and n is the number of compoundings per year. For options A and B, this formula becomes:

$$A(x) = 5{,}000\left(1+\frac{0.045}{1}\right)^{1x}$$

$$B(x) = 5{,}000\left(1+\frac{0.046}{4}\right)^{4x}$$

Substitute $x = 6$ into both equations to find how much money each option will have after 6 years:

$$A(6) = 5{,}000\left(1+\frac{0.045}{1}\right)^{1 \cdot 6} \approx 6{,}511.30$$

$$B(6) = 5{,}000\left(1+\frac{0.046}{4}\right)^{4 \cdot 6} \approx 6{,}578.87$$

Find the difference of the two options:
$$6{,}578.87 - 6{,}511.30 = 67.57$$

Option B will earn \$67.57 more than option A.

To find the amount of time it takes for the money to double in option B, solve the equation $10{,}000 = 5{,}000\left(1+\dfrac{0.046}{4}\right)^{4x}$ using logarithms:

$$10{,}000 = 5{,}000\left(1+\dfrac{0.046}{4}\right)^{4x}$$

$$\dfrac{10{,}000}{5{,}000} = \dfrac{5{,}000\left(1+\dfrac{0.046}{4}\right)^{4x}}{5{,}000}$$

$$2 = \left(1+\dfrac{0.046}{4}\right)^{4x}$$

$$2 = 1.0115^{4x}$$

$$\log_{1.0115} 2 = 4x$$

$$\dfrac{\log_{1.0115} 2}{4} = \dfrac{4x}{4}$$

$$\dfrac{\log_{1.0115} 2}{4} = x$$

Enter this equation into the calculator to solve it. The solution is $x \approx 15.15$, which rounds to the *nearest tenth of a year* to 15.2 years.

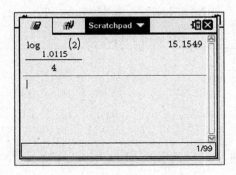

290   SELF-ANALYSIS CHART August 2016

Topic	Question Numbers	Number of Points	Your Points	Your Percentage
1. Polynomial Expressions and Equations	5, 11, 15, 20, 21, 33	2 + 2 + 2 + 2 + 2 + 4 = 14		
2. Complex Numbers	1, 27	2 +2 = 4		
3. Exponential Expressions and Equations	13, 22, 24, 34, 37	2 + 2 + 2 + 4 + 6 = 16		
4. Rational Expressions and Equations	17	2		
5. Radical Expressions and Equations	26, 35	2 + 4 = 16		
6. Trigonometric Expressions and Equations	10, 16, 25, 28	2 + 2 + 2 + 2 = 8		
7. Graphing	3, 6, 19	2 + 2 + 2 = 6		
8. Functions	14, 30, 31	2 + 2 + 2 = 6		
9. Systems of Equations	23	2		
10. Sequences and Series	8, 9, 18	2 + 2 + 2 = 6		
11. Probability	7, 32	2 + 2 = 4		
12. Statistics	2, 4, 12, 29, 36	2 + 2 + 2 + 2 + 4 = 12		

SELF-ANALYSIS CHART August 2016

HOW TO CONVERT YOUR RAW SCORE TO YOUR ALGEBRA II REGENTS EXAMINATION SCORE

The accompanying conversion chart must be used to determine your final score on the August 2016 Regents Examination in Algebra II. To find your final exam score, locate in the column labeled "Raw Score" the total number of points you scored out of a possible 86 points. Since partial credit is allowed in Parts II, III, and IV of the test, you may need to approximate the credit you would receive for a solution that is not completely correct. Then locate in the adjacent column to the right the scale score that corresponds to your raw score. The scale score is your final Algebra II Regents Examination score.

SELF-ANALYSIS CHART August 2016

Regents Examination in Algebra II—August 2016
Chart for Converting Total Test Raw Scores to Final
Examination Scores (Scaled Scores)

Raw Score	Scale Score	Performance Level	Raw Score	Scale Score	Performance Level	Raw Score	Scale Score	Performance Level
86	100	5	57	82	4	28	66	3
85	99	5	56	82	4	27	65	3
84	98	5	55	81	4	26	64	2
83	98	5	54	81	4	25	63	2
82	97	5	53	80	4	24	62	2
81	96	5	52	80	4	23	61	2
80	95	5	51	80	4	22	59	2
79	95	5	50	79	4	21	58	2
78	94	5	49	79	4	20	55	2
77	93	5	48	78	4	19	54	1
76	93	5	47	78	4	18	53	1
75	92	5	46	77	3	17	52	1
74	91	5	45	77	3	16	50	1
73	91	5	44	77	3	15	48	1
72	90	5	43	76	3	14	46	1
71	89	5	42	76	3	13	44	1
70	89	5	41	75	3	12	42	1
69	88	5	40	75	3	11	39	1
68	88	5	39	74	3	10	37	1
67	87	5	38	74	3	9	34	1
66	86	5	37	73	3	8	31	1
65	86	5	36	72	3	7	28	1
64	86	5	35	72	3	6	25	1
63	85	5	34	71	3	5	21	1
62	84	4	33	70	3	4	18	1
61	84	4	32	70	3	3	14	1
60	83	4	31	69	3	2	10	1
59	83	4	30	68	3	1	5	1
58	82	4	29	67	3	0	0	1

293

Examination
June 2017
Algebra II

HIGH SCHOOL MATH REFERENCE SHEET

Conversions

1 inch = 2.54 centimeters

1 meter = 39.37 inches

1 mile = 5280 feet

1 mile = 1760 yards

1 mile = 1.609 kilometers

1 kilometer = 0.62 mile

1 pound = 16 ounces

1 pound = 0.454 kilogram

1 kilogram = 2.2 pounds

1 ton = 2000 pounds

1 cup = 8 fluid ounces

1 pint = 2 cups

1 quart = 2 pints

1 gallon = 4 quarts

1 gallon = 3.785 liters

1 liter = 0.264 gallon

1 liter = 1000 cubic centimeters

Formulas

Triangle	$A = \dfrac{1}{2}bh$
Parallelogram	$A = bh$
Circle	$A = \pi r^2$
Circle	$C = \pi d$ or $C = 2\pi r$

EXAMINATION June 2017

Formulas (continued)

General Prisms	$V = Bh$
Cylinder	$V = \pi r^2 h$
Sphere	$V = \dfrac{4}{3}\pi r^3$
Cone	$V = \dfrac{1}{3}\pi r^2 h$
Pyramid	$V = \dfrac{1}{3}Bh$
Pythagorean Theorem	$a^2 + b^2 = c^2$
Quadratic Formula	$x = \dfrac{-b \pm \sqrt{b^2 - 4ac}}{2a}$
Arithmetic Sequence	$a_n = a_1 + (n-1)d$
Geometric Sequence	$a_n = a_1 r^{n-1}$
Geometric Series	$S_n = \dfrac{a_1 - a_1 r^n}{1 - r}$ where $r \neq 1$
Radians	$1 \text{ radian} = \dfrac{180}{\pi} \text{ degrees}$
Degrees	$1 \text{ degree} = \dfrac{\pi}{180} \text{ radians}$
Exponential Growth/Decay	$A = A_0 e^{k(t - t_0)} + B_0$

EXAMINATION June 2017

PART I

Answer all 24 questions in this part. Each correct answer will receive 2 credits. No partial credit will be allowed. For each statement or question, write in the space provided the numeral preceding the word or expression that best completes the statement or answers the question. [48 credits]

1 The graph of the function $p(x)$ is sketched below.

Which equation could represent $p(x)$?

(1) $p(x) = (x^2 - 9)(x - 2)$
(2) $p(x) = x^3 - 2x^2 + 9x + 18$
(3) $p(x) = (x^2 + 9)(x - 2)$
(4) $p(x) = x^3 - 2x^2 - 9x - 18$

1_____

2 What is the solution to $8(2^{x+3}) = 48$?

(1) $x = \dfrac{\ln 6}{\ln 2} - 3$

(2) $x = 0$

(3) $x = \dfrac{\ln 48}{\ln 16} - 3$

(4) $x = \ln 4 - 3$

2_____

296   **EXAMINATION June 2017**

3 Cheap and Fast gas station is conducting a consumer satisfaction survey. Which method of collecting data would most likely lead to a biased sample?

  (1) interviewing every 5th customer to come into the station

  (2) interviewing customers chosen at random by a computer at the checkout

  (3) interviewing customers who call an 800 number posted on the customers' receipts

  (4) interviewing every customer who comes into the station on a day of the week chosen at random out of a hat   3 _____

4 The expression $6xi^3(-4xi + 5)$ is equivalent to

  (1) $2x - 5i$ (3) $-24x^2 + 30x - i$

  (2) $-24x^2 - 30xi$ (4) $26x - 24x^2i - 5i$ 4 _____

5 If $f(x) = 3|x| - 1$ and $g(x) = 0.03x^3 - x + 1$, an approximate solution for the equation $f(x) = g(x)$ is

  (1) 1.96   (3) (−0.99, 1.96)

  (2) 11.29   (4) (11.29, 32.87)   5 _____

6 Given the parent function $p(x) = \cos x$, which phrase best describes the transformation used to obtain the graph of $g(x) = \cos(x + a) - b$, if a and b are positive constants?

  (1) right a units, up b units

  (2) right a units, down b units

  (3) left a units, up b units

  (4) left a units, down b units   6 _____

7 The solution to the equation $4x^2 + 98 = 0$ is

(1) ± 7

(3) $\pm \dfrac{7\sqrt{2}}{2}$

(2) $\pm 7i$

(4) $\pm \dfrac{7i\sqrt{2}}{2}$

7 _____

8 Which equation is represented by the graph shown below?

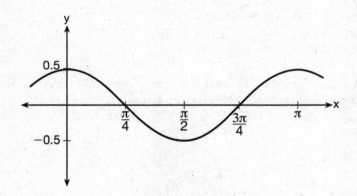

(1) $y = \dfrac{1}{2} \cos 2x$

(3) $y = \dfrac{1}{2} \cos x$

(2) $y = \cos x$

(4) $y = 2 \cos \dfrac{1}{2} x$

8 _____

9 A manufacturing company has developed a cost model, $C(x) = 0.15x^3 + 0.01x^2 + 2x + 120$, where x is the number of items sold, in thousands. The sales price can be modeled by $S(x) = 30 - 0.01x$. Therefore, revenue is modeled by $R(x) = x \cdot S(x)$.

The company's profit, $P(x) = R(x) - C(x)$, could be modeled by

(1) $0.15x^3 + 0.02x^2 - 28x + 120$
(2) $-0.15x^3 - 0.02x^2 + 28x - 120$
(3) $-0.15x^3 + 0.01x^2 - 2.01x - 120$
(4) $-0.15x^3 + 32x + 120$

9 _____

10 A game spinner is divided into 6 equally sized regions, as shown in the diagram below.

For Miles to win, the spinner must land on the number 6. After spinning the spinner 10 times, and losing all 10 times, Miles complained that the spinner is unfair. At home, his dad ran 100 simulations of spinning the spinner 10 times, assuming the probability of winning each spin is $\frac{1}{6}$. The output of the simulation is shown in the following diagram.

Which explanation is appropriate for Miles and his dad to make?

(1) The spinner was likely unfair, since the number 6 failed to occur in about 20% of the simulations.
(2) The spinner was likely unfair, since the spinner should have landed on the number 6 by the sixth spin.
(3) The spinner was likely not unfair, since the number 6 failed to occur in about 20% of the simulations.
(4) The spinner was likely not unfair, since in the output the player wins once or twice in the majority of the simulations.

10 _____

300　　　　**EXAMINATION June 2017**

11 Which binomial is a factor of $x^4 - 4x^2 - 4x + 8$?

(1) $x - 2$　　　　　　　(3) $x - 4$
(2) $x + 2$　　　　　　　(4) $x + 4$　　　　　11 _____

12 Given that $\sin^2 \theta + \cos^2 \theta = 1$ and $\sin \theta = -\dfrac{\sqrt{2}}{5}$, what is a possible value of $\cos \theta$?

(1) $\dfrac{5 + \sqrt{2}}{5}$　　　　　(3) $\dfrac{3\sqrt{3}}{5}$

(2) $\dfrac{\sqrt{23}}{5}$　　　　　　(4) $\dfrac{\sqrt{35}}{5}$　　　　　12 _____

13 A student studying public policy created a model for the population of Detroit, where the population decreased 25% over a decade. He used the model $P = 714(0.75)^d$, where P is the population, in thousands, d decades after 2010. Another student, Suzanne, wants to use a model that would predict the population after y years. Suzanne's model is best represented by

(1) $P = 714(0.6500)^y$　　　(3) $P = 714(0.9716)^y$
(2) $P = 714(0.8500)^y$　　　(4) $P = 714(0.9750)^y$　　13 _____

14 The probability that Gary and Jane have a child with blue eyes is 0.25, and the probability that they have a child with blond hair is 0.5. The probability that they have a child with both blue eyes and blond hair is 0.125. Given this information, the events blue eyes and blond hair are

　　　　I:　dependent
　　　　II:　independent
　　　　III:　mutually exclusive

(1) I, only　　　　　　　(3) I and III
(2) II, only　　　　　　　(4) II and III　　　　　14 _____

EXAMINATION June 2017   301

15 Based on climate data that have been collected in Bar Harbor, Maine, the average monthly temperature, in degrees F, can be modeled by the equation $B(x) = 23.914\sin(0.508x - 2.116) + 55.300$. The same governmental agency collected average monthly temperature data for Phoenix, Arizona, and found the temperatures could be modeled by the equation $P(x) = 20.238\sin(0.525x - 2.148) + 86.729$.

Which statement can *not* be concluded based on the average monthly temperature models x months after starting data collection?

(1) The average monthly temperature variation is more in Bar Harbor than in Phoenix.

(2) The midline average monthly temperature for Bar Harbor is lower than the midline temperature for Phoenix.

(3) The maximum average monthly temperature for Bar Harbor is 79°F, to the nearest degree.

(4) The minimum average monthly temperature for Phoenix is 20°F, to the nearest degree.   15 _____

16 For $x \neq 0$, which expressions are equivalent to one divided by the sixth root of x?

$$\text{I. } \frac{\sqrt[6]{x}}{\sqrt[3]{x}} \qquad \text{II. } \frac{x^{\frac{1}{6}}}{x^{\frac{1}{3}}} \qquad \text{III. } x^{-\frac{1}{6}}$$

(1) I and II, only   (3) II and III, only
(2) I and III, only   (4) I, II, and III   16 _____

17 A parabola has its focus at $(1, 2)$ and its directrix is $y = -2$. The equation of this parabola could be

(1) $y = 8(x + 1)^2$ (3) $y = 8(x - 1)^2$

(2) $y = \frac{1}{8}(x + 1)^2$ (4) $y = \frac{1}{8}(x - 1)^2$ 17 _____

302   **EXAMINATION June 2017**

18 The function $p(t) = 110e^{0.03922t}$ models the population of a city, in millions, t years after 2010. As of today, consider the following two statements:

      I.   The current population is 110 million.
     II.   The population increases continuously by approximately 3.9% per year.

This model supports

(1) I, only   (3) both I and II
(2) II, only   (4) neither I nor II   18 _____

19 To solve $\dfrac{2x}{x-2} - \dfrac{11}{x} = \dfrac{8}{x^2 - 2x}$, Ren multiplied both sides by the least common denominator. Which statement is true?

(1) 2 is an extraneous solution.

(2) $\dfrac{7}{2}$ is an extraneous solution.

(3) 0 and 2 are extraneous solutions.

(4) This equation does not contain any extraneous solutions.   19 _____

20 Given $f(9) = -2$, which function can be used to generate the sequence $-8, -7.25, -6.5, -5.75, \ldots$?

(1) $f(n) = -8 + 0.75n$
(2) $f(n) = -8 - 0.75(n - 1)$
(3) $f(n) = -8.75 + 0.75n$
(4) $f(n) = -0.75 + 8(n - 1)$ 20 _____

EXAMINATION June 2017   303

21 The function $f(x) = 2^{-0.25x} \cdot \sin\left(\dfrac{\pi}{2}x\right)$ represents a damped sound wave function. What is the average rate of change for this function on the interval $[-7, 7]$, to the *nearest hundredth*?

(1) -3.66 (3) -0.26

(2) -0.30 (4) 3.36 21 _____

22 Mallory wants to buy a new window air conditioning unit. The cost for the unit is \$329.99. If she plans to run the unit three months out of the year for an annual operating cost of \$108.78, which function models the cost per year over the lifetime of the unit, $C(n)$, in terms of the number of years, n, that she owns the air conditioner?

(1) $C(n) = 329.99 + 108.78n$

(2) $C(n) = 329.99 + 326.34n$

(3) $C(n) = \dfrac{329.99 + 108.78n}{n}$

(4) $C(n) = \dfrac{329.99 + 326.34n}{n}$ 22 _____

23 The expression $\dfrac{-3x^2 - 5x + 2}{x^3 + 2x^2}$ can be rewritten as

(1) $\dfrac{-3x - 3}{x^2 + 2x}$ (3) $-3x^{-1} + 1$

(2) $\dfrac{-3x - 1}{x^2}$ (4) $-3x^{-1} + x^{-2}$ 23 _____

304   **EXAMINATION June 2017**

24 Jasmine decides to put \$100 in a savings account each month. The account pays 3% annual interest, compounded monthly. How much money, S, will Jasmine have after one year?

(1) $S = 100(1.03)^{12}$

(2) $S = \dfrac{100 - 100(1.0025)^{12}}{1 - 1.0025}$

(3) $S = 100(1.0025)^{12}$

(4) $S = \dfrac{100 - 100(1.03)^{12}}{1 - 1.03}$

24 _____

EXAMINATION June 2017   305

PART II

Answer all 8 questions in this part. Each correct answer will receive 2 credits. Clearly indicate the necessary steps, including appropriate formula substitutions, diagrams, graphs, charts, etc. For all questions in this part, a correct numerical answer with no work shown will receive only 1 credit. [16 credits]

25 Given $r(x) = x^3 - 4x^2 + 4x - 6$, find the value of $r(2)$.

What does your answer tell you about $x - 2$ as a factor of $r(x)$? Explain.

306　　　　**EXAMINATION June 2017**

26 The weight of a bag of pears at the local market averages 8 pounds with a standard deviation of 0.5 pound. The weights of all the bags of pears at the market closely follow a normal distribution. Determine what percentage of bags, to the *nearest integer*, weighed *less* than 8.25 pounds.

27 Over the set of integers, factor the expression $4x^3 - x^2 + 16x - 4$ completely.

28 The graph below represents the height above the ground, h, in inches, of a point on a triathlete's bike wheel during a training ride in terms of time, t, in seconds.

Identify the period of the graph and describe what the period represents in this context.

29 Graph $y = 400(.85)^{2x} - 6$ on the set of axes below.

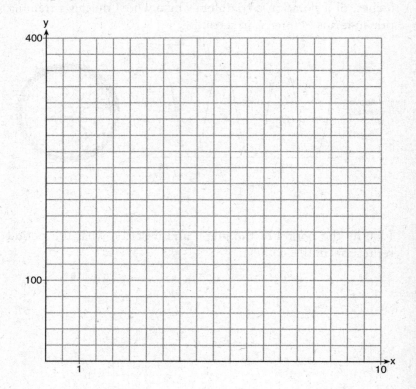

EXAMINATION June 2017

30 Solve algebraically for all values of x:

$$\sqrt{x-4} + x = 6$$

31 Write $\sqrt[3]{x} \cdot \sqrt{x}$ as a single term with a rational exponent.

310 **EXAMINATION June 2017**

32 Data collected about jogging from students with two older siblings are shown in the table below.

	Neither Sibling Jogs	One Sibling Jogs	Both Siblings Jog
Student Does Not Jog	1168	1823	1380
Student Jogs	188	416	400

Using these data, determine whether a student with two older siblings is more likely to jog if one sibling jogs or if both siblings jog. Justify your answer.

EXAMINATION June 2017   311

PART III

Answer all 4 questions in this part. Each correct answer will receive 4 credits. Clearly indicate the necessary steps, including appropriate formula substitutions, diagrams, graphs, charts, etc. For all questions in this part, a correct numerical answer with no work shown will receive only 1 credit. [16 credits]

33 Solve the following system of equations algebraically for all values of x, y, and z:

$$x + y + z = 1$$
$$2x + 4y + 6z = 2$$
$$-x + 3y - 5z = 11$$

312   **EXAMINATION June 2017**

34 Jim is looking to buy a vacation home for \$172,600 near his favorite southern beach. The formula to compute a mortgage payment, M, is $M = P \cdot \dfrac{r(1+r)^N}{(1+r)^N - 1}$ where P is the principal amount of the loan, r is the monthly interest rate, and N is the number of monthly payments. Jim's bank offers a monthly interest rate of 0.305% for a 15-year mortgage.

With no down payment, determine Jim's mortgage payment, rounded to the *nearest dollar*.

Algebraically determine and state the down payment, rounded to the *nearest dollar*, that Jim needs to make in order for his mortgage payment to be \$1100.

35 Graph $y = \log_2(x + 3) - 5$ on the set of axes below. Use an appropriate scale to include *both* intercepts.

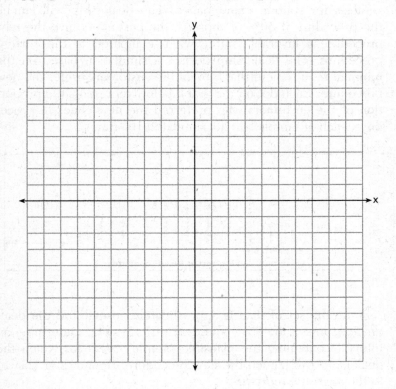

Describe the behavior of the given function as x approaches -3 and as x approaches positive infinity.

36 Charlie's Automotive Dealership is considering implementing a new check-in procedure for customers who are bringing their vehicles for routine maintenance. The dealership will launch the procedure if 50% or more of the customers give the new procedure a favorable rating when compared to the current procedure. The dealership devises a simulation based on the minimal requirement that 50% of the customers prefer the new procedure. Each dot on the graph below represents the proportion of the customers who preferred the new check-in procedure, each of sample size 40, simulated 100 times.

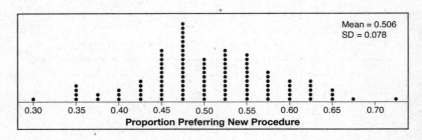

Assume the set of data is approximately normal and the dealership wants to be 95% confident of its results. Determine an interval containing the plausible sample values for which the dealership will launch the new procedure. Round your answer to the *nearest hundredth*.

Forty customers are selected randomly to undergo the new check-in procedure and the proportion of customers who prefer the new procedure is 32.5%. The dealership decides *not* to implement the new check-in procedure based on the results of the study. Use statistical evidence to explain this decision.

EXAMINATION June 2017

PART IV

Answer the question in this part. A correct answer will receive 6 credits. Clearly indicate the necessary steps, including appropriate formula substitutions, diagrams, graphs, charts, etc. A correct numerical answer with no work shown will receive only 1 credit. [6 credits]

37 A radioactive substance has a mass of 140 g at 3 p.m. and 100 g at 8 p.m. Write an equation in the form $A = A_0 \left(\frac{1}{2}\right)^{\frac{t}{h}}$ that models this situation, where h is the constant representing the number of hours in the half-life, A_0 is the initial mass, and A is the mass t hours after 3 p.m.

Using this equation, solve for h, to the *nearest ten thousandth*.

Determine when the mass of the radioactive substance will be 40 g. Round your answer to the *nearest tenth of an hour*.

Answers June 2017
Algebra II

Answer Key

PART I
1. (1)
2. (1)
3. (3)
4. (2)
5. (2)
6. (4)
7. (4)
8. (1)
9. (2)
10. (3)
11. (1)
12. (2)
13. (3)
14. (2)
15. (4)
16. (4)
17. (4)
18. (2)
19. (1)
20. (3)
21. (3)
22. (3)
23. (4)
24. (2)

PART II
25. -6, $(x - 2)$ is not a factor of $r(x)$
26. 69%
27. $(x^2 + 4)(4x - 1)$
28. Period is $\frac{2}{3}$. The wheel takes $\frac{2}{3}$ seconds to make one rotation.
29.
30. 5
31. $x^{\frac{5}{6}}$
32. More likely if both siblings jog

PART III
33. $(0, 2, -1)$
34. $1247; $20,407
35. y approaches negative infinity as x approaches -3. y approaches positive infinity as x approaches positive infinity.
36. The confidence interval is between 0.35 and 0.66; 0.325 is not between 0.35 and 0.66.

PART IV
37. $100 = 140\left(\frac{1}{2}\right)^{\frac{5}{h}}$; $h = 10.3002$; $x = 18.6$

In **Parts II–IV**, you are required to show how you arrived at your answers. For sample methods of solutions, see the *Answers Explained* section.

Answers Explained

PART I

1. When a graph has an x-intercept at $(a, 0)$, there is a factor of $(x - a)$. Since the graph has x-intercepts at -3, 2, and 3, the function could be:

$$p(x) = (x - (-3))(x - 2)(x - 3)$$
$$= (x + 3)(x - 2)(x - 3)$$

None of the answer choices looks exactly like this. Since

$x^2 - 9 = (x - 3)(x + 3)$, choice (1) is equivalent.

Another way to answer this is to graph the four answer choices on a graphing calculator to see which one most resembles the question.

When the function in choice (1) is graphed, it looks like the following.

For the TI-84:   For the TI-Nspire:

The correct choice is **(1)**.

2. To isolate the x, follow these four steps:

    Step 1: Divide both sides by 8.

    $$\frac{8(2^{x+3})}{8} = \frac{48}{8}$$
    $$2^{x+3} = 6$$

Step 2: Rearrange this exponential equation as a log equation. An equation of the form $a^b = c$ can be rewritten as $\log_a c = b$.

$$\log_2 6 = x + 3$$

Step 3: Subtract 3 from both sides of the equation.

$$\log_2 6 = x + 3$$
$$-3 = -3$$
$$\log_2 6 - 3 = x$$

Step 4: Since this does not perfectly match any of the choices, you can use the change of base formula, $\log_b a = \dfrac{\log_c a}{\log_c b}$. Use $c = e$.

$$x = \frac{\log_e 6}{\log_e 2} - 3$$
$$= \frac{\ln 6}{\ln 2} - 3$$

Another way to answer this question is to test each of the four answer choices to see which makes the left side of the equation equal to 48.

For the TI-84:   For the TI-Nspire:

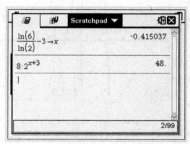

The correct choice is **(1)**.

3. A self-selection bias is when the subjects get to choose whether or not they want to participate in a survey. This is not as random as picking every 5th customer, using a random number generator to decide who participates, or interviewing every customer on a random day. People who volunteer to be in a survey by calling a phone number may be the people who are either very satisfied or very unsatisfied. However, they may not represent the average customer.

The correct choice is **(3)**.

4. First use the distributive property.

$$6xi^3(-4xi + 5) = 6xi^3 \cdot -4xi + 6xi^3 \cdot 5$$
$$= -24x^2i^4 + 30xi^3$$

Next, simplify the i-terms using the fact that $i^2 = -1$, $i^3 = -i$, and $i^4 = 1$. You could also use the graphing calculator to simplify the i-terms.

For the TI-84:   For the TI-Nspire:

 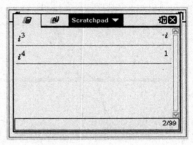

$$-24x^2i^4 + 30xi^3 = -24x^2(1) + 30x(-i)$$
$$= -24x^2 - 30xi$$

The correct choice is **(2)**.

5. There are three values for x that satisfy the equation $f(x) = g(x)$. It is difficult to solve the equation $3|x| - 1 = 0.03x^3 - x + 1$ with algebra. So the best way to solve this is with the graphing calculator.

On the same set of axes, graph the following equations:

$$y = 3|x| - 1 \text{ and } y = 0.03x^3 - x + 1$$

Then locate the x-coordinates of the intersection points.

For the TI-84:   For the TI-Nspire:

Based on this portion of the graph, two of the solutions are $x = -0.99$ and $x = 0.5$. It is tempting to answer choice (3) here. However, choice (3) is not correct because a solution to the equation is a value of x and not an ordered pair describing an intersection point.

If you zoom out on the graph, you will find another intersection point at $x = 11.29$.

For the TI-84:   For the TI-Nspire:

Since this is a multiple-choice question, the easiest way to solve it is to eliminate choices (3) and (4) first since they are ordered pairs instead of numbers. Then substitute the other choices into the two functions to see which results in the same output value for $f(x)$ and $g(x)$. Since $f(11.29) = 32.87$ and $g(11.29) = 32.87$, one solution is $x = 11.29$.

The correct choice is **(2)**.

6. If you pick values for a and b, like $a = 30$ and $b = 2$, and set your calculator to degree mode, you can use the graphing calculator to graph $p(x)$ and $g(x)$.

For the TI-84:   For the TI-Nspire:

The graph of $g(x)$ is like the graph of $p(x)$ but shifted 30 to the left and 2 down.

ANSWERS June 2017   **321**

In general, the graph of $p(x + a)$ will be the graph of $p(x)$ shifted a units to the left and the graph of $p(x) - b$ will be the graph of $p(x)$ shifted b units down.

The correct choice is **(4)**.

7. Use algebra to isolate the x.

$$4x^2 = -98$$

$$\frac{4x^2}{4} = \frac{-98}{4}$$

$$x^2 = \frac{-98}{4}$$

$$\sqrt{x^2} = \pm\sqrt{-\frac{98}{4}}$$

$$x = \pm\frac{\sqrt{-98}}{\sqrt{4}}$$

$$= \pm\frac{\sqrt{2 \cdot 49 \cdot (-1)}}{2}$$

$$= \pm\frac{7i\sqrt{2}}{2}$$

The correct choice is **(4)**.

8. A cosine curve with amplitude a and frequency b has the equation $y = a \cos(bx)$. In this curve, the amplitude is 0.5 since the highest point of the cosine curve is 0.5 above the middle of the curve. The period of the cosine curve is the difference between two consecutive high points. For this curve, the period is π because there is a high point at $(0, 0.5)$ and the next high point is at $(\pi, 0.5)$. The frequency can be calculated with the formula $b = \dfrac{2\pi}{\text{period}} = \dfrac{2\pi}{\pi} = 2$. Therefore, the equation is $y = 0.5 \cos(2x)$.

This could also be done by graphing the four choices on the graphing calculator to see which looks most like the graph. You would have to set your calculator to radians and use a window that matches the graph.

For the TI-84:

For the TI-Nspire:

The correct choice is **(1)**.

9. Subtract the polynomials to find the answer.

$$R(x) - C(x)$$
$$x \cdot S(x) - C(x)$$
$$x(30 - 0.01x) - (0.15x^3 + 0.01x^2 + 2x + 120)$$
$$30x - 0.01x^2 - 0.15x^3 - 0.01x^2 - 2x - 120$$
$$-0.15x^3 - 0.02x^2 + 28x - 120$$

The correct choice is **(2)**.

10. If something is very unlikely, it will not happen very much in a computer simulation. Since 6 failed to occur about 20% of the time in a simulation of a fair spinner, it is common for a fair spinner not to come up with 6 in any of 10 tries. This spinner is likely to be a fair spinner (also described as "likely not unfair").

Had the simulation shown that a fair spinner would not land on 6 only 1% of the time, Miles would have had some evidence that the spinner he used was an unfair spinner.

The correct choice is **(3)**.

ANSWERS June 2017   323

11. The quickest way to check if something like $x - a$ is a factor of a polynomial is to use the factor theorem. To check if $(x - a)$ is a factor, substitute $x = a$ into the polynomial. If the polynomial evaluates to zero, then $(x - a)$ is a factor. If it evaluates to something other than zero, then $(x - a)$ is not a factor. To check if $(x + a)$ is a factor, substitute $x = -a$ into the polynomial. If the polynomial evaluates to zero, then $(x + a)$ is a factor. If it evaluates to something other than zero, then $(x + a)$ is not a factor. Test each choice.

Choice (1): Substitute $x = 2$ into the polynomial.

$$2^4 - 4 \cdot 2^2 - 4 \cdot 2 + 8 =$$
$$16 - 16 - 8 + 8 = 0$$

Since substituting 2 for x into the polynomial makes it evaluate to zero, $(x - 2)$ is a factor.

Choice (2): Substitute $x = -2$ into the polynomial.

$$(-2)^4 - 4 \cdot (-2)^2 - 4 \cdot (-2) + 8 =$$
$$16 - 16 + 8 + 8 = 16$$

Since substituting -2 for x into the polynomial makes it evaluate to something other than zero, $(x + 2)$ is a not a factor.

Choice (3): Substitute $x = 4$ into the polynomial.

$$(4)^4 - 4 \cdot (4)^2 - 4 \cdot (4) + 8 =$$
$$256 - 64 - 16 + 8 = 184$$

Since substituting 4 for x into the polynomial makes it evaluate to something other than zero, $(x - 4)$ is not a factor.

Choice (4): Substitute $x = -4$ into the polynomial

$$(-4)^4 - 4 \cdot (-4)^2 - 4 \cdot (-4) + 8 =$$
$$256 - 64 + 16 + 8 = 216$$

Since substituting -4 for x into the polynomial makes it evaluate to something other than zero, $(x + 4)$ is not a factor.

The correct choice is **(1)**.

324 **ANSWERS June 2017**

12. Substitute the given value and solve for $\cos\theta$.

$$\sin^2\theta + \cos^2\theta = 1$$

$$\left(-\frac{\sqrt{2}}{5}\right)^2 + \cos^2\theta = 1$$

$$\frac{2}{25} + \cos^2\theta = 1$$

$$\frac{-\dfrac{2}{25} = -\dfrac{2}{25}}{\cos^2\theta = \dfrac{23}{25}}$$

$$\sqrt{\cos^2\theta} = \pm\sqrt{\frac{23}{25}}$$

$$\cos\theta = \pm\frac{\sqrt{23}}{5}$$

The correct choice is (**2**).

13. Since there are 10 years in a decade, $d = \frac{y}{10}$. Substitute for d in the equation.

$$P = 714(0.75)^{\frac{y}{10}}$$

To make this look more like the answer choices, rewrite $\frac{y}{10}$ as $\frac{1}{10} \cdot y$ and then use the rule of exponents, $x^{a \cdot b} = (x^a)^b$.

$$P = 714(0.75)^{\frac{1}{10} \cdot y}$$

$$= 714\left(0.75^{\frac{1}{10}}\right)^y$$

Then since $0.75^{\frac{1}{10}} \approx 0.9716$, the equation becomes $P = 714(0.9716)^y$.

Another way to solve this question is to pick a number for d such as $d = 5$ and calculate the value of $P = 714(0.75)^5 \approx 169$. Since 5 decades is the same as 50 years, check each of the answer choices with $y = 50$ to see which one gives the same answer.

For choice (3), $P = 714(0.9716)^{50} \approx 169$.

The correct choice is (**3**).

14. Two events, A and B, are independent if $P(A \text{ and } B) = P(A) \cdot P(B)$. In this problem, event A is "child with blue eyes" and event B is "child with blond hair." It is given that $P(A) = 0.25$, $P(B) = 0.5$, and $P(A \text{ and } B) = 0.125$.

So $P(A) \cdot P(B) = 0.25 \cdot 0.5 = 0.125$ and $P(A \text{ and } B)$ also equals 0.125. When this happens, the events are independent.

Analyze each of the four statements.

Analyze I: When events are independent, they are not dependent.

Analyze II: These are independent events because

$$P(A) \cdot P(B) = P(A \text{ and } B)$$

Analyze III: Mutually exclusive occurs when $P(A \text{ and } B) = 0$, which is not the case in this problem since it is given that $P(A \text{ and } B) = 0.125$.

The correct choice is **(2)**.

15. The graphs of $B(x)$ and $P(x)$ can be used to analyze the answer choices.

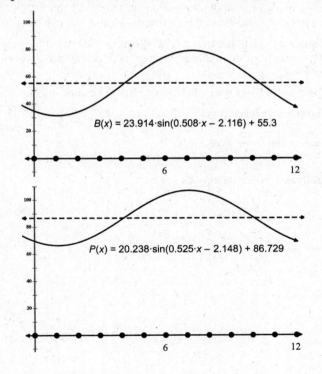

326 ANSWERS June 2017

Choice (1): The city with a bigger difference between its high temperature and its low temperature has a higher average monthly temperature variation. Since $B(x)$ has a difference of around 48 between its high point and its low point while $P(x)$ has a difference of around 40 between its high point and its low point, $B(x)$ has a higher average monthly temperature variation.

Even without the graphs, it can be determined that the graph of $B(x)$ has a larger amplitude. The amplitude is the coefficient before the sine function. For $B(x)$, the amplitude is 23.914. For $P(x)$, the amplitude is 20.238.

Choice (2): The midline average monthly temperature for Bar Harbor is 55.3. The midline average monthly temperature for Phoenix is 86.729. So for Bar Harbor, the midline temperature is lower.

Without the graph, the midline can be determined by looking at the constant that is added to the sine term. For $B(x)$, this is 55.300. For $P(x)$, this is 86.729.

Choice (3): The maximum of $B(x)$ can be calculated by adding the amplitude to the midline constant, $23.914 + 55.300 = 79.214 \approx 79$. On the graphing calculator, this can be done using the maximum function.

Choice (4): The minimum of $P(x)$ can be calculated by subtracting the amplitude from the midline constant, $86.729 - 20.238 = 66.491$, which is not 20.

The question asks which statement can *not* be concluded.

The correct choice is **(4)**.

16. Analyze the three statements.

Analyze I: Use the rule $\sqrt[n]{x} = x^{\frac{1}{n}}$.

$$\frac{\sqrt[6]{x}}{\sqrt[3]{x}} = \frac{x^{\frac{1}{6}}}{x^{\frac{1}{3}}}$$

Then use the rule $\dfrac{x^a}{x^b} = x^{a-b}$.

$$\frac{x^{\frac{1}{6}}}{x^{\frac{1}{3}}} = x^{\frac{1}{6}-\frac{1}{3}}$$

$$= x^{\frac{1}{6}-\frac{2}{6}}$$

$$= x^{-\frac{1}{6}}$$

Finally use the rule $x^{-a} = \dfrac{1}{x^a}$.

$$x^{-\frac{1}{6}} = \frac{1}{x^{\frac{1}{6}}}$$

This is one divided by the sixth root of x. So statement I is an answer.

Analyze II: Use the rule $\dfrac{x^a}{x^b} = x^{a-b}$.

$$\frac{x^{\frac{1}{6}}}{x^{\frac{1}{3}}} = x^{\frac{1}{6}-\frac{1}{3}}$$

$$= x^{\frac{1}{6}-\frac{2}{6}}$$

$$= x^{-\frac{1}{6}}$$

This is identical to one of the steps when analyzing statement I. So statement II is an answer.

Analyze III: Use the rule $x^{-a} = \dfrac{1}{x^a}$.

$$x^{-\frac{1}{6}} = \frac{1}{x^{\frac{1}{6}}}$$

This is identical to one of the steps when analyzing statement I. So statement III is an answer.

So all three expressions are equivalent to one divided by the sixth root of x.

The correct choice is **(4)**.

17. A parabola with its focus at (1, 2) and directrix at $y = -2$ looks like this.

The vertex of a parabola is halfway between the vertex and the directrix. For this problem, the point halfway between (1, 2) and $y = -2$ is (1, 0).

The vertex form of the equation of a parabola is $y = a(x - h)^2 + k$, where the vertex is (h, k). Since the vertex is (1, 0), the equation for this parabola will have the form $y = a(x - 1)^2 + 0 = y = a(x - 1)^2$. Only choices (3) and (4) have this form.

To calculate a, use the relationship $a = \dfrac{1}{4p}$, where p is half the distance between the focus and the directrix. Since the distance between the focus and the directrix is 4, p is 2. So $a = \dfrac{1}{4p} = \dfrac{1}{4 \cdot 2} = \dfrac{1}{8}$. The complete equation is $y = \dfrac{1}{8}(x-1)^2$.

The correct choice is **(4)**.

ANSWERS June 2017   329

18. In an exponential equation of the form $p(t) = Pe^{rt}$, the P is the initial value and the r is the annual growth rate compounded continuously. Since P is 110 in this problem, 110 million is the number of people at time $t = 0$, which is 2010. Statement I is incorrect since the current year is not 2010. Statement II is correct. If the word *continuously* was not included, statement II would not be correct since the annual growth rate is actually $e^{0.03922} \approx 1.04$.

The correct choice is **(2)**.

19. To solve a rational equation, first factor all the denominators.

$$\frac{2x}{x-2} - \frac{11}{x} = \frac{8}{x(x-2)}$$

The least common denominator is $x(x - 2)$. Multiply both sides of the equation by the least common denominator and simplify.

$$x(x-2)\left(\frac{2x}{x-2} - \frac{11}{x}\right) = x(x-2)\left(\frac{8}{x(x-2)}\right)$$

$$\frac{x(x-2)2x}{x-2} - \frac{x(x-2)11}{x} = \frac{x(x-2)8}{x(x-2)}$$

$$2x^2 - (x-2)11 = 8$$

$$2x^2 - (11x - 22) = 8$$

$$2x^2 - 11x + 22 = 8$$

$$\underline{ -8 = -8}$$

$$2x^2 - 11x + 14 = 0$$

$$(2x - 7)(x - 2) = 0$$

Solving for x seems to give two answers.

$$2x - 7 = 0$$
$$\underline{+7 = +7}$$
$$2x = 7$$
$$\frac{2x}{2} = \frac{7}{2}$$
$$x = \frac{7}{2}$$

and

$$x - 2 = 0$$
$$\underline{+2 = +2}$$
$$x = 2$$

However, the $x = 2$ solution is extraneous. If you substitute it into the original equation, then two of the denominators evaluate to zero.

The correct choice is **(1)**.

ANSWERS June 2017

20. The easiest way to solve this question is to test the four choices to see which of them evaluates to $f(1) = -8$ and $f(9) = -2$.

Choice (3):

$$f(1) = -8.75 + 0.75 \cdot 1$$
$$= -8.75 + 0.75$$
$$= -8$$

$$f(9) = -8.75 + 0.75 \cdot 9$$
$$= -8.75 + 6.75$$
$$= -2$$

None of the other choices have $f(1) = -8$ and $f(9) = -2$.

Another way to answer this question is to look at the sequence and notice that the first term is -8, and the common difference is 0.75, since each term is 0.75 greater than the previous term. The formula for the nth term of an arithmetic sequence given on the reference sheet is $a_n = a_1 + (n - 1)d$, where d is the common difference and a_1 is the first term.

For this problem, it becomes

$$a_n = -8 + (n - 1)0.75 = -8 + 0.75n - 0.75 = -8.75 + 0.75n$$

The fact is that $f(9) = -2$ is not really needed, although it could have been used to calculate the value of d with the equation $a_n = a_1 + (n - 1)d$.

$$-2 = -8 + (9 - 1)d$$
$$-2 = -8 + 8d$$
$$\underline{+8 = +8}$$
$$6 = 8d$$
$$\frac{6}{8} = \frac{8d}{8}$$
$$\frac{3}{4} = 0.75 = d$$

Since it is simpler to calculate d by just looking at the first two terms of the sequence, $f(9) = -2$ is not a necessary detail.

The correct choice is **(3)**.

ANSWERS June 2017   **331**

21. The average rate of change of a function f on an interval $[a, b]$ can be calculated with the average rate of change formula.

$$\text{average rate of change} = \frac{f(b) - f(a)}{b - a}$$

For this problem, use the following equation:

$$\frac{f(b) - f(a)}{b - a} = \frac{f(7) - f(-7)}{7 - (-7)}$$

To calculate the terms in the numerator, put your calculator into radian mode.

$$f(7) = 2^{-0.25 \cdot 7} \cdot \sin\left(\frac{\pi}{2} \cdot 7\right)$$

$$= -0.2973$$

$$f(-7) = 2^{-0.25 \cdot (-7)} \cdot \sin\left(\frac{\pi}{2} \cdot (-7)\right)$$

$$= 3.36359$$

$$\frac{f(7) - f(-7)}{7 - (-7)} = \frac{-0.2973 - 3.36359}{7 - (-7)}$$

$$= \frac{-3.6609}{14}$$

$$\approx -0.26$$

The correct choice is **(3)**.

22. The cost of having the air conditioner for one year is 329.99 + 108.78. For two years, the cost is 329.99 + 108.78 · 2. For three years, it is 329.99 + 108.78 · 3. How many months in the year Mallory uses the air conditioner does not matter since \$108.78 is the cost per year, not the cost per month. In general, the cost for n years is 329.99 + 108.78n. The question, though, asks for an expression for the cost per year over the lifetime of the unit, not the cost during any particular year. Therefore, divide the expression by n to get $C(n)$.

$$C(n) = \frac{329.99 + 108.78n}{n}$$

The correct choice is **(3)**.

332   **ANSWERS June 2017**

23. To simplify this expression, first factor the numerator and the denominator. To make factoring the numerator easier, start by factoring out the -1. The resulting quadratic polynomial will be simpler to factor.

$$\frac{-3x^2 - 5x + 2}{x^3 + 2x^2} = \frac{-(3x^2 + 5x - 2)}{x^2(x + 2)}$$

$$= \frac{-(3x - 1)(x + 2)}{x^2(x + 2)}$$

$$= \frac{-(3x - 1)}{x^2}$$

$$= \frac{-3x + 1}{x^2}$$

$$= \frac{-3x}{x^2} + \frac{1}{x^2}$$

$$= \frac{-3}{x} + \frac{1}{x^2}$$

$$= -3x^{-1} + x^{-2}$$

The correct choice is **(4)**.

24. The monthly interest rate is $\dfrac{0.03}{12}$ = 0.0025. If you assume that the money is put into the bank on the first of every month starting on January 1 and that the final value will be calculated right after the twelfth deposit is made on December 1, the first \$100 will be worth \$100(1.0025)^{11} since it will be in the bank for 11 months. The second \$100 will be worth \$100(1.0025)^{10} on December 1. The third \$100 will be worth \$100(1.0025)^9 on December 1, and so on. The twelfth deposit, made on December 1, will still be worth \$100 on that day. There are a total of twelve \$100 deposits. At the end of the year, they will have a combined value of 100(1.0025)^{11} + 100(1.0025)^{10} + 100(1.0025)^9 + $\cdots$ + 100. This is a geometric series. The following shows how this can be rewritten:

$$100 + 100(1.0025)^1 + 100(1.0025)^2 + \cdots + 100(1.0025)^{11}$$

ANSWERS June 2017   **333**

On the reference sheet is the formula for the sum of a geometric series, $S_n = \dfrac{a_1 - a_1 r^n}{1-r}$, where a_1 is the first term, r is the common ratio, and n is the number of terms.

For this problem, the formula becomes $S_{12} = \dfrac{100 - 100(1.0025)^{12}}{1-1.0025}$.

Even without having this formula, it is possible to eliminate two choices by noticing that the total value must be more than \$1,200 since that's how much would be in the account even with no interest. Choice (1) evaluates to only \$142.58. Choice (3) evaluates to only \$103.04.

The correct choice is **(2)**.

PART II

25. To find $r(2)$, substitute 2 for every x in the equation $r(x)$.

$$r(2) = 2^3 - 4 \cdot 2^2 + 4 \cdot 2 - 6$$
$$= 8 - 16 + 8 - 6$$
$$= -6$$

The factor theorem says that if $f(a) = 0$ then $(x - a)$ is a factor of $f(x)$. The theorem also says that if $f(a) \neq 0$, then $(x - a)$ is not a factor of $f(x)$. Since $r(2) \neq 0$, $(x - 2)$ is not a factor of $r(x)$.

26. This can be calculated by using the "normal cdf" function of the graphing calculator. This function determines the percent that can be expected between two numbers. When the question asks about the percent that can be expected below some number, make the lower bound something negative and very large since there is no way to make a negative infinity as the lower bound. Make the "lower" -9999, the "upper" 8.25, the "μ" 8, and the "σ" 0.5.

For the TI-84:

For the TI-Nspire:

When rounded to the nearest integer, the percentage of bags that weighed less than 8.25 pounds is 69%.

ANSWERS June 2017   335

27. A cubic equation can sometimes be factored by grouping. Start by factoring out a common factor from the first two terms. Then factor out a common factor from the last two terms.

$$4x^3 - x^2 + 16x - 4$$
$$x^2(4x - 1) + 4(4x - 1)$$

If the two terms have a common factor (which they do in this case), factor that out too.

$$(4x - 1)(x^2 + 4)$$

These two factors cannot be factored anymore without using imaginary numbers, so this is considered completely factored over the set of integers.

28. The period of a cosine curve is the horizontal distance between one high point and the next high point. Since the first high point is at $(0, 26)$ and the second is at $(\frac{2}{3}, 26)$ the period of the graph is $\frac{2}{3}$.

In this context, the $\frac{2}{3}$ is the amount of time the wheel takes to make one rotation.

29. Make a chart with x-values from 0 to 10.

x	y
0	394
1	283
2	203
3	145
4	103
5	73
6	51
7	35
8	24
9	15
10	10

This makes an exponential graph. Graphs of exponential equations of the form $y = a \cdot b^x + c$ have a horizontal asymptote at $y = c$, so this graph has a horizontal asymptote at $y = 6$.

30. Isolate the radical by subtracting x from both sides of the equation. Then square both sides and solve the resulting quadratic equation.

$$\sqrt{x-4} + x = 6$$
$$\underline{-x = -x}$$
$$\sqrt{x-4} = 6 - x$$
$$\left(\sqrt{x-4}\right)^2 = (6-x)^2$$
$$x - 4 = 36 - 12x + x^2$$
$$\underline{-x = -x}$$
$$-4 = x^2 - 13x + 36$$
$$\underline{+4 = +4}$$
$$0 = x^2 - 13x + 40$$
$$0 = (x-5)(x-8)$$

$x - 5 = 0$ or $x - 8 = 0$
$x = 5$ or $x = 0$

ANSWERS June 2017

Check both answers to see if either or both are extraneous solutions.

Check $x = 5$:

$$\sqrt{5-4} + 5 \overset{?}{=} 6$$

$$\sqrt{1} + 5 \overset{?}{=} 6$$

$$1 + 5 = 6$$

Check $x = 8$:

$$\sqrt{8-4} + 8 \overset{?}{=} 6$$

$$\sqrt{4} + 8 \overset{?}{=} 6$$

$$2 + 8 \overset{?}{=} 6$$

$$10 \neq 6$$

So 8 is an extraneous solution. The only solution is $x = 5$.

31. First use the rule $\sqrt[n]{x} = x^{\frac{1}{n}}$.

$$\sqrt[3]{x} \cdot \sqrt{x} = \sqrt[3]{x} \cdot \sqrt[2]{x} = x^{\frac{1}{3}} \cdot x^{\frac{1}{2}}$$

Then use the rule $x^a \cdot x^b = x^{a+b}$.

$$x^{\frac{1}{3}} \cdot x^{\frac{1}{2}} = x^{\frac{1}{3} + \frac{1}{2}}$$

$$= x^{\frac{2}{6} + \frac{3}{6}}$$

$$= x^{\frac{5}{6}}$$

338   **ANSWERS June 2017**

32. There are a total of 416 + 1823 = 2239 students who have one sibling who jogs. Of those students, only 416 of them also jog. So the probability that a student with one sibling who jogs also jogs is $\frac{416}{2239} \approx 18.6\%$.

There are a total of 400 + 1380 = 1780 students who have two siblings who jog. Of those students, only 400 of them also jog. The probability that a student with two siblings who jog also jogs is $\frac{400}{1780} \approx 22.5\%$.

So it is more likely for a student who has two siblings who jog to jog than it is for a student who has only one sibling who jogs.

ANSWERS June 2017

PART III

33. To solve a system of three equations with three unknowns, first find a way to combine two of the equations to eliminate one of the variables. Since the first equation has an x and the third equation has a $-x$, add the first and third equations. The result is a new equation that only has the variables y and z.

$$\begin{array}{r} x + y + z = 1 \\ -x + 3y - 5z = 11 \\ \hline 4y - 4z = 12 \end{array}$$

Next choose another pair of equations (either the first and second or the second and third). Eliminate the same variable, which is the x in this case. Eliminate the x from the second and third equations.

$$\begin{array}{r} 2x + 4y + 6z = 2 \\ -x + 3y - 5z = 11 \end{array}$$

$$\begin{array}{r} 2x + 4y + 6z = 2 \\ 2(-x + 3y - 5z) = 2(11) \end{array}$$

$$\begin{array}{r} 2x + 4y + 6z = 2 \\ + \quad -2x + 6y - 10z = 22 \\ \hline 10y - 4z = 24 \end{array}$$

Now solve the system of two equations.

$$\begin{array}{r} 4y - 4z = 12 \\ 10y - 4z = 24 \end{array}$$

$$\begin{array}{r} 4y - 4z = 12 \\ -1(10y - 4z) = -1(24) \end{array}$$

$$\begin{array}{r} 4y - 4z = 12 \\ + \quad -10y + 4z = -24 \\ \hline \dfrac{-6y}{-6} = \dfrac{-12}{-6} \end{array}$$

$$y = 2$$

340   **ANSWERS June 2017**

Substitute $y = 2$ into one of the two equations that have just y and z in them and solve for z.

$$4(2) - 4z = 12$$
$$8 - 4z = 12$$
$$\underline{-8 = -8}$$
$$\frac{-4z}{-4} = \frac{4}{-4}$$
$$z = -1$$

Substitute $y = 2$ and $z = -1$ into one of the original three equations and solve for x.

$$x + 2 + (-1) = 1$$
$$x + 1 = 1$$
$$\underline{-1 = -1}$$
$$x = 0$$

The solution to the system is $x = 0$, $y = 2$, and $z = -1$, $(0, 2, -1)$.

34. Start by finding N. Since there are 12 months in a year, $12 \cdot 15 = 180$. Now substitute $P = 172{,}600$, $r = 0.00305$, and $N = 180$ into the equation, and solve for M.

$$M = 172{,}600 \cdot \frac{0.00305(1+0.00305)^{180}}{(1+0.00305)^{180} - 1}$$
$$= 1247.49$$

Rounded to the nearest dollar, Jim's mortgage payment is $1247 with no down payment.

If Jim makes a down payment of x, the amount of his loan will be $172{,}600 - x$. Substitute $P = 172{,}600 - x$, $M = 1100$, $r = 0.00305$, and $N = 180$ into the equation, and solve for x.

$$1100 = (172{,}600 - x) \cdot \frac{0.00305(1+0.00305)^{180}}{(1+0.00305^{180}) - 1}$$

$$1100 = (172{,}600 - x) \cdot \frac{0.00305(1.00305)^{180}}{(1.00305^{180}) - 1}$$

$$1100 = (172{,}600 - x) \cdot 0.0072276558$$

$$\frac{1100}{0.0072276558} = \frac{(172{,}600 - x) \cdot 0.0072276558}{0.0072276558}$$

$$152{,}193.19 = 172{,}600 - x$$
$$\underline{-172{,}600 = -172{,}600}$$
$$\overline{-20{,}406.81 = -x}$$
$$20{,}406.81 = x$$

Rounded to the nearest dollar, Jim needs to make a down payment of $20,407 in order for his mortgage payment to be $1100.

35. Before selecting the scale, determine the intercepts.

For the y-intercept, set $x = 0$ and solve for y.

$$y = \log_2(0 + 3) - 5$$
$$y = \log_2 3 - 5$$
$$y \approx -3.4$$

The y-intercept is $(0, -3.4)$.

For the x-intercept, set $y = 0$ and solve for x.

$$0 = \log_2(x + 3) - 5$$
$$+5 = +5$$
$$5 = \log_2(x + 3)$$
$$2^5 = x + 3$$
$$32 = x + 3$$
$$-3 = -3$$
$$29 = x$$

The x-intercept is $(29, 0)$.

By making each box 3 units, both intercepts will appear on the graph. Plot the two intercepts on the graph.

This can also be graphed on the graphing calculator. Set the window so that the x- and y-values go from -30 to $+30$.

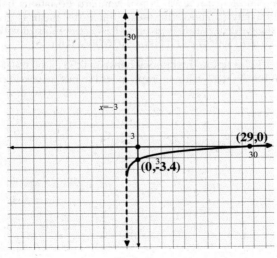

For the TI-84:   For the TI-Nspire:

For x-values very close to -3 like -2.99999, the value of

$$\log_2(-2.9999 + 3) = \log_2(0.0001)$$

becomes a very large negative number. Since log is undefined for negative numbers, there is a vertical asymptote at $x = -3$. The behavior of the function as x approaches -3 is that the y-value approaches negative infinity.

Although it looks like there might be a horizontal asymptote, there is not one. For very large numbers like $x = 10,000,000$, the value of $\log_2(10,000,000 + 3) = \log_2(10,000,003)$ is well over 10. As the x-value increases, the y-value also increases. The range of a log graph is all real numbers. So as x approaches positive infinity, y also approaches positive infinity.

36. The 95% confidence interval is two standard deviations below and two standard deviations above the mean. Since the mean is given as 0.506 and the standard deviation is given as 0.078, the interval will be between $0.506 - 2(0.078) = 0.35$ and $0.506 + 2(0.078) = 0.66$.

Because 0.325 is not in the interval between 0.35 and 0.66, it is too low of a percent for the dealership to implement the new check-in procedure.

ANSWERS June 2017   **343**

PART IV

37. Substitute $A_0 = 140$, $A = 100$, and $t = 5$ (the interval from 3 p.m. to 8 p.m.) into the equation.

$$100 = 140\left(\frac{1}{2}\right)^{\frac{5}{h}}$$

Solve this equation for h.

$$\frac{100}{140} = \frac{140\left(\frac{1}{2}\right)^{\frac{5}{h}}}{140}$$

$$\frac{5}{7} = \left(\frac{1}{2}\right)^{\frac{5}{h}}$$

$$\log_{\frac{1}{2}}\left(\frac{5}{7}\right) = \frac{5}{h}$$

$$0.4854 = \frac{5}{h}$$

$$0.4854h = 5$$

$$\frac{0.4854h}{0.4854} = \frac{5}{0.4854}$$

$$h = 10.3002$$

Substitute $A_0 = 140$, $A = 40$, and $h = 10.3002$ into the equation, and solve for t.

$$40 = 140\left(\frac{1}{2}\right)^{\frac{t}{10.3002}}$$

$$\frac{40}{140} = \frac{140\left(\frac{1}{2}\right)^{\frac{t}{10.3002}}}{140}$$

$$\frac{2}{7} = \left(\frac{1}{2}\right)^{\frac{t}{10.3002}}$$

$$\log_{\frac{1}{2}}\left(\frac{2}{7}\right) = \frac{t}{10.3002}$$

$$1.8073 = \frac{t}{10.3002}$$

$$18.6 = t$$

344   SELF-ANALYSIS CHART June 2017

Topic	Question Numbers	Number of Points	Your Points	Your Percentage
1. Polynomial Expressions and Equations	9, 11, 17, 22, 23, 25, 27	2 + 2 + 2 + 2 + 2 + 2 + 2 = 14		
2. Complex Numbers	4, 7	2 + 2 = 4		
3. Exponential Expressions and Equations	2, 13, 16, 18, 24, 31, 34, 37	2 + 2 + 2 + 2 + 2 + 2 + 4 + 6 = 22		
4. Rational Expressions and Equations	19	2		
5. Radical Expressions and Equations	30	2		
6. Trigonometric Expressions and Equations	6, 8, 12, 15, 28	2 + 2 + 2 + 2 + 2 = 10		
7. Graphing	1, 5, 29, 35	2 + 2 + 2 + 4 = 10		
8. Functions	21	2		
9. Systems of Equations	33	4		
10. Sequences and Series	20	2		
11. Probability	14, 32	2 + 2 = 4		
12. Statistics	3, 10, 26, 36	2 + 2 + 2 + 4 = 10		

SELF-ANALYSIS CHART June 2017   **345**

HOW TO CONVERT YOUR RAW SCORE TO YOUR ALGEBRA II REGENTS EXAMINATION SCORE

The accompanying conversion chart must be used to determine your final score on the June 2017 Regents Examination in Algebra II. To find your final exam score, locate in the column labeled "Raw Score" the total number of points you scored out of a possible 86 points. Since partial credit is allowed in Parts II, III, and IV of the test, you may need to approximate the credit you would receive for a solution that is not completely correct. Then locate in the adjacent column to the right the scale score that corresponds to your raw score. The scale score is your final Algebra II Regents Examination score.

346   SELF-ANALYSIS CHART June 2017

Regents Examination in Algebra II—June 2017
Chart for Converting Total Test Raw Scores to Final
Examination Scores (Scaled Scores)

Raw Score	Scale Score	Performance Level	Raw Score	Scale Score	Performance Level	Raw Score	Scale Score	Performance Level
86	100	5	57	82	4	28	67	3
85	99	5	56	82	4	27	66	3
84	98	5	55	82	4	26	65	3
83	97	5	54	81	4	25	64	2
82	97	5	53	81	4	24	63	2
81	96	5	52	80	4	23	62	2
80	95	5	51	80	4	22	60	2
79	94	5	50	80	4	21	59	2
78	94	5	49	79	4	20	56	2
77	93	5	48	79	4	19	55	2
76	92	5	47	79	4	18	53	1
75	92	5	46	78	4	17	52	1
74	91	5	45	78	4	16	50	1
73	90	5	44	77	3	15	48	1
72	90	5	43	77	3	14	45	1
71	89	5	42	77	3	13	43	1
70	89	5	41	76	3	12	41	1
69	88	5	40	76	3	11	38	1
68	88	5	39	75	3	10	35	1
67	87	5	38	75	3	9	32	1
66	87	5	37	74	3	8	29	1
65	86	5	36	74	3	7	26	1
64	86	5	35	73	3	6	23	1
63	86	5	34	72	3	5	19	1
62	85	5	33	72	3	4	16	1
61	84	4	32	71	3	3	12	1
60	84	4	31	70	3	2	8	1
59	83	4	30	69	3	1	4	1
58	83	4	29	68	3	0	0	1

347

Examination
August 2017
Algebra II

HIGH SCHOOL MATH REFERENCE SHEET

Conversions

1 inch = 2.54 centimeters

1 meter = 39.37 inches

1 mile = 5280 feet

1 mile = 1760 yards

1 mile = 1.609 kilometers

1 kilometer = 0.62 mile

1 pound = 16 ounces

1 pound = 0.454 kilogram

1 kilogram = 2.2 pounds

1 ton = 2000 pounds

1 cup = 8 fluid ounces

1 pint = 2 cups

1 quart = 2 pints

1 gallon = 4 quarts

1 gallon = 3.785 liters

1 liter = 0.264 gallon

1 liter = 1000 cubic centimeters

Formulas

Triangle $A = \dfrac{1}{2}bh$

Parallelogram $A = bh$

Circle $A = \pi r^2$

Circle $C = \pi d$ or $C = 2\pi r$

EXAMINATION August 2017

Formulas (continued)

General Prisms $\qquad V = Bh$

Cylinder $\qquad V = \pi r^2 h$

Sphere $\qquad V = \dfrac{4}{3}\pi r^3$

Cone $\qquad V = \dfrac{1}{3}\pi r^2 h$

Pyramid $\qquad V = \dfrac{1}{3}Bh$

Pythagorean Theorem $\qquad a^2 + b^2 = c^2$

Quadratic Formula $\qquad x = \dfrac{-b \pm \sqrt{b^2 - 4ac}}{2a}$

Arithmetic Sequence $\qquad a_n = a_1 + (n-1)d$

Geometric Sequence $\qquad a_n = a_1 r^{n-1}$

Geometric Series $\qquad S_n = \dfrac{a_1 - a_1 r^n}{1 - r}$ where $r \neq 1$

Radians $\qquad 1 \text{ radian} = \dfrac{180}{\pi} \text{ degrees}$

Degrees $\qquad 1 \text{ degree} = \dfrac{\pi}{180} \text{ radians}$

Exponential Growth/Decay $\qquad A = A_0 e^{k(t - t_0)} + B_0$

EXAMINATION August 2017

PART I

Answer all 24 questions in this part. Each correct answer will receive 2 credits. No partial credit will be allowed. For each statement or question, write in the space provided the numeral preceding the word or expression that best completes the statement or answers the question. [48 credits]

1 The function $f(x) = \dfrac{x-3}{x^2 + 2x - 8}$ is undefined when x equals

   (1) 2 or –4   (3) 3, only
   (2) 4 or –2   (4) 2, only   1 _____

2 Which expression is equivalent to $(3k - 2i)^2$, where i is the imaginary unit?

   (1) $9k^2 - 4$ (3) $9k^2 - 12ki - 4$
   (2) $9k^2 + 4$ (4) $9k^2 - 12ki + 4$ 2 _____

3 The roots of the equation $x^2 + 2x + 5 = 0$ are

   (1) –3 and 1   (3) $-1 + 2i$ and $-1 - 2i$
   (2) –1, only   (4) $-1 + 4i$ and $-1 - 4i$ 3 _____

4 The solution set for the equation $\sqrt{x+14} - \sqrt{2x+5} = 1$ is

   (1) {–6}   (3) {18}
   (2) {2}   (4) {2, 22}   4 _____

5 As x increases from 0 to $\dfrac{\pi}{2}$, the graph of the equation $y = 2 \tan x$ will

   (1) increase from 0 to 2
   (2) decrease from 0 to –2
   (3) increase without limit
   (4) decrease without limit   5 _____

6 Which equation represents a parabola with the focus at (0, −1) and the directrix $y = 1$?
(1) $x^2 = -8y$
(2) $x^2 = -4y$
(3) $x^2 = 8y$
(4) $x^2 = 4y$

6_____

7 Which diagram represents an angle, α, measuring $\frac{13\pi}{20}$ radians drawn in standard position, and its reference angle, θ?

7_____

EXAMINATION August 2017   351

8 What are the zeros of $P(m) = (m^2 - 4)(m^2 + 1)$?

(1) 2 and −2, only   (3) −4, i, and −i

(2) 2, −2, and −4   (4) 2, −2, i, and −i 8 _____

9 The value of a new car depreciates over time. Greg purchased a new car in June 2011. The value, V, of his car after t years can be modeled by the equation

$$\log_{0.8}\left(\frac{V}{17000}\right) = t.$$

What is the average decreasing rate of change per year of the value of the car from June 2012 to June 2014, to the *nearest ten dollars per year*?

(1) 1960   (3) 2450

(2) 2180   (4) 2770   9 _____

10 Iridium-192 is an isotope of iridium and has a half-life of 73.83 days. If a laboratory experiment begins with 100 grams of Iridium-192, the number of grams, A, of Iridium-192 present after t days would be

$$A = 100\left(\frac{1}{2}\right)^{\frac{t}{73.83}}.$$

Which equation approximates the amount of Iridium-192 present after t days?

(1) $A = 100\left(\dfrac{73.83}{2}\right)^t$ (3) $A = 100(0.990656)^t$

(2) $A = 100\left(\dfrac{1}{147.66}\right)^t$ (4) $A = 100(0.116381)^t$ 10 _____

352   **EXAMINATION August 2017**

11 The distribution of the diameters of ball bearings made under a given manufacturing process is normally distributed with a mean of 4 cm and a standard deviation of 0.2 cm. What proportion of the ball bearings will have a diameter less than 3.7 cm?

(1) 0.0668   (3) 0.8664

(2) 0.4332   (4) 0.9500   11 _____

12 A polynomial equation of degree three, $p(x)$, is used to model the volume of a rectangular box. The graph of $p(x)$ has x intercepts at –2, 10, and 14. Which statements regarding $p(x)$ could be true?

A. The equation of $p(x) = (x - 2)(x + 10)(x + 14)$.
B. The equation of $p(x) = -(x + 2)(x - 10)(x - 14)$.
C. The maximum volume occurs when $x = 10$.
D. The maximum volume of the box is approximately 56.

(1) A and C (3) B and C

(2) A and D (4) B and D 12 _____

13 Which expression is equivalent to $\dfrac{4x^3 + 9x - 5}{2x - 1}$, where $x \neq \dfrac{1}{2}$?

(1) $2x^2 + x + 5$ (3) $2x^2 - x + 5$

(2) $2x^2 + \dfrac{11}{2} + \dfrac{1}{2(2x - 1)}$ (4) $2x^2 - x + 4 + \dfrac{1}{2x - 1}$ 13 _____

14 The inverse of the function $f(x) = \dfrac{x + 1}{x - 2}$ is

(1) $f^{-1}(x) = \dfrac{x + 1}{x + 2}$ (3) $f^{-1}(x) = \dfrac{x + 1}{x - 2}$

(2) $f^{-1}(x) = \dfrac{2x + 1}{x - 1}$ (4) $f^{-1}(x) = \dfrac{x - 1}{x + 1}$ 14 _____

15 Which expression has been rewritten correctly to form a true statement?

(1) $(x + 2)^2 + 2(x + 2) - 8 = (x + 6)x$
(2) $x^4 + 4x^2 + 9x^2y^2 - 36y^2 = (x + 3y)^2(x - 2)^2$
(3) $x^3 + 3x^2 - 4xy^2 - 12y^2 = (x - 2y)(x + 3)^2$
(4) $(x^2 - 4)^2 - 5(x^2 - 4) - 6 = (x^2 - 7)(x^2 - 6)$ 15 _____

16 A study conducted in 2004 in New York City found that 212 out of 1334 participants had hypertension. Kim ran a simulation of 100 studies based on these data. The output of the simulation is shown in the diagram below.

Proportion of Successes

At a 95% confidence level, the proportion of New York City residents with hypertension and the margin of error are closest to

(1) proportion ≈ .16; margin of error ≈ .01
(2) proportion ≈ .16; margin of error ≈ .02
(3) proportion ≈ .01; margin of error ≈ .16
(4) proportion ≈ .02; margin of error ≈ .16   16 _____

354   **EXAMINATION August 2017**

17 Which scenario is best described as an observational study?

  (1) For a class project, students in Health class ask every tenth student entering the school if they eat breakfast in the morning.

  (2) A social researcher wants to learn whether or not there is a link between attendance and grades. She gathers data from 15 school districts.

  (3) A researcher wants to learn whether or not there is a link between children's daily amount of physical activity and their overall energy level. During lunch at the local high school, she distributed a short questionnaire to students in the cafeteria.

  (4) Sixty seniors taking a course in Advanced Algebra Concepts are randomly divided into two classes. One class uses a graphing calculator all the time, and the other class never uses graphing calculators. A guidance counselor wants to determine whether there is a link between graphing calculator use and students' final exam grades.

            17 _____

18 Which sinusoid has the greatest amplitude?

(1)   (3)

$y = 3\sin(\theta - 3) + 5$ $y = -5\sin(\theta - 1) - 3$

(2)   (4)   18 _____

19 Consider the system shown below.
$$2x - y = 4$$
$$(x + 3)^2 + y^2 = 8$$

The two solutions of the system can be described as
(1) both imaginary
(2) both irrational
(3) both rational
(4) one rational and one irrational   19 _____

20 Which binomial is *not* a factor of the expression $x^3 - 11x^2 + 16x + 84$?
(1) $x + 2$ (3) $x - 6$
(2) $x + 4$ (4) $x - 7$ 20 _____

356   **EXAMINATION August 2017**

21 A ball is dropped from a height of 32 feet. It bounces
and rebounds 80% of the height from which it was
falling. What is the total downward distance, in feet,
the ball traveled up to the 12th bounce?

(1) 29   (3) 120

(2) 58   (4) 149   21 _____

22 A public opinion poll was conducted on behalf of
Mayor Ortega's reelection campaign shortly before
the election. 264 out of 550 likely voters said they
would vote for Mayor Ortega; the rest said they
would vote for his opponent.

Which statement is *least* appropriate to make,
according to the results of the poll?

(1) There is a 48% chance that Mayor Ortega will
win the election.

(2) The point estimate ($\hat{p}$) of voters who will vote for
Mayor Ortega is 48%.

(3) It is most likely that between 44% and 52% of
voters will vote for Mayor Ortega.

(4) Due to the margin of error, an inference cannot
be made regarding whether Mayor Ortega or his
opponent is most likely to win the election.   22 _____

23 What does $\left(\dfrac{-54x^9}{y^4} \right)^{\frac{2}{3}}$ equal?

(1) $\dfrac{9ix^6 \sqrt[3]{4}}{y \sqrt[3]{y^2}}$ (3) $\dfrac{9x^6 \sqrt[3]{4}}{y \sqrt[3]{y}}$

(2) $\dfrac{9ix^6 \sqrt[3]{4}}{y^2 \sqrt[3]{y^2}}$ (4) $\dfrac{9x^6 \sqrt[3]{4}}{y^2 \sqrt[3]{y^2}}$ 23 _____

EXAMINATION August 2017   **357**

24 The Rickerts decided to set up an account for their daughter to pay for her college education. The day their daughter was born, they deposited $1000 in an account that pays 1.8% compounded annually. Beginning with her first birthday, they deposit an additional $750 into the account on each of her birthdays. Which expression correctly represents the amount of money in the account n years after their daughter was born?

(1) $a_n = 1000(1.018)^n + 750$

(2) $a_n = 1000(1.018)^n + 750n$

(3) $a_0 = 1000$

$a_n = a_{n-1}(1.018) + 750$

(4) $a_0 = 1000$

$a_n = a_{n-1}(1.018) + 750n$ 24 _____

358   **EXAMINATION August 2017**

PART II

Answer all 8 questions in this part. Each correct answer will receive 2 credits. Clearly indicate the necessary steps, including appropriate formula substitutions, diagrams, graphs, charts, etc. For all questions in this part, a correct numerical answer with no work shown will receive only 1 credit. [16 credits]

25 Explain how $(-8)^{\frac{4}{3}}$ can be evaluated using properties of rational exponents to result in an integer answer.

26 A study was designed to test the effectiveness of a new drug. Half of the volunteers received the drug. The other half received a sugar pill. The probability of a volunteer receiving the drug and getting well was 40%. What is the probability of a volunteer getting well, given that the volunteer received the drug?

EXAMINATION August 2017   **359**

27 Verify the following Pythagorean identity for all values of x and y:

$$(x^2 + y^2)^2 = (x^2 - y^2)^2 + (2xy)^2$$

28 Mrs. Jones had hundreds of jelly beans in a bag that contained equal numbers of six different flavors. Her student randomly selected four jelly beans and they were all black licorice. Her student complained and said "What are the odds I got all of that kind?" Mrs. Jones replied, "simulate rolling a die 250 times and tell me if four black licorice jelly beans is unusual."

Explain how this simulation could be used to solve the problem.

EXAMINATION August 2017

29 While experimenting with her calculator, Candy creates the sequence 4, 9, 19, 39, 79, ...

Write a recursive formula for Candy's sequence.

Determine the eighth term in Candy's sequence.

EXAMINATION August 2017   361

30 In New York State, the minimum wage has grown exponentially. In 1966, the minimum wage was \$1.25 an hour and in 2015, it was \$8.75. Algebraically determine the rate of growth to the *nearest percent*.

31 Algebraically determine whether the function $j(x) = x^4 - 3x^2 - 4$ is odd, even, or neither.

32 On the axes below, sketch a possible function $p(x) = (x - a)(x - b)(x + c)$, where a, b, and c are positive, $a > b$, and $p(x)$ has a positive y-intercept of d. Label all intercepts.

EXAMINATION August 2017   363

PART III

Answer all 4 questions in this part. Each correct answer will receive 4 credits. Clearly indicate the necessary steps, including appropriate formula substitutions, diagrams, graphs, charts, etc. For all questions in this part, a correct numerical answer with no work shown will receive only 1 credit. [16 credits]

33 Solve for all values of p: $\dfrac{3p}{p-5} - \dfrac{2}{p+3} = \dfrac{p}{p+3}$

364   **EXAMINATION August 2017**

34 Simon lost his library card and has an overdue library book. When the book was 5 days late, he owed $2.25 to replace his library card and pay the fine for the overdue book. When the book was 21 days late, he owed $6.25 to replace his library card and pay the fine for the overdue book.

Suppose the total amount Simon owes when the book is n days late can be determined by an arithmetic sequence. Determine a formula for a_n, the nth term of this sequence.

Use the formula to determine the amount of money, in dollars, Simon needs to pay when the book is 60 days late.

35 a) On the axes below, sketch *at least one* cycle of a sine curve with an amplitude of 2, a midline at $y = -\frac{3}{2}$, and a period of 2π.

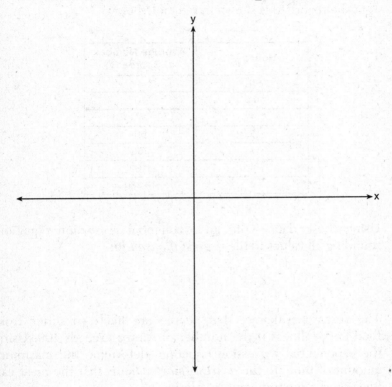

b) Explain any differences between a sketch of $y = 2\sin\left(x - \frac{\pi}{3}\right) - \frac{3}{2}$ and the sketch from part *a*.

366   **EXAMINATION August 2017**

36 Using a microscope, a researcher observed and recorded the number of bacteria spores on a large sample of uniformly sized pieces of meat kept at room temperature. A summary of the data she recorded is shown in the table below.

Hours (x)	Average Number of Spores (y)
0	4
0.5	10
1	15
2	60
3	260
4	1130
6	16,380

Using these data, write an exponential regression equation, rounding all values to the *nearest thousandth*.

The researcher knows that people are likely to suffer from food-borne illness if the number of spores exceeds 100. Using the exponential regression equation, determine the maximum amount of time, to the *nearest quarter hour*, that the meat can be kept at room temperature safely.

PART IV

Answer the question in this part. **A correct answer will receive 6 credits.** Clearly indicate the necessary steps, including appropriate formula substitutions, diagrams, graphs, charts, etc. A correct numerical answer with no work shown will receive only 1 credit. [6 credits]

37 The value of a certain small passenger car based on its use in years is modeled by $V(t) = 28482.698(0.684)^t$, where $V(t)$ is the value in dollars and t is the time in years. Zach had to take out a loan to purchase the small passenger car. The function $Z(t) = 22151.327(0.778)^t$, where $Z(t)$ is measured in dollars, and t is the time in years, models the unpaid amount of Zach's loan over time.

Graph $V(t)$ and $Z(t)$ over the interval $0 \le t \le 5$, on the set of axes below.

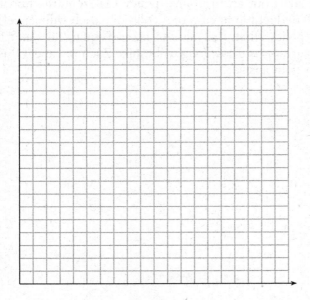

Question 37 is continued on the next page.

368   **EXAMINATION August 2017**

Question 37 continued.

State when $V(t) = Z(t)$, to *the nearest hundredth*, and interpret its meaning in the context of the problem.

Zach takes out an insurance policy that requires him to pay a $3000 deductible in case of a collision. Zach will cancel the collision policy when the value of his car equals his deductible. To the *nearest year*, how long will it take Zach to cancel this policy? Justify your answer.

369

Answers
August 2017
Algebra II

Answer Key

PART I

1. (1)	**5.** (3)	**9.** (3)	**13.** (1)	**17.** (2)	**21.** (4)
2. (3)	**6.** (2)	**10.** (3)	**14.** (2)	**18.** (4)	**22.** (1)
3. (3)	**7.** (4)	**11.** (1)	**15.** (1)	**19.** (1)	**23.** (4)
4. (2)	**8.** (4)	**12.** (4)	**16.** (2)	**20.** (2)	**24.** (3)

PART II

25. 16

26. 0.8

27. Both sides equal $x^4 + 2x^2y^2 + y^4$

28. See explanation

29. $a_1 = 4$
$a_n = 2a_{n-1} + 1$ for $n > 1$, 639

30. 4%

31. Even

32. See explanation for graph

PART III

33. $\{-5, -1\}$

34. $a_n = 1.25 + (n - 1) \cdot 0.25$, 16

35. See explanation for graph. The graph of the equation is like the original graph but shifted to the right.

36. $y = 4.168 \cdot 3.981^x$, 2.25 hours.

PART IV

37. See explanation for graph, 1.95 years, 6 years

In **Parts II–IV**, you are required to show how you arrived at your answers. For sample methods of solutions, see the *Answers Explained* section.

370 ANSWERS August 2017

Answers Explained

PART I

1. A fraction is undefined when the denominator is 0. Since this function has a denominator, the values of x that make that denominator equal 0 are the ones that make the function undefined. To find these values, solve the equation:

$$x^2 + 2x - 8 = 0$$
$$(x + 4)(x - 2) = 0$$

$$
\begin{array}{ccc}
x + 4 = 0 & & x - 2 = 0 \\
\underline{-4 = -4} & \text{or} & \underline{+2 = +2} \\
x = -4 & & x = 2
\end{array}
$$

The correct choice is **(1)**.

2. Start by multiplying and combining like terms:

$$(3k - 2i)(3k - 2i) =$$
$$(3k)(3k) + (3k)(-2i) - (2i)(3k) + (-2i)(-2i) =$$
$$9k^2 - 6ki - 6ki + 4i^2 =$$
$$9k^2 - 12ki + 4i^2$$

Since i is the imaginary unit, $i = \sqrt{-1}$ and $i^2 = -1$. Replace i^2 with -1 in the expression:

$$9k^2 - 12ki + 4i^2 =$$
$$9k^2 - 12ki + 4(-1) =$$
$$9k^2 - 12ki - 4$$

The correct choice is **(3)**.

ANSWERS August 2017

3. Use the quadratic formula to find the values of x that satisfy this equation. Use $a = 1$, $b = 2$, and $c = 5$:

$$x = \frac{-b \pm \sqrt{b^2 - 4ac}}{2a}$$

$$x = \frac{-2 \pm \sqrt{2^2 - 4 \cdot 1 \cdot 5}}{2 \cdot 1}$$

$$x = \frac{-2 \pm \sqrt{4 - 20}}{2} = \frac{-2 \pm \sqrt{-16}}{2} = \frac{-2 \pm \sqrt{16(-1)}}{2}$$

$$= \frac{-2 \pm \sqrt{16}\sqrt{-1}}{2} = \frac{-2 \pm 4i}{2} = -1 \pm 2i$$

The correct choice is **(3)**.

4. To solve this equation using algebra, first isolate one of the radicals and then square both sides of the equation:

$$\sqrt{x+14} - \sqrt{2x+5} = 1$$
$$\underline{\phantom{\sqrt{x+14}} + \sqrt{2x+5} = +\sqrt{2x+5}}$$
$$\sqrt{x+14} = 1 + \sqrt{2x+5}$$
$$\left(\sqrt{x+14}\right)^2 = \left(1 + \sqrt{2x+5}\right)^2$$
$$x + 14 = \left(1 + \sqrt{2x+5}\right)\left(1 + \sqrt{2x+5}\right)$$
$$x + 14 = 1 + 1\sqrt{2x+5} + 1\sqrt{2x+5} + 2x + 5$$
$$x + 14 = 1 + 2\sqrt{2x+5} + 2x + 5$$
$$x + 14 = 2x + 6 + 2\sqrt{2x+5}$$

Now isolate the remaining radical, and square both sides of the equation:

$$x + 14 = 2x + 6 + 2\sqrt{2x+5}$$
$$\underline{-2x - 6 = -2x - 6}$$
$$-x + 8 = 2\sqrt{2x+5}$$
$$(-x+8)^2 = \left(2\sqrt{2x+5}\right)^2$$
$$(-x+8)(-x+8) = \left(2\sqrt{2x+5}\right)^2$$
$$(-x)(-x) + (-x)8 + 8(-x) + 8 \cdot 8 = 2^2(2x+5)$$
$$x^2 - 16x + 64 = 4(2x+5)$$
$$x^2 - 16x + 64 = 8x + 20$$

ANSWERS August 2017

Simplify and solve the quadratic equation by moving all terms to one side of the equation and factoring:

$$x^2 - 16x + 64 = 8x + 20$$
$$\underline{-8x - 20 = -8x - 20}$$
$$x^2 - 24x + 44 = 0$$
$$(x - 22)(x - 2) = 0$$

$$\begin{array}{ccc}
x - 22 = 0 & & x - 2 = 0 \\
\underline{+22 = +22} & \text{or} & \underline{+2 = +2} \\
x = 22 & & x = 2
\end{array}$$

At this point, it seems like the answer is choice (4). When solving radical equations, though, there are often "extraneous" solutions. This means some solutions to the quadratic equation are not solutions to the original equation. To see if either of these solutions need to be rejected, test both to see if they make the original equation true.

Test $x = 2$:

$$\sqrt{2 + 14} - \sqrt{2 \cdot 2 + 5} \overset{?}{=} 1$$
$$\sqrt{16} - \sqrt{9} \overset{?}{=} 1$$
$$4 - 3 \overset{?}{=} 1$$
$$1 = 1$$
✔

Test $x = 22$:

$$\sqrt{22 + 14} - \sqrt{2 \cdot 22 + 5} \overset{?}{=} 1$$
$$\sqrt{36} - \sqrt{49} \overset{?}{=} 1$$
$$6 - 7 \overset{?}{=} 1$$
$$-1 \neq 1$$

Since $x = 22$ is not a solution to the original equation, the complete solution set is {2}.

Solving this question with algebra is very time consuming and difficult to do without making some careless error. If this was a Part II, III, or IV question, you would have to do it that way. Since this is a multiple-choice question, you could just test the four values used in the answer choices to see which of them make the equation true. Since only 2 makes the equation true, it must be the solution set.

The correct choice is **(2)**.

5. The graph of $y = 2 \tan x$ looks like this:

This graph can be produced on the graphing calculator if you put the calculator into radian mode.

For the TI-84:   For the TI-Nspire:

There is a vertical asymptote at $x = \frac{\pi}{2}$, so the curve gets closer and closer to the asymptote without ever reaching it. The closer the curve gets, the higher the y-coordinate is. So we can say that the graph increases without limit.

The correct choice is **(3)**.

6. If you are given the focus and directrix of a parabola, the vertex of the parabola will be halfway between the focus and the directrix. The parabola curves around the focus away from the directrix.

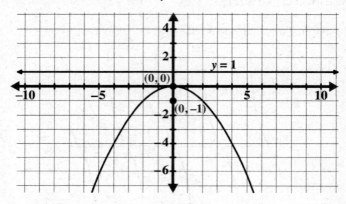

For this focus and directrix, the point halfway between them is (0, 0).

The vertex form of a parabola with vertex (h, k) has the equation:

$$y = \pm a(x - h)^2 + k$$

For this example, h and k are both 0. When the focus is above the directrix, a will be positive. In this case, the focus is below the directrix, so a will be negative. The value of a can be calculated with the formula $a = \dfrac{1}{4p}$, where p is the distance between the vertex and the focus.

Since the vertex is 1 unit from the focus, the value of p is 1, so:

$$a = \dfrac{1}{4p} = \dfrac{1}{4 \cdot 1} = \dfrac{1}{4}$$

The equation of the parabola is $y = -\dfrac{1}{4}(x-0)^2 + 0 = -\dfrac{1}{4}x^2$. After multiplying both sides of the equation by -4, it becomes $-4y = x^2$.

The correct choice is **(2)**.

ANSWERS August 2017   375

7. An angle in standard position has the positive x-axis as its initial ray and another ray rotated counterclockwise as its terminal ray. Choice (2) is not in standard position, so it can be eliminated. In degrees, angle α can be calculated by multiplying the radian measure by $\dfrac{180°}{\pi}$:

$$\alpha = \frac{13\pi}{20} \times \frac{180°}{\pi} = 117°$$

The reference angle θ is always an acute angle, so choice (1) can also be eliminated.

When an angle terminates in quadrant II, the reference angle is the acute angle formed by the negative x-axis and the terminal ray. This is represented in choice (4).

The measure of the quadrant II reference angle can also be found by subtracting the original angle from 180°:

$$\begin{aligned} \theta &= 180° - \alpha \\ &= 180° - 117° \\ &= 63° \end{aligned}$$

The measure of θ in choice (3) appears to be less than 63°.

The correct choice is **(4)**.

8. The zeros of a function are the input values that will make the function evaluate to zero. To find the zeros of P, solve the equation:

$$\begin{aligned} 0 &= (m^2 - 4)(m^2 + 1) \\ 0 &= (m - 2)(m + 2)(m^2 + 1) \end{aligned}$$

If any of these factors is equal to zero, the product will be zero:

$$
\begin{array}{ccccc}
m - 2 = 0 & & m + 2 = 0 & & m^2 + 1 = 0 \\
\underline{+2 = +2} & \text{or} & \underline{-2 = -2} & \text{or} & \underline{-1 = -1} \\
m = 2 & & m = -2 & & m^2 = -1 \\
& & & & \sqrt{m^2} = \pm\sqrt{-1} \\
& & & & m = \pm i
\end{array}
$$

The correct choice is **(4)**.

ANSWERS August 2017

9. The average rate of change can be calculated using the following formula:

$$\text{Average Rate of Change} = \frac{\text{Value in 2014} - \text{Value in 2012}}{2014 - 2012}$$

Calculate the values in the numerator using the property that $\log_b a = c$ can be rewritten as $b^c = a$.

Calculate the value in 2012 ($t = 1$):

$$\log_{0.8}\left(\frac{V}{17000}\right) = 1$$

$$0.8^1 = \frac{V}{17000}$$

$$17000 \cdot 0.8^1 = 17000 \cdot \frac{V}{17000}$$

$$13600 = V$$

Calculate the value in 2014 ($t = 3$):

$$\log_{0.8}\left(\frac{V}{17000}\right) = 3$$

$$0.8^3 = \frac{V}{17000}$$

$$17000 \cdot 0.8^3 = 17000 \cdot \frac{V}{17000}$$

$$8704 = V$$

$$\text{Average Rate of Change} = \frac{\text{Value in 2014} - \text{Value in 2012}}{2014 - 2012}$$

$$= \frac{8704 - 13600}{2014 - 2012}$$

$$= \frac{-4896}{2}$$

$$\approx -2450$$

Since the question asks for the average decreasing rate of change to the *nearest ten dollars*, the answer is +2450 instead of –2450.

The correct choice is **(3)**.

ANSWERS August 2017

10. A property of exponents is that $x^{ab} = (x^a)^b$. This can be used to write the given equation as an equivalent equation:

$$A = 100\left(\frac{1}{2}\right)^{\frac{t}{73.83}}$$

$$= 100\left(\frac{1}{2}\right)^{\frac{1}{73.83} \cdot t}$$

$$= 100\left(\frac{1}{2}^{\frac{1}{73.83}}\right)^{t}$$

$$\approx 100(0.990656)^t$$

The easiest way to solve this question, however, would just be to pick a number like $t = 10$ and use the given equation to calculate the value of A:

$$A = 100\left(\frac{1}{2}\right)^{\frac{10}{73.83}} \approx 91$$

Then check each of the answer choices to see which of them also produces a value of A equal to 91 for $t = 10$.

Test choice (3):

$$A = 100(0.990656)^{10} \approx 91$$

The correct choice is **(3)**.

11. Use the normal cdf feature of the graphing calculator. The upper bound is 3.7, the standard deviation σ is 0.2, and the mean μ is 4. For the lower bound (since there is no lower bound), use a large negative number like −9999.

For the TI-84:

Press [2ND] [VARS] [2] to access the normal cdf function.

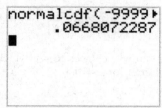

For the TI-Nspire:

Press [menu] [5] [5] [2] to access the Normal Cdf function.

0.0668 is the percent of ball bearings with a diameter less than 3.7 cm in this normal distribution.

The correct choice is **(1)**.

12. When the graph of a polynomial function has an x-intercept at a, the function has a factor of $(x - a)$. This function will have as three of its factors $(x - (-2)) = (x + 2)$, $(x - 10)$, and $(x - 14)$.

    Statement A does not have these factors, so it cannot be true.

    Statement B does have the correct factors and also an additional factor of -1 because of the negative sign in front of the factors. Therefore this statement can be true.

    At $x = 10$, the value of the function is 0, which is not the maximum. Statement (C) cannot be true.

    If you graph $p(x) = -(x + 2)(x - 10)(x - 14)$ and use the maximum feature of the graphing calculator, it will tell you that the maximum value will occur at $x = 12.14$ and the maximum volume will be approximately 56.

    For the TI-84:

    For the TI-Nspire:

    Both statements B and D could be true, while statements A and C could not be true.

    The correct choice is **(4)**.

380   ANSWERS August 2017

13. Divide $4x^3 + 9x - 5$ by $2x - 1$ using polynomial long division:

$$\begin{array}{r}
2x^2 + 1x + 5 \\
2x - 1 \overline{\smash{\big)}\, 4x^3 + 0x^2 + 9x - 5} \\
\underline{-(4x^3 - 2x^2)} \\
2x^2 + 9x - 5 \\
\underline{-(2x^2 - x)} \\
10x - 5 \\
\underline{-(10x - 5)} \\
0
\end{array}$$

The correct choice is (**1**).

14. Finding the inverse can be done in three steps:

Step 1: Change the $f(x)$ into a y.

$$y = \frac{x+1}{x-2}$$

Step 2: Change each x into a y and each y into an x.

$$x = \frac{y+1}{y-2}$$

Step 3: Solve for y in terms of x.

For this question, this last step is very involved because y occurs two times. Cross multiply, and then simplify. Move all the terms that have a y in them to one side of the equation and all other terms to the other side. Finally, factor out the y and divide both sides to isolate the y. This will be the answer for the inverse.

ANSWERS August 2017

$$x = \frac{y+1}{y-2}$$

$$(y-2)x = y+1$$

$$xy - 2x = y + 1$$

$$\underline{+2x = +2x}$$

$$xy = y + 1 + 2x$$

$$\underline{-y = -y}$$

$$xy - y = 1 + 2x$$

$$y(x-1) = 1 + 2x$$

$$\frac{y(x-1)}{x-1} = \frac{1+2x}{x-1}$$

$$y = \frac{1+2x}{x-1}$$

$$y = \frac{2x+1}{x-1}$$

$$f^{-1}(x) = \frac{2x+1}{x-1}$$

The correct choice is **(2)**.

15. Test the choices until you find one that is true.

Choice (1):

$$(x+2)^2 + 2(x+2) - 8 \overset{?}{=} (x+6)x$$

$$(x+2)(x+2) + 2(x+2) - 8 \overset{?}{=} x^2 + 6x$$

$$x^2 + 2x + 2x + 4 + 2x + 4 - 8 \overset{?}{=} x^2 + 6x$$

$$x^2 + 6x = x^2 + 6x$$

✔

The correct choice is **(1)**.

382　　　　**ANSWERS August 2017**

16. A plot of a sampling distribution of sample proportion has a proportion close to the median number of the data points. For this example, that is 0.16. The margin of error is the difference between the median and the highest and lowest numbers. Since the lowest numbers are approximately 0.02 away from the median and the highest numbers are also approximately 0.02 away from the median, the margin of error is approximately 0.02.

    The correct choice is **(2)**.

17. In an observational study, the person doing the study must simply observe what the subjects are doing and not interfere or interact with the subjects in any way.

    Choice (1) is a survey where the subjects answer questions rather than get observed.

    Choice (2) is an observational study since the researcher is gathering the data without any interaction with the actual subjects.

    Choice (3) is also a survey.

    Choice (4) is called an experimental study since the researcher decides which of the subjects use calculators and which ones do not.

    The correct choice is **(2)**.

18. The amplitude of a sine curve is the distance from the midline to the maximum of the curve. It can also be thought of as half the distance between the y-coordinates of the maximum point and the minimum point.

    Choice (1): The amplitude is 2. The minimum has a y-coordinate of 3, and the maximum has a y-coordinate of 7. Half the distance between 3 and 7 is 2.

    Choice (2): The amplitude is 4. The minimum has a y-coordinate of –5, and the maximum has a y-coordinate of 3. Half the distance between –5 and 3 is 4.

    Choice (3): When the equation is given, the amplitude is the absolute value of the coefficient before the sine function. For this choice, the amplitude is 3.

    Choice (4): The coefficient before the sine function is –5, so the amplitude is 5.

    Of the four choices, 5 is the greatest amplitude.

    The correct choice is **(4)**.

ANSWERS August 2017   383

19. The best way to solve a system of a linear and quadratic equation is to isolate one of the variables in the linear equation, substitute that variable into the quadratic equation, and then solve.

Isolate the y-term in the linear equation:

$$2x - y = 4$$
$$\underline{-2x = -2x}$$
$$-y = -2x + 4$$
$$\frac{-y}{-1} = \frac{-2x + 4}{-1}$$
$$y = 2x - 4$$

Substitute into the quadratic equation:

$$(x + 3)^2 + y^2 = 8$$
$$(x + 3)^2 + (2x - 4)^2 = 8$$
$$x^2 + 6x + 9 + 4x^2 - 16x + 16 = 8$$
$$5x^2 - 10x + 25 = 8$$
$$\underline{-8 = -8}$$
$$5x^2 - 10x + 17 = 0$$

Since the question does not ask for the solutions but for what type of solutions there will be, this can be completed by calculating the discriminant of the quadratic equation:

$$D = b^2 - 4ac$$

When the discriminant is negative, the equation has two imaginary solutions.

When the discriminant is zero, the equation has two equal rational solutions.

When the discriminant is a positive perfect square, the equation has two unequal rational solutions.

When the discriminant is positive but not a perfect square, the equation has two unequal irrational roots.

$$D = (-10)^2 - 4 \cdot 5 \cdot 17$$
$$= 100 - 340$$
$$= -240$$

384   **ANSWERS August 2017**

Since the discriminant is negative, this equation—and also the original system of equations—has two imaginary solutions.

The correct choice is **(1)**.

20. The quickest way to check if an expression of the form $x + a$ (or $x - a$) is a factor of a polynomial is to use the factor theorem. The factor theorem says that if in a polynomial function $f(a) = 0$, then $(x - a)$ is a factor of $f(x)$. Test the four choices.

Choice (1): To check if $x + 2$ is a factor, substitute -2 for x into the polynomial:

$$(-2)^3 - 11(-2)^2 + 16(-2) + 84 = -8 - 44 - 32 + 84 = 0$$

Since $f(-2) = 0$, $x + 2$ is a factor.

Choice (2): To check if $x + 4$ is a factor, substitute -4 for x into the polynomial:

$$(-4)^3 - 11(-4)^2 + 16(-4) + 84 = -64 - 176 - 64 + 84 = -220 \neq 0$$

Since $f(-4) \neq 0$, $x + 4$ is *not* a factor.

Choice (3): To check if $x - 6$ is a factor, substitute $+6$ for x into the polynomial:

$$6^3 - 11 \cdot 6^2 + 16 \cdot 6 + 84 = 216 - 396 + 96 + 84 = 0$$

Since $f(6) = 0$, $x - 6$ is a factor.

Choice (4): To check if $x - 7$ is a factor, substitute $+7$ for x into the polynomial:

$$7^3 - 11 \cdot 7^2 + 16 \cdot 7 + 84 = 343 - 539 + 112 + 84 = 0$$

Since $f(7) = 0$, $x - 7$ is a factor.

Only $x + 4$ is not a factor.

The correct choice is **(2)**.

21. After the ball drops 32 feet, it bounces up to a height of $32 \times 0.8 = 25.6$ feet and then down again. By the second bounce, the ball has a total downward distance (which does not include the distance that it travels on the way back up) of $32 + 25.6$ feet, which is already about 58 feet. There are still 10 more bounces to go so choices (1) and (2) can be eliminated.

ANSWERS August 2017 385

Without using a formula, the total downward distance can be calculated by making a chart with the downward distance for each of the twelve bounces and then adding them together. Each distance will be the previous distance multiplied by 0.8.

Bounce	Distance
1	32
2	25.6
3	20.48
4	16.38
5	13.11
6	10.49
7	8.39
8	6.71
9	5.37
10	4.29
11	3.44
12	2.75
Total	**149**

The "official" way to solve this question, however, is to recognize the pattern as a geometric series.

The downward distance for the first bounce is 32 feet.

The downward distance for the second bounce is $32 \cdot 0.8$ feet.

The downward distance for the third bounce is $32 \cdot 0.8^2$ feet.

The downward distance for the fourth bounce is $32 \cdot 0.8^3$ feet.

Notice that the exponent is always one less than the bounce number.

This pattern continues.

The downward distance for the twelfth bounce is $32 \cdot 0.8^{11}$ feet.

The sum of all the downward distances, then, is:

$$32 + 32 \cdot 0.8 + 32 \cdot 0.8^2 + 32 \cdot 0.8^3 + \cdots + 32 \cdot 0.8^{11}$$

386   **ANSWERS August 2017**

This is a finite geometric series with $a_1 = 32$, $r = 0.8$, and $n = 12$. The formula for the sum of a geometric series is given in the reference sheet provided on the Regents.

$$S_n = \frac{a_1 - a_1 r^n}{1 - r}$$

$$= \frac{32 - 32 \cdot 0.8^{12}}{1 - 0.8}$$

$$\approx 149 \text{ feet}$$

The correct choice is **(4)**.

22. Of the voters who were polled, 264 out of 550 (approximately 48%) said they would vote for Mayor Ortega. Each of the choices needs to be analyzed separately to see which is least appropriate.

Choice (1): The percent of people who vote for Mayor Ortega is not the same as the chance that Mayor Ortega will win. For example, if only 10% of the sample said they were going to vote for Mayor Ortega, there would be closer to a 0% chance that Mayor Ortega would win. Although there is some relationship between the two percentages, they are not the same thing.

Choice (2): The percent based on just the 550 people polled is known as the point estimate, so this is accurate.

Choice (3): The 48% found in the poll is not likely to be the exact percent of people who vote for Mayor Ortega in a full election. However, assuming that the sample size is a reasonable size, the percent is likely be near 48%. There really isn't enough information to say that the actual voter range from 48% is "likely" to be ±4%, but this is still more of an appropriate conclusion than the one from choice (1).

Choice (4): Since so little information is given in the problem—we don't know the total number of voters, for example—it is really hard to make any conclusions based on the poll. It is then appropriate to say that no inference can be made about who will ultimately win the election.

The correct choice is **(1)**.

23. First use the property of exponents that $\left(\dfrac{a}{b}\right)^x = \dfrac{a^x}{b^x}$:

$$\left(\frac{-54x^9}{y^4}\right)^{\frac{2}{3}} = \frac{(-54x^9)^{\frac{2}{3}}}{\left(y^4\right)^{\frac{2}{3}}}$$

ANSWERS August 2017   387

Next use the property of exponents that $(ab)^x = a^x b^x$:

$$\frac{(-54x^9)^{\frac{2}{3}}}{(y^4)^{\frac{2}{3}}} = \frac{(-54)^{\frac{2}{3}}(x^9)^{\frac{2}{3}}}{(y^4)^{\frac{2}{3}}}$$

Next use the property of exponents that $\left(a^x\right)^y = a^{xy}$:

$$\frac{(-54)^{\frac{2}{3}}(x^9)^{\frac{2}{3}}}{(y^4)^{\frac{2}{3}}} = \frac{(-54)^{\frac{2}{3}}x^{9\cdot\frac{2}{3}}}{y^{4\cdot\frac{2}{3}}} = \frac{(-54)^{\frac{2}{3}}x^6}{y^{\frac{8}{3}}}$$

Next use the property of exponents that $a^{\frac{x}{y}} = \left(\sqrt[y]{a}\right)^x$ or $a^{\frac{x}{y}} = \left(\sqrt[y]{a^x}\right)$:

$$\frac{(-54)^{\frac{2}{3}}x^6}{y^{\frac{8}{3}}} = \frac{\left(\sqrt[3]{-54}\right)^2 x^6}{\sqrt[3]{y^8}}$$

Finally, simplify the radicals by factoring the perfect cubes out of the expressions inside the cube roots, and then simplify:

$$\frac{\left(\sqrt[3]{-54}\right)^2 x^6}{\sqrt[3]{y^8}} = \frac{\left(\sqrt[3]{-27\cdot 2}\right)^2 x^6}{\sqrt[3]{y^6 y^2}}$$

$$= \frac{\left(-3\sqrt[3]{2}\right)^2 x^6}{y^2 \sqrt[3]{y^2}}$$

$$= \frac{(-3)^2 \sqrt[3]{2}^2 x^6}{y^2 \sqrt[3]{y^2}}$$

$$= \frac{(-3)^2 \sqrt[3]{2^2} x^6}{y^2 \sqrt[3]{y^2}}$$

$$= \frac{9x^6 \sqrt[3]{4}}{y^2 \sqrt[3]{y^2}}$$

The correct choice is (4).

388   **ANSWERS August 2017**

24. The money in the account the day their daughter was born is represented by $a_0 = 1000$.

    After one year, since the interest is compounded annually, the Rickerts receive the 1.8% interest and they deposit an additional $750. Calculating 1.8% interest added to the $1000 can be done two ways. First, calculate $1000 + \$1000(0.018) = \$1018$. Alternatively, just multiply $\$1000(1.018) = \$1018$. Using this second approach for calculating the amount of money after the interest is earned, the amount of money after one year—including the interest and the new deposit—can be represented as $a_1 = 1000(1.018) + 750 = \$1768$.

    After two years, the Rickerts receive 1.8% interest on the money that was in the account after one year and the additional $750 deposit. One way to represent this is with the equation $a_2 = 1768(1.018) + 750 \approx \$2550$. It could also be expressed in terms of a_1 as $a_2 = a_1(1.018) + 750$.

    Likewise, after three years, the amount of money in the account can be expressed as $a_3 = a_2(1.018) + 750$.

    In general when $n > 1$, the recursive expression for calculating a_n is $a_n = a_{n-1}(1.018) + 750$.

    The correct choice is **(3)**.

PART II

25. Use the property of fractional exponents that says $x^{\frac{n}{d}} = \left(\sqrt[d]{x}\right)^n$:

$$(-8)^{\frac{4}{3}} = \left(\sqrt[3]{-8}\right)^4$$
$$= (-2)^4$$
$$= 16$$

26. Use the property of probability that says $P(A \text{ and } B) = P(A) \cdot P(B \text{ given } A)$. The event A is the volunteer receiving the drug. The event B is the patient getting well.

$P(A \text{ and } B)$ is given as $40\% = 0.4$.

$P(A)$ is given as $50\% = 0.5$.

$P(A \text{ and } B) = P(A) \cdot P(B \text{ given } A)$.

$$0.4 = 0.5 \cdot P(B \text{ given } A)$$
$$\frac{0.4}{0.5} = \frac{0.5 \cdot P(B \text{ given } A)}{0.5}$$
$$\frac{0.4}{0.5} = P(B \text{ given } A)$$
$$0.8 = P(B \text{ given } A)$$

390   **ANSWERS August 2017**

27. Simplify each side of the equation, and compare the results:

$$(x^2 + y^2)^2 = (x^2 - y^2)^2 + (2xy)^2$$

For the left-hand side:

$$\left(x^2 + y^2\right)^2 =$$
$$\left(x^2 + y^2\right)\left(x^2 + y^2\right) =$$
$$x^4 + x^2 y^2 + x^2 y^2 + y^4 =$$
$$x^4 + 2x^2 y^2 + y^4 =$$

For the right-hand side:

$$= \left(x^2 - y^2\right)^2 + \left(2xy\right)^2$$
$$= \left(x^2 - y^2\right)\left(x^2 - y^2\right) + \left(2xy\right)^2$$
$$= x^4 - x^2 y^2 - x^2 y^2 + y^4 + 4x^2 y^2$$
$$= x^4 - 2x^2 y^2 + y^4 + 4x^2 y^2$$
$$= x^4 + 2x^2 y^2 + y^4$$

Both sides simplify to the same expression, making the identity true.

28. Since there are six different flavors of jelly beans and there are six sides on the die, you could simulate taking a jelly bean out of the bag by rolling one die. Assign one of the possible numbers, like number 1, to correspond to picking a black jelly bean from the bag.

To see how likely it is to pick four jelly beans at random and have all of them be black, you can roll the die four times and record how many 1s there were. Sometimes there will be no 1s. Sometimes there will be one 1. Sometimes there will be two 1s. Sometimes there will be three 1s. Sometimes there will be four 1s.

Do this experiment 62 times, which will require 248 dice rolls. Find the percent of times out of the 62 experiments that there were four 1s. That will be approximately the probability of getting four black jelly beans from the bag.

ANSWERS August 2017

29. In this sequence, each term after the first one is one more than double the previous term.

As a recursive formula, this would be:

$$a_1 = 4$$
$$a_n = 2a_{n-1} + 1 \text{ for } n > 1$$

To find the eighth term, the sixth and seventh terms need to be found:

$$
\begin{aligned}
a_6 &= 2a_{6-1} + 1 \\
&= 2a_5 + 1 \\
&= 2 \cdot 79 + 1 \\
&= 159
\end{aligned}
$$

$$
\begin{aligned}
a_7 &= 2a_{7-1} + 1 \\
&= 2a_6 + 1 \\
&= 2 \cdot 159 + 1 \\
&= 319
\end{aligned}
$$

$$
\begin{aligned}
a_8 &= 2a_{8-1} + 1 \\
&= 2a_7 + 1 \\
&= 2 \cdot 319 + 1 \\
&= 639
\end{aligned}
$$

30. The formula for exponential growth is $A_n = A_0 (1 + r)^n$, where A_0 is the starting value, r is the rate of growth, n is the amount of time, and A_n is the ending value.

For this problem, use the following:

$$8.75 = 1.25(1+r)^{49}$$
$$\frac{8.75}{1.25} = \frac{1.25(1+r)^{49}}{1.25}$$
$$7 = (1+r)^{49}$$
$$\sqrt[49]{7} = \sqrt[49]{(1+r)^{49}}$$
$$1.0405 \approx 1+r$$
$$-1 = -1$$
$$0.040 \approx r$$

The rate of growth is approximately 4%.

31. An odd function is one where $f(-x) = -f(x)$. An even function is one where $f(-x) = f(x)$.

Calculate $j(-x)$, and compare it to $j(x)$ and $-j(x)$:

$$j(-x) = (-x)^4 - 3(-x)^2 - 4$$
$$= x^4 - 3x^2 - 4$$

$$j(x) = x^4 - 3x^2 - 4$$

$$-j(x) = -(x^4 - 3x^2 - 4)$$
$$= -x^4 - 3x^2 + 4$$

Since $j(-x) = j(x)$, the function is even.

Since $j(-x) = j(x)$, the function is not odd.

32. If a polynomial function has a factor of $(x - a)$, the graph of that function will have an x-intercept at $(a, 0)$. If it has a factor of $(x - b)$, the graph will have an x-intercept at $(b, 0)$. If the polynomial function has a factor of $(x + c)$, the graph will have an x-intercept at $(-c, 0)$.

Make a curve that passes through these three points. It will look a little like an "N" and will have a y-intercept $(0, d)$ where d is also positive.

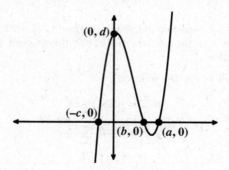

A simpler way to solve this is to pick positive values for a, b, and c with $a > b$, such as $a = 3$, $b = 2$, and $c = 1$. Then use the graphing calculator to make a graph of $p(x) = (x - 3)(x - 2)(x + 1)$.

For the TI-84:   For the TI-Nspire:

PART III

33. Multiply both sides of the equation by the LCM of all the denominators. In this case, the LCM is $(p-5)(p+3)$:

$$\frac{3p}{p-5} - \frac{2}{p+3} = \frac{p}{p+3}$$

$$\frac{\cancel{(p-5)}(p+3)3p}{\cancel{p-5}} - \frac{(p-5)\cancel{(p+3)}2}{\cancel{p+3}} = \frac{(p-5)\cancel{(p+3)}p}{\cancel{p+3}}$$

$$(p+3)3p - (p-5)2 = (p-5)p$$

$$3p^2 + 9p - 2p + 10 = p^2 - 5p$$

$$3p^2 + 7p + 10 = p^2 - 5p$$

To solve this quadratic equation, move all the terms to one side of the equation and try to factor:

$$3p^2 + 7p + 10 = p^2 - 5p$$

$$\underline{-p^2 + 5p = -p^2 + 5p}$$

$$2p^2 + 12p + 10 = 0$$

$$2\left(p^2 + 6p + 5\right) = 0$$

$$2(p+1)(p+5) = 0$$

$$\begin{array}{ccc} p+1=0 & & p+5=0 \\ \underline{-1=-1} & \text{or} & \underline{-5=-5} \\ p=-1 & & p=-5 \end{array}$$

This equation has two solutions: $p = -1$ and $p = -5$.

ANSWERS August 2017   395

34. In this arithmetic sequence, the position in the sequence is the number of days late and the number in that position is the amount of money owed.

It is given that $a_5 = 2.25$ and $a_{21} = 6.25$. To find the value of d, the constant difference from one term to the next, divide the change from term 5 to term 21 by 16 because the common difference must be added 16 times to get from term 5 to term 21:

$$d = \frac{6.25 - 2.25}{21 - 16}$$
$$= \frac{4}{16}$$
$$= 0.25$$

The formula for the nth term of an arithmetic sequence, which can be found in the reference sheet in the back of the Regents, is $a_n = a_1 + (n - 1)d$, where a_n is the nth term, a_1 is the first term, and d is the constant difference.

You can use the given information that $a_5 = 2.25$ together with the value of d that was just calculated to find the value of a_1, which is needed for the formula:

$$a_n = a_1 + (n - 1)d$$
$$a_5 = a_1 + (5 - 1)d$$
$$2.25 = a_1 + 4 \cdot 0.25$$
$$2.25 = a_1 + 1$$
$$\underline{-1 = \quad -1}$$
$$1.25 = a_1$$

So the formula for the nth term is $a_n = 1.25 + (n - 1) \cdot 0.25$.

Use the formula to find the 60th term:

$$a_{60} = 1.25 + (60 - 1) \cdot 0.25$$
$$= 1.25 + 59 \cdot 0.25$$
$$= 1.25 + 14.75$$
$$= 16$$

When the book is 60 days late, the total amount Simon owes will be $16.

35. Start by graphing the horizontal midline $y = -\dfrac{3}{2}$.

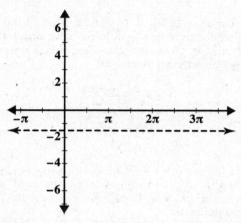

Since the amplitude is given as 2, the highest point will be 2 units above the midline and the lowest point will be 2 units below the midline. Draw horizontal guidelines 2 units above and two units below the midline at $y = \dfrac{1}{2}$ and $y = -\dfrac{7}{2}$. The curve will fit between these two lines.

The initial point of one cycle of a sine curve is always on the midline. The simplest place to start is at the point $\left(0, -\frac{3}{2}\right)$. The period is the horizontal distance between the initial point of one cycle of the sine curve and the ending point of that cycle. Since the period is given as 2π, graph the ending point of the cycle at $\left(2\pi, -\frac{3}{2}\right)$.

The sine curve passes through the midline halfway between the initial and the ending points. The curve reaches a maximum point at $\frac{1}{4}$ and a minimum point at $\frac{3}{4}$ of the way through the complete cycle. The complete curve looks like this:

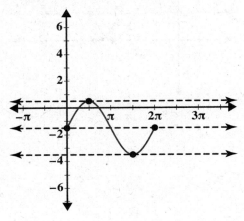

The graph of an equation with the structure $y = A \sin[B(x - C)] + D$ is a sine curve with an amplitude of A, a period of $\frac{2\pi}{B}$, a midline of $y = D$, and a horizontal shift to the right of C.

The graph of $y = 2\sin\left(x - \frac{\pi}{3}\right) - \frac{3}{2}$ has an amplitude of 2, a midline of $y = -\frac{3}{2}$, a period of $\frac{2\pi}{1} = 2\pi$, and a shift of $\frac{\pi}{3}$ to the right compared to $y = 2\sin(x) - \frac{3}{2}$.

The original graph also has an amplitude of 2, a midline of $y = -\frac{3}{2}$, and a period of 2π. So the only difference between a sketch of the equation $y = 2\sin\left(x - \frac{\pi}{3}\right) - \frac{3}{2}$ and the sketch from part (a) is that the sketch of the equation in part (b) is shifted to the right by $\frac{\pi}{3}$.

Here is a graph of one cycle of this equation on the same graph as the sketch from part (a):

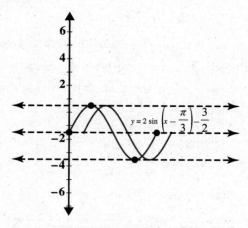

36. On your graphing calculator, enter the data and use the exponential regression feature.

For the TI-84:

Press [STAT] [1] and enter the x-values into L1 and the y-values into L2.
Press [STAT] [RIGHT] [0], then select the "Calculate" option.

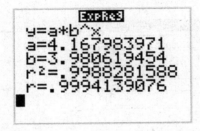

For the TI-Nspire:

From the home screen, select "Lists & Spreadsheet." Enter the data into columns A and B. Press [menu] [4] [A] to select the Exponential Regression function. For X List, enter "x," and for Y List enter "y." Press the "OK" button.

The exponential regression equation is $y = 4.168(3.981)^x$.

Substitute $y = 100$ into the equation, and solve for x using logarithms:

$$100 = 4.168(3.981)^x$$

$$\frac{100}{4.168} = \frac{4.168(3.981)^x}{4.168}$$

$$24 = 3.981^x$$

$$x = \log_{3.981} 24$$

$$\approx 2.3$$

Since the question asks you to round to the nearest quarter hour and 2.3 is closer to 2.25 than it is to 2.5, the answer is 2.25 hours.

PART IV

37. On the x-axis, make every box equal to $\frac{1}{4}$ units so the largest value is 5.

On the y-axis, make every box equal to 1500 units so the largest value is 30,000.

To plot the graph, find the values of V(t) and Z(t) for t = 0, 1, 2, 3, 4, and 5.

t	V(t)	Z(t)
0	28,483	22,151
1	19,482	17,234
2	13,326	13,408
3	9115	10,431
4	6235	8116
5	4264	6314

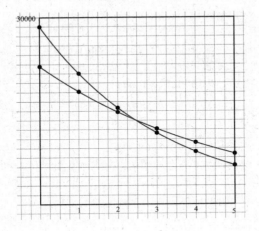

To find the t value for which $V(t) = Z(t)$, use the graphing calculator to find the x-coordinate of the intersection point of the two curves.

For the TI-84:   For the TI-Nspire:

Rounded to the nearest hundredth, $V(t) = Z(t)$ when $t \approx 1.95$ (or on the TI-Nspire because of rounding issues, 1.92). This means that after approximately 1.95 years (or with the TI-Nspire, 1.92 years) the value of the car is the same as the amount Zach still owes on the loan.

Find the value of t for which $V(t) = 3000$:

$$3000 = 28482.698(0.684)^t$$
$$\frac{3000}{28482.698} = \frac{28482.698(0.684)^t}{28482.698}$$
$$0.1053271 = 0.684^t$$
$$t = \log_{0.684} 0.1053271$$
$$= 5.92$$
$$\approx 6$$

It will take approximately 6 years for Zach to cancel this policy.

SELF-ANALYSIS CHART August 2017

Topic	Question Numbers	Number of Points	Your Points	Your Percentage
1. Polynomial Expressions and Equations	6, 8, 12, 13, 15, 20, 27, 32	2 + 2 + 2 + 2 + 2 + 2 + 2 + 2 = 16		
2. Complex Numbers	2, 3	2 + 2 = 4		
3. Exponential Expressions and Equations	9, 10, 30, 36, 37	2 + 2 + 2 + 4 + 6 = 16		
4. Rational Expressions and Equations	33	4		
5. Radical Expresions and Equations	4, 23, 25	2 + 2 + 2 = 6		
6. Trigonometric Expressions and Equations	5, 7, 18, 35	2 + 2 + 2 + 4 = 10		
7. Functions	1, 14, 31	2 + 2 + 2 = 6		
8. Systems of Equations	19	2		
9. Sequences and Series	21, 24, 29, 34	2 + 2 + 2 + 4 = 10		
10. Probability	26	2		
11. Statistics	11, 16, 17, 22, 28	2 + 2 + 2 + 2 + 2 = 10		

404   **SELF-ANALYSIS CHART August 2017**

HOW TO CONVERT YOUR RAW SCORE TO YOUR ALGEBRA II REGENTS EXAMINATION SCORE

The accompanying conversion chart must be used to determine your final score on the August 2017 Regents Examination in Algebra II. To find your final exam score, locate in the column labeled "Raw Score" the total number of points you scored out of a possible 86 points. Since partial credit is allowed in Parts II, III, and IV of the test, you may need to approximate the credit you would receive for a solution that is not completely correct. Then locate in the adjacent column to the right the scale score that corresponds to your raw score. The scale score is your final Algebra II Regents Examination score.

SELF-ANALYSIS CHART August 2017   405

Regents Examination in Algebra II—August 2017
Chart for Converting Total Test Raw Scores to Final
Examination Scores (Scaled Scores)

Raw Score	Scale Score	Performance Level	Raw Score	Scale Score	Performance Level	Raw Score	Scale Score	Performance Level
86	100	5	57	82	4	28	67	3
85	99	5	56	81	4	27	66	3
84	98	5	55	81	4	26	66	3
83	97	5	54	80	4	25	65	3
82	96	5	53	80	4	24	63	2
81	96	5	52	80	4	23	62	2
80	95	5	51	79	4	22	61	2
79	94	5	50	79	4	21	60	2
78	93	5	49	79	4	20	58	2
77	93	5	48	78	4	19	55	2
76	92	5	47	78	4	18	54	1
75	91	5	46	78	4	17	53	1
74	91	5	45	77	3	16	51	1
73	90	5	44	77	3	15	49	1
72	89	5	43	76	3	14	47	1
71	89	5	42	76	3	13	45	1
70	88	5	41	76	3	12	43	1
69	87	5	40	75	3	11	40	1
68	87	5	39	75	3	10	37	1
67	86	5	38	74	3	9	34	1
66	86	5	37	74	3	8	31	1
65	86	5	36	73	3	7	28	1
64	85	5	35	73	3	6	25	1
63	84	4	34	72	3	5	21	1
62	84	4	33	71	3	4	17	1
61	83	4	32	71	3	3	13	1
60	83	4	31	70	3	2	9	1
59	82	4	30	69	3	1	5	1
58	82	4	29	68	3	0	0	1

It's finally here—
online Regents exams from the experts!

- Take complete practice tests by date, or choose questions by topic
- All questions answered with detailed explanations
- Instant test results let you know where you need the most practice

Online Regents exams are available in the following subjects:

- Biology–The Living Environment
- Chemistry–The Physical Setting
- Earth Science–The Physical Setting
- English (Common Core)
- Global History and Geography
- Algebra I (Common Core)
- Geometry (Common Core)
- Algebra II (Common Core)
- Physics–The Physical Setting
- U.S. History and Government

Getting started is a point and a click away!

Only $24.99 for a subscription...includes all of the subjects listed above!

Teacher and classroom rates also available.

For more subscription information and to try our demo visit our site at *www.barronsregents.com*.

Prices subject to change without notice.

Join the half-million students each year who rely on Barron's Redbooks for Regents Exams success

Algebra I
ISBN 978-1-4380-0665-9

Algebra II
ISBN 978-1-4380-0843-1

Biology—The Living Environment
ISBN 978-0-8120-3197-3

Chemistry—The Physical Setting
ISBN 978-0-8120-3163-8

Earth Science—The Physical Setting
ISBN 978-0-8120-3165-2

English
ISBN 978-0-8120-3191-1

Geometry
ISBN 978-1-4380-0763-2

Global History and Geography
ISBN 978-0-8120-4344-0

Physics—The Physical Setting
ISBN 978-0-8120-3349-6

U.S. History and Government
ISBN 978-0-8120-3344-1

Each book: Paperback, $8.99, Can$11.50

Available at your local bookstore or visit **www.barronseduc.com**

Barron's Educational Series, Inc.
250 Wireless Blvd.
Hauppauge, NY 11788
Order toll-free: 1-800-645-3476
Order by fax: 1-631-434-3217

In Canada:
Georgetown Book Warehouse
34 Armstrong Ave.
Georgetown, Ontario L7G 4R9
Canadian orders: 1-800-247-7160
Order by fax: 1-800-887-1594

For extra Regents review visit our website: **www.barronsregents.com**

(#224) R10/17

Barron's Let's Review—Textbook companions to help you ace your Regents Exams

Let's Review: Algebra I
ISBN 978-1-4380-0604-8

Let's Review: Algebra II
ISBN 978-1-4380-0844-8

Let's Review: Biology—The Living Environment, 6th Ed.
ISBN 978-1-4380-0216-3

Let's Review: Chemistry—The Physical Setting, 6th Ed.
ISBN 978-1-4380-0959-9

Let's Review: Earth Science—The Physical Setting, 5th Ed.
ISBN 978-1-4380-0911-7

Let's Review: English, 5th Ed.
ISBN 978-1-4380-0626-0

Let's Review: Geometry
ISBN 978-1-4380-0702-1

Let's Review: Global History and Geography, 6th Ed.
ISBN 978-1-4380-1131-8

Let's Review: Physics, 5th Ed.
ISBN 978-1-4380-0630-7

Let's Review: U.S. History and Government, 6th Ed.
ISBN 978-1-4380-0962-9

Each book: Paperback, $14.99, Can18.50

Available at your local bookstore or visit **www.barronseduc.com**

Barron's Educational Series, Inc.
250 Wireless Blvd.
Hauppauge, N.Y. 11788
Order toll-free: 1-800-645-3476
Order by fax: 1-631-434-3217

In Canada:
Georgetown Book Warehouse
34 Armstrong Ave.
Georgetown, Ontario L7G 4R9
Canadian orders: 1-800-247-7160
Order by fax: 1-800-887-1594

For extra Regents review visit our website: **www.barronsregents.com**

Prices subject to change without notice.

#225 R10/17

ADD REAL POWER
to your preparation for Regents Exams
...with BARRON'S REGENTS POWER PACKS

They're concentrated, high-power help for students preparing for the New York State Regents exams. They also offer terrific extra classroom help to students taking final exams in the other 49 states. Combined in one money-saving package is a copy of the latest *Barron's Regents Exams and Answers* and a copy of *Barron's Let's Review* in each of the subjects listed below. *Let's Review* Books feature in-depth reviews of their subjects plus additional Regents exams to help students practice for the real thing. Purchase of *Barron's Regents Power Packs* represents an average savings of $3.99 off the total list price of both books.

Algebra I Power Pack
ISBN 978-1-4380-7574-7

Algebra II Power Pack
ISBN 978-1-4380-7659-1

Biology–Living Environment Power Pack
ISBN 978-1-4380-7434-4

Chemistry–The Physical Setting Power Pack
ISBN 978-1-4380-7838-0

Earth Science–Physical Setting Power Pack
ISBN 978-1-4380-7839-7

English Power Pack
ISBN 978-1-4380-7599-0

Geometry Power Pack
ISBN 978-1-4380-7619-5

Global History and Geography Power Pack
ISBN 978-1-4380-7296-8

Physics Power Pack
ISBN 978-1-4380-7598-3

U.S. History and Gov't. Power Pack
ISBN 978-1-4380-7840-3

Each package: $19.99, Can$24.99

Available at your local bookstore or visit **www.barronseduc.com**

Barron's Educational Series, Inc.
250 Wireless Blvd.
Hauppauge, NY 11788
Order toll-free: 1-800-645-3476
Order by fax: 1-631-434-3217

In Canada:
Georgetown Book Warehouse
34 Armstrong Ave.
Georgetown, Ontario L7G 4R9
Canadian orders: 1-800-247-7160
Order by fax: 1-800-887-1594

For extra Regents review visit our website: **www.barronsregents.com**

Barron's Regents Exams and Answers

FREE

Algebra II

Latest Examination

- A full-length New York State Regents Examination
- Complete Answers

Barron's Educational Series, Inc.

CHECK OUR WEBSITE FOR THE LATEST REGENTS EXAM IN ALGEBRA II

To ensure that you are always up-to-date we will post a PDF of the most current Regents exam with answers and explanations on our website:

http://barronsregents.com/supps/rha294.html

(#267) R10/16

Barron's Review Course Series
Let's Review:
Algebra II

Gary M. Rubinstein
B.A. Mathematics
Tufts University
M.S. Computer Science
University of Colorado

Barron's Educational Series, Inc.

About the Author

Gary Rubinstein has been teaching math for 25 years. He is a three-time recipient of the Math for America Master Teacher Fellowship. Gary lives with his wife Erica and his two children, Sarah and Sam. He has a YouTube channel at nymathteacher where students can find extra test tips and strategies for learning Algebra II.

Dedication

To Erica, Sarah, and Sam

© Copyright 2017 by Barron's Educational Series, Inc.

All rights reserved.
No part of this publication may be reproduced in any form or by any means without the written permission of the copyright owner.

All inquiries should be addressed to:
Barron's Educational Series, Inc.
250 Wireless Boulevard
Hauppauge, New York 11788
www.barronseduc.com

ISBN: 978-1-4380-0844-8

ISSN: 2472-9167

PRINTED IN THE UNITED STATES OF AMERICA
9 8 7 6 5 4 3 2 1

10%
POST-CONSUMER WASTE
Paper contains a minimum of 10% post-consumer waste (PCW). Paper used in this book was derived from certified, sustainable forestlands.

TABLE OF CONTENTS

PREFACE		**v**
CHAPTER 1	**POLYNOMIAL EXPRESSIONS AND EQUATIONS**	**1**
	1.1 Polynomial Arithmetic	1
	1.2 Polynomial Factoring	8
	1.3 The Remainder Theorem and the Factor Theorem	15
	1.4 Polynomial Equations	18
	1.5 Quadratic Graphs	23
	1.6 Polynomial Graphs	41
CHAPTER 2	**RATIONAL EXPRESSIONS AND EQUATIONS**	**57**
	2.1 Arithmetic with Rational Expressions	57
	2.2 Solving Rational Equations	68
	2.3 Graphing Rational Functions	75
CHAPTER 3	**EXPONENTIAL AND LOGARITHMIC EXPRESSIONS AND EQUATIONS**	**82**
	3.1 Properties of Exponents	82
	3.2 Solving Exponential Equations by Guess and Check or by Graphing	88
	3.3 Logarithms	99
	3.4 Transformed Graphs of Exponential and Logarithmic Functions	111
	3.5 Using Exponential or Logarithmic Equations in Real-World Scenarios	117
CHAPTER 4	**RADICAL EXPRESSIONS AND EQUATIONS**	**126**
	4.1 Simplifying Radicals	126
	4.2 Imaginary and Complex Numbers	130
	4.3 Solving Radical Equations	137
	4.4 Graphs of Radical Functions	144
CHAPTER 5	**TRIGONOMETRIC EXPRESSIONS AND EQUATIONS**	**157**
	5.1 Unit Circle Trigonometry	157
	5.2 Trigonometry Equations	176
	5.3 Radian Measure	183
	5.4 Graphs of the Sine and Cosine Functions	189
	5.5 Graphically Solving Trig Equations	212
	5.6 Modeling Real-World Scenarios with Trig Functions	215
	5.7 Trigonometry Identities	221

Table of Contents

CHAPTER 6	**SYSTEMS OF EQUATIONS**	**228**
	6.1 Two Equations with Two Unknowns	228
	6.2 The Elimination Method for Three Equations with Three Unknowns	238
	6.3 Using a Graphing Calculator to Solve Systems of Equations	244
CHAPTER 7	**FUNCTIONS**	**248**
	7.1 Composite Functions	248
	7.2 Inverse Functions	256
CHAPTER 8	**SEQUENCES**	**269**
	8.1 Sequences and Series	269
CHAPTER 9	**PROBABILITY**	**276**
	9.1 Sample Spaces	276
	9.2 Calculating Probabilities Involving Independent Events	286
	9.3 Calculating Probabilities Involving Dependent Events	292
CHAPTER 10	**NORMAL DISTRIBUTION**	**301**
	10.1 Standard Deviation	301
	10.2 Normal Distribution	309
CHAPTER 11	**STATISTICS**	**316**
	11.1 Types of Statistical Studies	316
	11.2 Inferential Statistics	318
CHAPTER 12	**TEST-TAKING STRATEGIES**	**331**
	12.1 Time Management	331
	12.2 Know How to Get Partial Credit	332
	12.3 Know Your Calculator	334
	12.4 Use the Reference Sheet	340
	12.5 How Many Points Do You Need to Pass?	341

ANSWERS AND SOLUTION HINTS TO PRACTICE EXERCISES	**342**
GLOSSARY OF ALGEBRA II TERMS	**363**
THE ALGEBRA II REGENTS EXAMINATION	**370**
June 2016 Examination	373
Answer Key	395
June 2017 Examination	397
Answer Key	419
INDEX	**421**

iv

PREFACE

In 2009, New York State adopted the Common Core Standards in order to qualify for President Obama's Race To The Top initiative. The Common Core math curriculum is more difficult than the previous math curriculum. The new state tests, including the Common Core Algebra II Regents, are more difficult as well.

Certain topics that had been in the Algebra II/Trigonometry curriculum for decades have been removed for not being rigorous enough. Other topics have been added with the goal of making 21st-century American students more career and college ready than their predecessors.

The main topics that have been cut from the curriculum are permutations, combinations, Bernoulli trials, binomial expansion, and the majority of the trigonometric identities like the sine sum, cosine sum, sine difference, cosine difference, and the different double-angle and half-angle formulas. Other topics have been moved into earlier grades like the law of sines and the law of cosines. More than half of the trigonometry that had once been part of Algebra II is no longer part of the course.

Other topics have been added to fill the gaps left by those topics now considered obsolete. Primarily, these new topics are often taught in AP Statistics as part of inferential statistics.

Aside from the change in topics, there is a change in the style of questions. Students now need to think more deeply about the topics because questions are intentionally phrased in a less straightforward way than they had been in the past.

Conquering the Algebra II test, something that was never an easy feat beforehand, has gotten much more difficult and will require more test preparation than before. Getting this book is a great first step toward that goal. In addition to reviewing all of the topics that can appear on this test, this book includes nearly 1,000 practice questions of various difficulty levels. This book can serve as a review or even as a way to learn the material for the first time. Teachers can also use this book to guide their pacing. They can focus on the types of questions that are most likely to appear on the test and spend less time on complicated aspects of the Common Core curriculum that are unlikely to be on the test.

The Common Core is part of a grand plan that is intended to propel our country to the top of the international rankings in math and reading. Good luck. We are all counting on you!

Gary Rubinstein
Math Teacher
2016

Chapter One
POLYNOMIAL EXPRESSIONS AND EQUATIONS

1.1 POLYNOMIAL ARITHMETIC

KEY IDEAS

A *polynomial* is an expression like $x^2 - 5x + 6$ that combines numbers and variables raised to different powers. Just as two numbers can be added, subtracted, multiplied, or divided, polynomials can be too.

Multiplying a Polynomial by a Constant

To multiply a polynomial by a constant, multiply the constant by each of the coefficients of the polynomial. This is sometimes called *distributing* the constant through the polynomial.

To multiply $3x^2 - 2x + 5$ by 4, multiply each of the coefficients by 4.

$$4 \cdot (3x^2 - 2x + 5) = 12x^2 - 8x + 20$$

Adding Polynomials

To add two polynomials, combine the *like terms*, which have the same variable raised to the same exponent.

To add the two polynomials $(x^2 - 5x + 6) + (2x^2 + 3x - 2)$, first combine the two x^2-terms, $x^2 + 2x^2 = 3x^2$. Then combine the two x-terms, $-5x + 3x = -2x$. Then combine the two constant terms $+6 - 2 = +4$. The sum is $3x^2 - 2x + 4$.

Subtracting Polynomials

Subtracting polynomials is more complicated than adding polynomials.

$$(3x^2 + 5x - 1) - (x^2 - 2x + 3)$$

Since the $-$ sign can be thought of as a negative 1 (-1) and the coefficient of the x^2 in the second polynomial is really a 1, this can be rewritten as:

$$(3x^2 + 5x - 1) - 1(1x^2 - 2x + 3)$$

Distribute the -1 through the second polynomial. The parentheses are no longer needed.

Polynomial Expressions and Equations

$$3x^2 + 5x - 1 - 1x^2 + 2x - 3$$

Combine like terms.

$$2x^2 + 7x - 4$$

Multiplying Polynomials

Multiplying two polynomials requires multiplying each combination of one term from the first polynomial with one term from the second polynomial and then combining all the products. The most common type of polynomial multiplication is when each of the polynomials has just two terms (called *binomials*). The four combinations can then be remembered with the word **FOIL**.

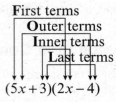

First terms in each expression: $5x \cdot 2x = +10x^2$

Outer terms in each expression: $5x \cdot -4 = -20x$

Inner terms in each expression: $+3 \cdot 2x = +6x$

Last terms in each expression: $+3 \cdot -4 = -12$

$$10x^2 - 20x + 6x - 12 = 10x^2 - 14x - 12$$

If one or both of the polynomials has more than two terms, then the **FOIL** method does not apply. Instead, get all the combinations by multiplying the first term in the polynomial on the left by all the terms in the polynomial on the right. Then multiply the second term in the polynomial on the left by all the terms in the polynomial on the right. Continue until the last term in the polynomial on the left has been multiplied by all the terms in the polynomial on the right.

Example 1

Multiply $(x + 3)(2x^2 - 4x + 5)$

Solution: First multiply the x in the binomial by each of the terms of $2x^2 - 4x + 5$. Then multiply the $+3$ in the binomial by each of the terms of $2x^2 - 4x + 5$. Combine the like terms.

$$(x \cdot 2x^2) + (x \cdot (-4x)) + (x \cdot 5) + (3 \cdot 2x^2) + (3 \cdot (-4x)) + (3 \cdot 5)$$
$$= 2x^3 - 4x^2 + 5x + 6x^2 - 12x + 15$$
$$= 2x^3 + 2x^2 - 7x + 15$$

Multiplication Patterns

Two useful patterns for multiplying binomials without using FOIL or the combination method are the perfect square multiplying pattern and the difference of perfect squares multiplying pattern.

Simplifying $(x + 5)^2$ with FOIL becomes $x^2 + 5x + 5x + (5 \cdot 5) = x^2 + 10x + 25$. The coefficient of the x in the solution is double the $+5$, whereas the constant in the solution is the square of $+5$. In general, $(x + a)^2 = x^2 + 2ax + a^2$.

Example 2

Use the perfect square multiplying pattern to simplify $(x - 3)^2$.

Solution: The coefficient will be $2 \cdot (-3) = -6$, and the constant will be $(-3)^2 = +9$.

$$x^2 - 6x + 9$$

When the only difference between two binomials is that one has a $+$ between the two terms and the other has a $-$ between the two terms, there is a shortcut for multiplying the binomials.

Simplifying $(x - 5)(x + 5)$ with FOIL becomes $x^2 + 5x - 5x - 25 = x^2 - 25$. There is no x-term in the answer, and the constant term is the negative square of the constant term of either of the binomials. In general, $(x - a)(x + a) = x^2 - a^2$.

Example 3

Use the difference of perfect squares multiplying pattern to simplify

$$(x - 3)(x + 3)$$

Solution: Since the only difference between the two binomials is the sign between the two terms, this pattern can be used. The answer is $x^2 - 3^2 = x^2 - 9$.

Dividing Polynomials

Dividing polynomials requires a process very similar to long division for numbers.

$$(2x^3 + x^2 - 11x + 12) \div (x + 3)$$

Step 1:

Set up for the long division process.

$$x+3\overline{)2x^3 + x^2 - 11x + 12}$$

3

Polynomial Expressions and Equations

Step 2:

Determine what you would need to multiply by the first term of the divisor (x in this example) to get the first term in the dividend ($2x^3$ in this example). Since you would need to multiply x by $2x^2$ to get $2x^3$, the first term of the solution is $2x^2$. Put that term over the x^2-term in the dividend.

$$x+3 \overline{)2x^3 + x^2 - 11x + 12} \quad \overset{2x^2}{}$$

Step 3:

Multiply the $2x^2$ by the $x + 3$ to get $2x^3 + 6x$. Put that product under the $2x^3 + x^2$. Subtract and bring down the $-11x$.

$$
\begin{array}{r}
2x^2 \\
x+3 \overline{)2x^3 + x^2 - 11x + 12} \\
-(2x^3 + 6x^2) \\
\hline
-5x^2 - 11x
\end{array}
$$

Step 4:

Determine what you would need to multiply by the first term of the divisor (x in this example) to get the first term in the expression at the bottom ($-5x^2$ in this example). Since you would need to multiply x by $-5x$ to get $-5x^2$, the second term of the solution is $-5x$. Put that term over the x-term in the dividend. Multiply the $-5x$ by the $x + 3$, and put the product under the $-5x^2 - 11x$. Then subtract and bring down the $+12$.

$$
\begin{array}{r}
2x^2 - 5x \\
x+3 \overline{)2x^3 + x^2 - 11x + 12} \\
-(2x^3 + 6x^2) \\
\hline
-5x^2 - 11x \\
-(-5x^2 - 15x) \\
\hline
4x + 12
\end{array}
$$

Step 5:

Determine what you would need to multiply by the first term of the divisor (x in this example) to get the first term in the expression at the bottom ($4x$ in this example). Since you would need to multiply x by 4 to get $4x$, the third term of the solution is $+4$. Put that term over the constant term in the dividend. Multiply the $+4$ by the $x + 3$, and put the product under the $4x + 2$. Then subtract. As there is nothing left to bring down, this final number at the bottom is the *remainder*. Since the remainder in this example is 0, we sometimes say there is no remainder and that $x + 3$ divides evenly into $2x^3 - x^2 - 11x + 12$.

$$
\begin{array}{r}
2x^2 - 5x + 4 \\
x+3{\overline{\smash{\big)}\,2x^3 +\ \ x^2 - 11x + 12}} \\
-(2x^3 + 6x^2) \\
\hline
-5x^2 - 11x \\
-(-5x^2 - 15x) \\
\hline
4x + 12 \\
-(4x + 12) \\
\hline
0
\end{array}
$$

Step 6:

You can check your answer by multiplying the solution by the divisor and then adding the remainder to see if the result is equal to the dividend.

$$(x + 3)(2x^2 - 5x + 4) + 0$$
$$= 2x^3 - 5x^2 + 4x + 6x^2 - 15x + 12 + 0$$
$$= 2x^3 + x^2 - 11x + 12$$

Polynomial Expressions and Equations

Example 4

What is the quotient and remainder (if any) of the following?

$$(3x^3 + 8x^2 - 14x + 13) \div (x + 4)$$

(1) $3x^2 + 4x - 2$, remainder 5
(2) $3x^2 - 4x + 2$, remainder 5
(3) $3x^2 + 4x + 2$, remainder 6
(4) $3x^2 - 4x - 2$, remainder 6

Solution:

$$
\require{enclose}
\begin{array}{r}
3x^2 - 4x + 2 \\[2pt]
x+4 \enclose{longdiv}{3x^3 + 8x^2 - 14x + 13} \\
\end{array}
$$

$$-(3x^3 + 12x^2)$$

$$-4x^2 - 14x$$
$$-(-4x^2 - 16x)$$

$$2x + 13$$
$$-(2x + 8)$$

$$5$$

The answer is choice (2).

Since Example 4 is a multiple-choice question, an alternative way to do this one would be to check each of the answer choices. Multiply each potential solution by $x + 4$, and add the potential remainder. The answer choice that gives you $3x^3 + 8x^2 - 14x + 13$ is the correct one.

Checking choice (2) would look like:

$$(x + 4)(3x^2 - 4x + 2) + 5$$
$$= 3x^3 - 4x^2 + 2x + 12x^2 - 16x + 8 + 5$$
$$= 3x^3 + 8x^2 - 14x + 13$$

The other choices all give different results. Admittedly, performing four multiplications to check the four choices might take longer than dividing. If on the test you forget how to divide polynomials and there is a multiple-choice question involving polynomial division, then this method would be a way to get the correct answer.

Check Your Understanding of Section 1.1

A. Multiple-Choice

1. What is $5 \cdot (2x^2 + 7x - 3)$?
   (1) $10x^2 + 35x - 15$
   (2) $10x^2 + 7x - 3$
   (3) $2x^2 + 7x - 15$
   (4) $2x^2 + 35x - 15$

2. What is $(-4) \cdot (2x^2 + 7x - 3)$?
   (1) $-8x^2 - 28x - 12$
   (2) $-8x^2 - 28x + 12$
   (3) $-8x^2 + 7x - 3$
   (4) $2x^2 + 7x + 12$

3. What is $(3x^2 - 5x + 7) + (2x^2 + 3x - 4)$?
   (1) $5x^2 - 8x + 3$
   (2) $5x^4 - 2x + 3$
   (3) $5x^2 - 2x + 11$
   (4) $5x^2 - 2x + 3$

4. What is $(5x^2 - 3x + 8) - (2x^2 + 4x - 2)$?
   (1) $3x^2 - 7x + 10$
   (2) $3x^2 + x + 10$
   (3) $3x^2 - 7x + 6$
   (4) $3x^2 + x + 6$

5. What is $2 \cdot (3x^2 - 4x + 7) - 3(x^2 - 2x - 5)$?
   (1) $3x^2 - 14x - 1$
   (2) $3x^2 - 14x + 29$
   (3) $3x^2 - 2x - 1$
   (4) $3x^2 - 2x + 29$

6. What is $(5x - 2)(2x + 7)$?
   (1) $10x^2 - 14$
   (2) $10x^2 + 35x - 14$
   (3) $10x^2 - 4x - 14$
   (4) $10x^2 + 31x - 14$

7. What is $(4x^2 + 2)(3x - 1)$?
   (1) $12x^3 - 4x^2 + 6x - 2$
   (2) $12x^3 - 2$
   (3) $12x^2 + 4x^2 - 6x + 2$
   (4) $12x^2 - 4x^2 - 6x - 2$

8. What is $(x + 2)(3x^2 + 2x - 7)$?
   (1) $3x^3 + 2x - 14$
   (2) $3x^3 + 8x^2 - 3x - 14$
   (3) $3x^3 + 8x^2 + 3x - 14$
   (4) $3x^3 - 8x^2 - 3x - 14$

9. What is $(x - 7)^2$?
   (1) $x^2 - 14x + 49$
   (2) $x^2 - 14x - 49$
   (3) $x^2 + 49$
   (4) $x^2 - 49$

10. What is $(3x^2 + 5)^2$?
    (1) $9x^4 + 25$
    (2) $9x^4 + 15x^2 + 25$
    (3) $9x^4 + 15x - 25$
    (4) $9x^4 + 30x^2 + 25$

7

Polynomial Expressions and Equations

11. What is $(x^3 + x^2 - 18x - 3) \div (x - 4)$?
    (1) $x^2 + 5x + 2$ R3
    (2) $x^2 + 5x + 2$ R4
    (3) $x^2 + 5x + 2$ R5
    (4) $x^2 + 5x + 2$ R6

B. *Show how you arrived at your answers.*

1. Simplify $(2x^2 - 3x + 4)^2$.

2. Zahra calculated $(5x^2 + 7x + 10) - (2x^2 + 3x + 4)$ as $5x^2 + 7x + 10 - 2x^2 + 3x + 4$ and got $3x^2 + 10x + 14$. What mistake did Zahra make?

3. If $(x + 5)(x + a) = x^2 + 7x + 10$, what is the value of a?

4. If $(x + a)^2 = x^2 + 10x + 25$, what is the value of a?

5. Simplify $(x + 3)^3$.

1.2 POLYNOMIAL FACTORING

KEY IDEAS

Factoring an integer means finding two other integers (other than 1) whose product is equal to the original integer. For example, the integer 15 has the *factors* 3 and 5 since $3 \cdot 5 = 15$. Likewise, a polynomial like $x^2 - 5x + 6$ can be factored into $(x - 2)(x - 3)$ because $(x - 2)(x - 3) = x^2 - 5x + 6$. When a polynomial is factored, the factors can provide useful information about the polynomial that was not apparent in the nonfactored form.

Just like some numbers can't be factored (for example, 7 and other prime numbers) some polynomials cannot be factored either. When a polynomial can be factored, there are several different methods of obtaining the factorization, depending on the polynomial.

Greatest Common Factor Factoring

Greatest common factor factoring is the first type of factoring you should always try. If all the terms of a polynomial have a common factor, that common factor can be "factored out." Often the only common factor is the number 1, in which case this type of factoring is not useful.

An example of using this method is factoring the polynomial $6x^3 - 4x^2 + 8x$. Each term of the polynomial $6x^3 - 4x^2 + 8x$ has a factor of $2x$. When the $2x$ is factored out, it is generally put on the left side and the remaining factor is put on the right in parentheses.

First write the $2x$ on the left of the parentheses

$$2x(\qquad\qquad) = 6x^3 - 4x^2 + 8x$$

Now imagine distributing the $2x$ through as if you were going to multiply the $2x$ by what needs to go inside the parentheses. For each term in the expression on the right, think "What does $2x$ need to be multiplied by to become this?" For the first term, $6x^3$, you have to multiply $2x$ by $3x^2$ ($2 \cdot 3 = 6$, and $x \cdot x^2 = x^3$). So the first term inside the parentheses is $3x^2$.

Do this for the other two terms to get

$$2x(3x^2 - 2x + 4) = 6x^3 - 4x + 8x$$

Example 1

Factor the polynomial $8x^4 - 12x^3 + 16x^2$.

Solution: Each of the three terms has a factor of $4x^2$. The factorization is

$$4x^2(2x^2 - 3x + 4)$$

Factoring a Quadratic Trinomial into the Product of Two Binomials

What's the opposite of FOIL? Is it LIOF? No. The opposite is factoring a quadratic trinomial like $x^2 + 5x + 6$ into the product of two binomials.

If the trinomial is of the form $x^2 + bx + c = 0$, find two numbers that have a sum of b and a product of c. For $x^2 + 5x + 6$, the two numbers that have a sum of 5 and a product of 6 are $+2$ and $+3$. Those numbers become the constants of the two binomials $(x + 2)$ and $(x + 3)$. So $x^2 + 5x + 6 = (x + 2)(x + 3)$.

Example 2

Factor the polynomial $x^2 + 3x - 10$ into the product of two binomials.

Solution: The two numbers that have a sum of $+3$ and a product of -10 are -2 and $+5$. So the two factors are $(x - 2)$ and $(x + 5)$.

$$x^2 + 3x - 10 = (x - 2)(x + 5)$$

Polynomial Expressions and Equations

Perfect Square Trinomial Factoring

The trinomials $x^2 + 6x + 9$, $x^2 + 8x + 16$, and $x^2 + 10x + 25$ are three examples of perfect square trinomials. These can be factored into $(x + 3)^2$, $(x + 4)^2$, and $(x + 5)^2$, respectively. The way to recognize a perfect square trinomial of the form $x^2 + bx + c$ is to compare c to $\left(\dfrac{b}{2}\right)^2$:

If $c = \left(\dfrac{b}{2}\right)^2$, the trinomial is a perfect square and can be factored as $\left(x + \dfrac{b}{2}\right)^2$.

For $x^2 + 6x + 9$, since $9 = \left(\dfrac{6}{2}\right)^2$, the trinomial is a perfect square. It can be factored as $\left(x + \dfrac{6}{2}\right)^2 = (x + 3)^2$.

Example 3

Which of the following is a perfect square trinomial?

(1) $x^2 + 8x + 25$
(2) $x^2 - 18x + 36$
(3) $x^2 + 10x - 25$
(4) $x^2 - 12x + 36$

Solution: Since $\left(-\dfrac{12}{2}\right)^2 = (-6)^2 = 36$, choice (4) is a perfect square trinomial. Notice that choice (3) would need to have +25 instead of −25 to be a perfect square trinomial.

Example 4

Factor $x^2 - 18x + 81$.

Solution: Since $\left(-\dfrac{18}{2}\right)^2 = (-9)^2 = 81$, the trinomial is a perfect square trinomial. It can be factored as $\left(x + \left(-\dfrac{18}{2}\right)\right)^2 = (x - 9)^2$.

Difference of Perfect Squares Factoring

A quadratic expression like $x^2 - 5^2$ is known as the difference between perfect squares since each of the terms is a perfect square and there is a subtraction sign between the two signs. In general, the expression $x^2 - a^2$ can be factored into $(x - a)(x + a)$. For the example $x^2 - 5^2$, it factors into $(x - 5)(x + 5)$.

Example 5

Factor $x^2 - 81$.

Solution: Since 81 can be written as 9^2, $x^2 - 81 = x^2 - 9^2 = (x - 9)(x + 9)$.

Factoring Cubic Expressions by Grouping

Polynomial expressions that have one of the variables raised to the third power are called cubic polynomials. Generally, they are very difficult to factor. Sometimes a technique called factor by grouping can be used to factor certain cubic polynomials.

Factor by grouping is when you have a four-term cubic expression and you factor a common factor from the first two terms and another common factor from the last two terms. Then you cross your fingers and hope that there will be a new common factor that you can then factor out.

For example, the polynomial $x^3 - 3x^2 + 4x - 12$ can be factored this way.

An x^2 can be factored out of the first two terms, and a 4 can be factored out of the last two terms.

$$x^2(x - 3) + 4(x - 3)$$

At this point, notice that both of the terms have a factor of $(x - 3)$. This can be factored out to become $(x - 3)(x^2 + 4)$ just like factoring an expression like $x^2 \cdot a + 4 \cdot a = a(x^2 + 4)$.

Example 6

Completely factor $x^3 + 2x^2 - 9x - 18$.

Solution: First factor an x^2 out of the first two terms.

$$x^2(x + 2) - 9x - 18$$

Now you can factor a -9 out of the second two terms so that the $(x + 2)$ factor matches the other one.

Polynomial Expressions and Equations

$$x^2(x + 2) - 9(x + 2)$$
$$(x + 2)(x^2 - 9)$$

Because the question says to factor "completely," check to see if any of the expressions can be factored further.

Since $x^2 - 9$ is a difference between perfect squares, it can be factored into $(x - 3)(x + 3)$. So the original expression can be completely factored to $(x + 2)(x - 3)(x + 3)$.

Factoring More Complicated Expressions

Polynomials that have exponents greater than 3 can sometimes be factored by rewriting the polynomials in an equivalent form that can be factored with other methods.

The expression $x^6 - 16$ has an exponent of 6. Since 6 is an even number, x^6 can be expressed as $(x^3)^2$. Since 16 is a perfect square, the original expression can be written as $(x^3)^2 - 4^2$, which can then be factored with the difference of perfect squares pattern: $(x^3 - 4)(x^3 + 4)$.

The trinomial $x^4 + 5x^2 - 14$ resembles the kind of quadratic trinomial from earlier in this chapter. It can be written as $(x^2)^2 + 5(x^2) - 14$. This looks a lot like $u^2 + 5u - 14$. If it were $u^2 + 5u - 14$, it would factor into $(u + 7)(u - 2)$. So the original trinomial factors instead into $(x^2 + 7)(x^2 - 2)$.

Example 7

Factor $x^8 + 10x^4 + 25$.

Solution: This can be expressed as $(x^4)^2 + 10(x^4) + 25$, which looks a lot like $u^2 + 10u + 25$. Since $u^2 + 10u + 25$ has the perfect square trinomial pattern, $25 = \left(\dfrac{10}{2}\right)^2$, it factors into $(u + 5)^2$. So the example factors into $(x^4 + 5)^2$.

Algebra Identities

When two algebraic expressions can be simplified to the same expression, it is called an identity. Proving that something is an identity requires simplifying one or both sides of the equation until the two sides are identical. Until an identity is proved, there will be a small question mark over the equal sign like $\overset{?}{=}$. After the identity is established, the question mark over the equals sign is replaced with a check $\overset{\checkmark}{=}$.

12

Example 8

Prove the following identity.

$$(x+y)^2 - 4xy \overset{?}{=} (x-y)^2$$

Solution:

$$(x+y)^2 - 4xy \overset{?}{=} (x-y)^2$$

$$x^2 + 2xy + y^2 - 4xy \overset{?}{=} x^2 - 2xy + y^2$$

$$x^2 - 2xy + y^2 \overset{\checkmark}{=} x^2 - 2xy + y^2$$

Number Theory Proofs

Sometimes a theorem about numbers can be proved by turning the theorem into an identity to be verified.

Example 9

The nth triangular number is $\dfrac{n(n+1)}{2}$. The nth square number is n^2.

Prove that the sum of the nth triangular number and the $(n + 1)$st triangular number is equal to the $(n + 1)$st square number.

Solution:

$$\frac{n(n+1)}{2} + \frac{(n+1)(n+2)}{2} \overset{?}{=} (n+1)^2$$

$$\frac{n^2 + n + n^2 + 3n + 2}{2} \overset{?}{=} (n+1)^2$$

$$\frac{2n^2 + 4n + 2}{2} \overset{?}{=} (n+1)^2$$

$$n^2 + 2n + 1 \overset{?}{=} (n+1)^2$$

$$(n+1)^2 \overset{\checkmark}{=} (n+1)^2$$

Polynomial Expressions and Equations

Check Your Understanding of Section 1.2

A. Multiple-Choice

1. Which shows $6x^3 - 15x^2 + 21x$ factored?
   (1) $3x(2x^2 - 5x + 7)$ (3) $2x(3x^2 - 15x^2 + 21x)$
   (2) $3x(2x^2 - 15x^2 + 21x)$ (4) $6x(x^2 - 2x^2 + 3x)$

2. Which shows $x^2 + 8x + 12$ factored?
   (1) $(x + 2)(x + 6)$ (3) $(x + 1)(x + 12)$
   (2) $(x + 3)(x + 4)$ (4) $(x + 3)(x + 5)$

3. Which shows $x^2 + 4x - 12$ factored?
   (1) $(x + 2)(x - 6)$ (3) $(x - 2)(x + 6)$
   (2) $(x - 3)(x + 4)$ (4) $(x - 1)(x - 12)$

4. Which shows $x^2 + 10x + 25$ factored?
   (1) $(x - 5)^2$ (3) $(x + 1)(x + 25)$
   (2) $(x + 5)(x - 5)$ (4) $(x + 5)^2$

5. Which shows $x^2 - 12x + 36$ factored?
   (1) $(x + 6)^2$ (3) $(x - 1)(x - 36)$
   (2) $(x + 6)(x - 6)$ (4) $(x - 6)^2$

6. Which shows $x^2 - 64$ factored?
   (1) $(x - 8)^2$ (3) $(x + 8)^2$
   (2) $(x - 8)(x + 8)$ (4) $(x - 8)(x - 8)$

7. Which shows $x^3 + 2x^2 + 6x + 12$ factored?
   (1) $(x + 6)(x^2 - 12)$ (3) $(x + 2)^2(x + 3)$
   (2) $(x + 2)(x^2 + 6)$ (4) $(x + 1)(x + 2)(x + 6)$

8. Which shows $x^3 - 5x^2 + 4x - 20$ factored?
   (1) $(x - 5)(x^2 + 4)$ (3) $(x - 2)^2(x + 5)$
   (2) $(x - 4)(x^2 + 5)$ (4) $(x - 1)(x + 4)(x + 5)$

9. Which shows $x^6 - 25$ factored?
   (1) $(x^3 - 5)(x^3 + 5)$ (3) $(x^3 - 5)^2$
   (2) $(x^6 - 5)(x^6 + 5)$ (4) $(x^3 + 5)^2$

10. Which shows $x^6 - 2x^3 - 24$ factored?
    (1) $(x^3 + 6)(x^3 - 4)$
    (2) $(x^3 - 8)(x^3 + 3)$
    (3) $(x^3 - 6)(x^3 + 4)$
    (4) $(x^3 + 8)(x^3 - 3)$

B. *Show how you arrived at your answers.*

1. Factor $2x^2 + 7x + 3$.

2. Factor $x^2 + (17 + 39)x + (17 \cdot 39)$.

3. What value for c can $x^2 - 18x + c$ be factored into $(x - 9)^2$?

4. How can the fact that $24^2 - 5^2 = 551$ be used to find the factors (not including 1 or 551) of 551?

5. Completely factor $x^4 - 13x + 36$.

1.3 THE REMAINDER THEOREM AND THE FACTOR THEOREM

KEY IDEAS

When something is a factor of a number, like 5 is a factor of 10, there will be no remainder when the number is divided by the factor. This is called *the factor theorem*. When something is not a factor of a number, like 3 is not a factor of 10, there will be some remainder when the number is divided by the factor. With polynomial division, there is a theorem called *the remainder theorem* that enables you to determine the remainder of some divisions without going through the long division process.

The Remainder Theorem

If you divide the polynomial function $f(x) = x^3 - 5x^2 + 6x - 3$ by $(x - 4)$ with the long division process, you get $x^2 - x + 2$ remainder 5. The remainder theorem says that you will also get the number 5 if you substitute +4 (the opposite of the −4 in the $(x - 4)$ divisor) into the $f(x) = x^3 - 5x^2 + 6x - 3$.

To check if the remainder theorem works for this example, evaluate $f(4) = 4^3 - (5 \cdot 4^2) + (6 \cdot 4) - 3 = 64 - 80 + 24 - 3 = 5$.

15

Polynomial Expressions and Equations

=== **MATH FACTS** ===

The remainder theorem says that the remainder when a polynomial equation is divided by $(x - a)$ is equivalent to the value of the polynomial when $+a$ is substituted for x. If the expression is $(x + a)$, then substitute $-a$ into the polynomial.

Example 1

What is the remainder when $f(x) = 4x^3 - 9x^2 + 7x + 1$ is divided by $(x + 5)$?

(1) 5
(2) −5
(3) $f(5)$
(4) $f(-5)$

Solution: The answer is choice (4). The remainder theorem states that the remainder will be the same as whatever the function evaluates to for the opposite of the constant in the divisor. For this example, $f(-5) = (4 \cdot (-5)^3) - (9 \cdot (-5)^2) + 7(-5) + 1 = -759$.

The Factor Theorem

When a binomial like $(x - 2)$ is a factor of a polynomial like $4x^3 - 3x^2 - 17x + 14$, it means that there will be a remainder of 0 when $4x^3 - 3x^2 - 17x + 14$ is divided by $x - 2$. If you evaluate the polynomial $4x^3 - 3x^2 - 17x + 14$ for $x = 2$, it becomes $(4 \cdot 2^3) - (3 \cdot 2^2) - (17 \cdot 2) + 14 = 32 - 12 - 34 + 14 = 0$, just as the remainder theorem predicted. From this, we get the factor theorem.

=== **MATH FACTS** ===

The factor theorem says that if $(x - a)$ is a factor of a polynomial, then the value of the polynomial when $+a$ is substituted for x will be 0.

Example 2

$(x - 3)$ is a factor of which polynomial?

(1) $f(x) = x^3 - 9x^2 + 25x - 19$
(2) $f(x) = x^3 - 9x^2 + 25x - 20$
(3) $f(x) = x^3 - 9x^2 + 25x - 21$
(4) $f(x) = x^3 - 9x^2 + 25x - 22$

Solution: Choice (3) is the answer. Evaluate each choice at $x = 3$. Choice (3) becomes $3^3 - (9 \cdot 3^2) + (25 \cdot 3) - 21 = 27 - 81 + 75 - 21 = 0$. The factor theorem says that if $f(3) = 0$, then $(x - 3)$ is a factor of $f(x)$.

Check Your Understanding of Section 1.3

A. Multiple-Choice

1. What is the remainder when $x^3 + x^2 - 9x - 5$ is divided by $x - 3$?
   (1) 4　　　　　(2) 5　　　　　(3) 6　　　　　(4) 7

2. What is the remainder when $2x^3 + 7x^2 - 10x + 28$ is divided by $x + 5$?
   (1) 2　　　　　(2) 3　　　　　(3) 4　　　　　(4) 5

3. If $f(x) = x^3 - 7x^2 + 5x - 9$, what is the remainder when $f(x)$ is divided by $(x + 4)$?
   (1) $f(4)$　　　(2) $f(-4)$　　　(3) $f(9)$　　　(4) $f(-9)$

4. If the remainder when $x^3 - 2x^2 + ax - 3$ is divided by $x - 2$ is 7, what is the value of a?
   (1) 2　　　　　(2) 3　　　　　(3) 4　　　　　(4) 5

5. If the remainder when $x^3 + ax^2 - 5x + 4$ is divided by $x - 3$ is 97, what is the value of a?
   (1) 7　　　　　(2) 8　　　　　(3) 9　　　　　(4) 10

6. If $(x + 4)(x - 5) = x^2 - x - 20$, what is the remainder when $x^2 - x - 20$ is divided by $(x + 4)$?
   (1) 3　　　　　(2) 2　　　　　(3) 1　　　　　(4) 0

7. Which of the following is a factor of $x^3 + 3x^2 - 10x - 24$?
   (1) $(x - 1)$　　(2) $(x - 2)$　　(3) $(x - 3)$　　(4) $(x - 4)$

8. Which of the following is a factor of $2x^4 - 9x^3 - 9x^2 + 46x + 24$?
   (1) $(x - 3)$　　(2) $(x + 4)$　　(3) $(x - 2)$　　(4) $(x - 1)$

9. If $(x - 5)$ is a factor of $x^3 - 4x^2 + 7x + a$, what is the value of a?
   (1) 20　　　　(2) −20　　　　(3) 60　　　　(4) −60

10. If $(x + 2)$ is a factor of $x^3 + ax^2 - 5x - 26$, what is the value of a?
    (1) 4　　　　　(2) 5　　　　　(3) 6　　　　　(4) 7

Polynomial Expressions and Equations

B. *Show how you arrived at your answers.*

1. What is the remainder when $x^3 + 5x^2 - 7x + 3$ is divided by $x - 4$?

2. $f(x) = x^4 - 3x^3 + 6x - 11$. If $f(3) = 7$, what is the remainder when $x^4 - 3x^3 + 6x - 11$ is divided by $(x - 3)$?

3. If the remainder when $x^3 + 4x^2 - 9x + 2$ is divided by $(x - 4)$ is 94, what is the remainder be when $x^3 + 4x^2 - 9x + 8$ is divided by $x - 4$?

4. Zoe and Jose tried to figure out if $x - 7$ is a factor of $x^3 - 12x^2 + 43x - 56$. Zoe did it by dividing, and Jose did it more quickly with the remainder theorem. How did Jose do it?

5. If $f(x) = x^4 + x^3 - x^2 + 11x - 12$, then $f(-3) = 0$. What is one factor of $f(x)$?

1.4 POLYNOMIAL EQUATIONS

A *polynomial equation* like $x^2 - 5x + 6 = 0$ involves an equal sign with a polynomial expression on one or both sides. The solution set of a polynomial equation is the set of numbers that make the left side of the equation equal to the right side of the equation. Polynomial equations usually have more than one solution.

Solving Quadratic Equations that Have No x-Term

A *quadratic equation* is one where the largest exponent is a 2. The simplest type of quadratic equation is when there is no x-term, such as the quadratic equation $x^2 = 9$.

To solve $x^2 = 9$, take the square root of each side. Be sure to put a ± on the square root on the right-hand side of the equal sign to account for the fact that $(+3)^2$ and $(-3)^2$ both equal +9.

$$x^2 = 9$$
$$\sqrt{x^2} = \pm\sqrt{9}$$
$$x = \pm 3$$

The solution set is $\{3, -3\}$.

Solving Factored Polynomial Equations

The equation $(x - 2)(x + 3) = 0$ is a quadratic equation. If the left side was simplified, the highest exponent would be 2.

Equations like this with two or more factors on the left side of the equation and a zero on the right side can be solved very quickly.

The only way $(x - 2)(x + 3)$ can equal zero is if one of the two factors is also equal to zero.

So $(x - 2)(x + 3)$ will equal zero if $x - 2$ is 0 or if $x + 3 = 0$.

$$(x - 2)(x - 3) = 0$$

$$
\begin{array}{ccc}
x - 2 = 0 & \text{or} & x - 3 = 0 \\
\underline{+2 = +2} & & \underline{+3 = +3} \\
x = 2 & \text{or} & x = 3
\end{array}
$$

The solution set is $\{2, 3\}$.

Example 1

What is the solution set to the equation $(x - 1)(x - 3) = 0$?

    Solution:

$$
\begin{array}{ccc}
x - 1 = 0 & \text{or} & x - 3 = 0 \\
x = 1 & \text{or} & x = 3
\end{array}
$$

The solution set is $\{1, 3\}$.

Example 2

What is the solution set to the equation $(x - 1)(x - 3)(x + 5) = 0$?

    Solution: With three factors, it is still true that if any one of them is equal to zero, the product of the three factors is also equal to zero. So the solution set can be found by using the three equations.

$$
\begin{array}{ccccc}
x - 1 = 0 & \text{or} & x - 3 = 0 & \text{or} & x + 5 = 0 \\
x = 1 & \text{or} & x = 3 & \text{or} & x = -5
\end{array}
$$

The solution set is $\{1, 3, -5\}$.

Solving Quadratic Equations by Factoring

Not all quadratic polynomials factor. If one does in an equation where there is a zero on the right-hand side of the equal sign, the solution set can be found very quickly.

Polynomial Expressions and Equations

$$x^2 - 5x + 6 = 0$$

If possible, start by factoring the left-hand side:

$$(x - 2)(x - 3) = 0$$

$$x - 2 = 0 \quad \text{or} \quad x - 3 = 0$$

$$\underline{+2 = +2} \qquad \underline{+3 = +3}$$

$$x = 2 \quad \text{or} \quad x = 3$$

The solution set is {2, 3}.

Example 3

Find the solution set to the equation $x^2 - 2x - 3 = 5$.

Solution: Even though the left-hand side of this equation can factor, do not factor until the right-hand side of the equation is a zero. Do this by subtracting 5 from both sides of the equation.

$$x^2 - 2x - 3 = 5$$

$$\underline{\qquad -5 = -5}$$

$$x^2 - 2x - 8 = 0$$

$$(x - 4)(x + 2) = 0$$

$$x - 4 = 0 \quad \text{or} \quad x + 2 = 0$$

$$x = 4 \quad \text{or} \quad x = -2$$

The solution set is {4, −2}.

The two solutions, 4 and −2, are the opposites of the two constants in the factors, $(x - 4)$ and $(x + 2)$. In general, if the *factors* of a quadratic expression are $(x - a)$ and $(x - b)$, then the *zeros* or *roots* of the quadratic expression are $x = a$ and $x = b$. This also works in reverse. If the zeros of a quadratic expression are $x = a$ and $x = b$, then the factors of the expression are $(x - a)$ and $(x - b)$.

Example 4

Find a quadratic expression whose roots are +5 and −3.

Solution: If the roots are +5 and −3, the factors are $(x - 5)$ and $(x + 3)$. So the quadratic expression could be $(x - 5)(x + 3)$, which can be simplified to $x^2 - 2x - 15$.

Example 5

What are the roots of a cubic expression that has the factors $(x + 6)$, $(x - 3)$, and $(x - 5)$?

Solution: Set each of the factors equal to zero to find the roots.

$(x + 6) = 0 \qquad (x - 3) = 0 \qquad (x - 5) = 0$

$\qquad x = -6 \qquad\qquad x = 3 \qquad\qquad x = 5$

The roots are $\{-6, 3, 5\}$.

Solving Quadratic Equations with the Quadratic Formula

When the quadratic expression does not factor, the equation has irrational roots and can be solved with the quadratic formula.

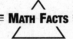 **MATH FACTS**

The two solutions to the quadratic equation $ax^2 + bx + c = 0$ can be found with the quadratic formula: $x = \dfrac{-b \pm \sqrt{b^2 - 4ac}}{2a}$

*Note: Simplifying radicals is explained in more detail in Chapter 4.

An example of a quadratic equation where the quadratic expression cannot factor is $x^2 - 4x + 1 = 0$, where $a = 1$, $b = -4$, and $c = 1$.

According to the quadratic formula:

$$x = \frac{-(-4) \pm \sqrt{(-4)^2 - 4(1)(1)}}{2(1)} = \frac{4 \pm \sqrt{16 - 4}}{2} = \frac{4 \pm \sqrt{12}}{2} = \frac{4 \pm 2\sqrt{3}}{2} = 2 \pm \sqrt{3}$$

The solution set is $\{2 + \sqrt{3}, 2 - \sqrt{3}\}$.

Example 6

Use the quadratic formula to find the two solutions to the equation $x^2 - 6x + 7 = 0$.

Solution:

$$x = \frac{-(-6) \pm \sqrt{(-6)^2 - 4(1)(7)}}{2(1)} = \frac{6 \pm \sqrt{36 - 28}}{2} = \frac{6 \pm \sqrt{8}}{2} = \frac{6 \pm 2\sqrt{2}}{2} = 3 \pm \sqrt{2}$$

Polynomial Expressions and Equations

Check Your Understanding of Section 1.4

A. *Multiple-Choice*

1. What is the solution set of $x^2 = 16$?
   (1) $\{4\}$ (2) $\{4, -4\}$ (3) $\{8\}$ (4) $\{8, -8\}$

2. What is the solution set of $(x - 4)(x - 7) = 0$?
   (1) $\{4, 7\}$ (2) $\{-4, -7\}$ (3) $\{-4, 7\}$ (4) $\{4, -7\}$

3. What is the solution set of $(x - 3)(x + 9) = 0$?
   (1) $\{-3, 9\}$ (2) $\{3, 9\}$ (3) $\{-3, -9\}$ (4) $\{3, -9\}$

4. What is the solution set of $(x - 2)(x - 5)(x + 1) = 0$?
   (1) $x = -2, x = -5, x = 1$ (3) $x = -2, x = 5, x = -1$
   (2) $x = 2, x = -5, x = 1$ (4) $x = 2, x = 5, x = -1$

5. What is the solution set of $x^2 + 10x + 24 = 0$?
   (1) $x = -4, x = -6$ (3) $x = -4, x = 6$
   (2) $x = 4, x = 6$ (4) $x = 4, x = -6$

6. What is the solution set of $x^2 + 3x + 2 = 12$?
   (1) $x = -5, x = 2$ (3) $x = 5, x = -2$
   (2) $x = -1, x = -2$ (4) $x = 1, x = 2$

7. Which equation has the solutions $x = 5, x = -2$?
   (1) $(x + 5)(x - 2) = 0$ (3) $(x - 5)(x + 2) = 0$
   (2) $(x + 5)(x + 2) = 0$ (4) $(x - 5)(x - 2) = 0$

8. Which equation has the solutions $x = -4, x = 3$?
   (1) $x^2 - x - 12 = 0$ (3) $x^2 - x + 12 = 0$
   (2) $x^2 + x + 12 = 0$ (4) $x^2 + x - 12 = 0$

9. What are the solutions to $x^2 - 6x + 4 = 0$?
   (1) $x = 5 \pm \sqrt{3}$ (3) $x = 3 \pm \sqrt{5}$
   (2) $x = 3 \pm \sqrt{2}$ (4) $x = 2 \pm \sqrt{5}$

10. What are the solutions to $x^2 - 10x + 23$?
    (1) $x = 2 \pm \sqrt{5}$ (3) $x = 3 \pm \sqrt{2}$
    (2) $x = 5 \pm \sqrt{2}$ (4) $x = 5 \pm \sqrt{3}$

B. *Show how you arrived at your answers.*

1. Lila solves the equation $x^2 - 25 = 0$ by first factoring $x^2 - 25$ into $(x - 5)(x + 5)$. Skylar solves the same equation by first adding 25 to both sides of the equation to get $x^2 = 25$. Who is right?

2. What are the three solutions to $x^3 + 4x^2 - 9x - 36 = 0$?

3. Create an equation that has the three solutions $x = 2, 4,$ and 5.

4. In terms of b, what is the solution to the quadratic equation $x^2 + bx - 4 = 0$?

5. Noelle used the quadratic formula to solve $x^2 + 10x + 16 = 0$. Delilah solved it without the quadratic formula. What did Delilah notice that enabled her to solve this equation without using the quadratic formula?

1.5 QUADRATIC GRAPHS

KEY IDEAS

A graph of the solution set of a two-variable quadratic equation, like $y = x^2 - 5x + 6$, has a shape called a *parabola*. To produce an accurate sketch of this graph, determine the *vertex* of the parabola and the *x*-intercepts of the parabola if any exist. Depending on the form of the equation, you can use several methods to produce this graph.

The most basic quadratic graph is the graph of $y = x^2$. Some of the ordered pairs in the solution set of this equation are (0, 0), (1, 1), (2, 4), and (3, 9). Since the square of a negative number is positive, ordered pairs with negative *x*-coordinates will have positive *y*-coordinates. So (−1, 1), (−2, 4), and (−3, 9) are three more ordered pairs. When these seven ordered pairs are graphed, they can be connected with a U-shaped curve called a *parabola*. The lowest point, which is (0, 0) for this parabola, is the parabola's *vertex*.

Polynomial Expressions and Equations

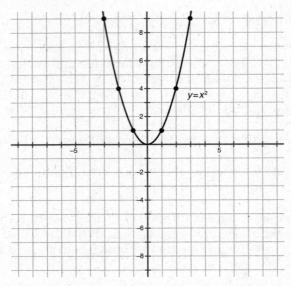

This graph is also the graph of the *quadratic function $f(x) = x^2$*.

Changing the *a* Coefficient

If a coefficient greater than 1 is in front of the x^2, such as $y = 2x^2$, the graph is still a parabola. However, it is narrower than the one created by $y = x^2$. The coefficient is usually called *a*, so the equation is $y = ax^2$. The vertex is still at (0, 0).

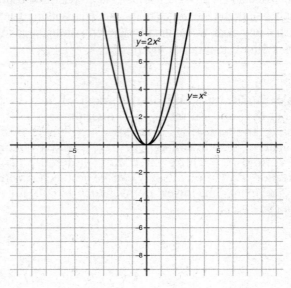

If *a* is negative, the parabola will be reflected over the *x*-axis and resemble an upside-down "U."

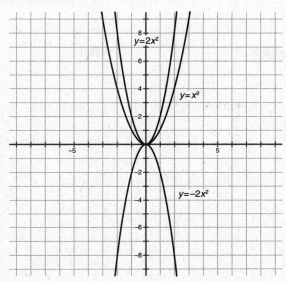

Vertex Form of a Quadratic Equation

A quadratic equation like $y = 1(x - 2)^2 - 5$ is said to be in *vertex form*. Vertex form is an expression like $y = a(x - h)^2 + k$, where *a*, *h*, and *k* are replaced with numbers. When a quadratic equation is in vertex form, the *x*-coordinate of the vertex is the opposite of the constant in the parentheses and the *y*-coordinate of the vertex is the constant outside the parentheses. For $y = 1(x - 2)^2 - 5$, the vertex is $(-(-2), -5) = (2, -5)$.

MATH FACTS

The vertex of the parabola defined by $y = a(x - h)^2 + k$ is (h, k). If *a* is greater than 1, the parabola will be narrow. If *a* is between 0 and 1, the parabola will be wide. If *a* is negative, the parabola will resemble an upside-down "U."

Polynomial Expressions and Equations

Example 1

What are the coordinates of the vertex of the parabola defined by $y = 3(x + 2)^2 - 5$?

- (1) (2, 5)
- (2) (2, −5)
- (3) (−2, 5)
- (4) (−2, −5)

Solution: Choice (4) is correct. The x-coordinate is $-(+2) = -2$, and the y-coordinate is −5.

Example 2

Which could be the equation for this parabola?

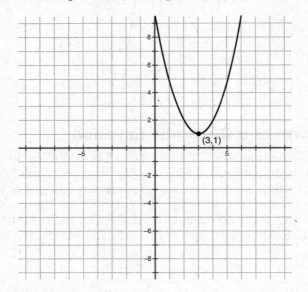

- (1) $y = 1(x + 3)^2 + 1$
- (2) $y = 1(x + 3)^2 - 1$
- (3) $y = 1(x - 3)^2 + 1$
- (4) $y = 1(x - 3)^2 - 1$

Solution: Choice (3) is correct. The constant inside the parentheses is the opposite of the x-coordinate of the vertex. The constant outside the parentheses is the y-coordinate.

Graphing Quadratic Equations

To sketch a more accurate graph, find the x-intercepts and y-intercept of the parabola. To find the x-intercepts, set $y = 0$ and solve for the values (if there are any) of x. The y-intercept can be found by setting $x = 0$ and solving for y.

Example 3

Sketch a graph, including the vertex and the x-intercepts (if any) and y-intercept, of $y = (x - 3)^2 - 1$.

Solution: The vertex is $(3, -1)$. To find the x-intercepts, solve the equation $0 = (x - 3)^2 - 1$:

$$\begin{array}{r} 0 = (x-3)^2 - 1 \\ \underline{+1 = +1} \\ 1 = (x-3)^2 \\ \pm\sqrt{1} = \sqrt{(x-3)^2} \\ \pm 1 = x - 3 \\ \underline{+3 = +3} \\ 3 \pm 1 = x \\ 4 = x \quad \text{or} \quad 2 = x \end{array}$$

The two x-intercepts are $(4, 0)$ and $(2, 0)$.

The y-intercept can be found with the equation $y = (0 - 3)^2 - 1 = 9 - 1 = 8$. So the y-intercept is $(0, 8)$.

Graph the vertex, the two x-intercepts, and the y-intercept. Sketch a parabola that passes through the four points.

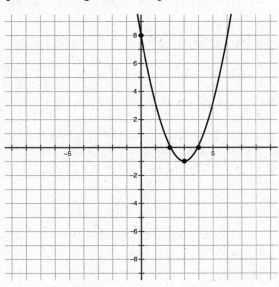

Polynomial Expressions and Equations

Vertex form is related to graphs of transformed functions. The quadratic function $g(x) = 1(x - 2)^2 - 5$ is related to the basic quadratic function $f(x) = x^2$. The $g(x)$ function can be expressed as $g(x) = f(x - 2) - 5$. The graph of a function defined this way is a translation of the original graph 2 units to the right and 5 units down. When the vertex of the graph of $f(x) = x^2$, which is $(0, 0)$, is translated 2 units to the right and 5 units down, the vertex of the transformed parabola will be $(2, -5)$.

Standard Form of a Quadratic Equation

Most quadratic functions and two-variable quadratic equations are not written in vertex form. Instead, they look like $f(x) = x^2 - 5x + 6$ or $y = x^2 - 5x + 6$. An equation in this simplified form is called *standard form*, $ax^2 + bx + c$. When the equation is in standard form, determining the vertex takes more work.

MATH FACTS

The x-coordinate of the vertex of the graph of $f(x) = ax^2 + bx + c$ is $\dfrac{-b}{2a}$. The y-coordinate of the vertex is the value of $f\left(\dfrac{-b}{2a}\right)$. The vertex of the parabola defined by $f(x) = x^2 - 6x + 8$, has an x-coordinate of $\dfrac{-(-6)}{2 \cdot 1} = 3$ and a y-intercept of $f(3) = -1$.

Example 4

What is the vertex of the graph of $f(x) = 2x^2 - 12x + 3$?

Solution: The x-coordinate of the vertex is $\dfrac{-b}{2a} = \dfrac{-(-12)}{2 \cdot 2} = \dfrac{12}{4} = 3$. The y-coordinate of the vertex is $f(3) = 2 \cdot 3^2 - 12 \cdot 3 + 3 = -15$. So the vertex is at $(3, -15)$.

Finding the x-intercepts of the graph of a quadratic function in standard form requires factoring. For the graph of $f(x) = x^2 - 6x + 8$, the x-intercepts can be found with the equation $0 = x^2 - 6x + 8$:

$$0 = x^2 - 6x + 8$$

$$0 = (x - 4)(x - 2)$$

$$x - 4 = 0 \text{ or } x - 2 = 0$$

$$x = 4 \text{ or } x = 2$$

So the x-intercepts are (4, 0) and (2, 0).

The y-intercept can be found by setting $x = 0$ and solving for y, $y = 0^2 - 6 \cdot 0 + 8 = 8$.

So the y-intercept is (0, 8).

Since the vertex was determined to be (3, −1) in a previous Math Fact, the graph looks like this:

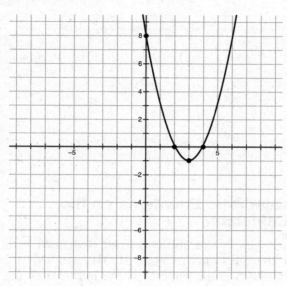

Example 5

Find the vertex and the x-intercepts of the graph of the function $f(x) = x^2 - 2x - 3$. Use these points to sketch the graph.

Solution: The x-coordinate of the vertex is $\dfrac{-(-2)}{2 \cdot 1} = +1$. The y-coordinate of the vertex is $f(1) = 1^2 - 2 \cdot 1 - 3 = -4$. The vetex is (1, −4). The x-intercepts can be found with the equation $0 = x^2 - 2x - 3$:

$$0 = x^2 - 2x - 3$$
$$0 = (x + 1)(x - 3)$$
$$0 = x + 1 \text{ or } 0 = x - 3$$
$$-1 = x \text{ or } 3 = x$$

The x-intercepts are (−1, 0) and (3, 0).

The graph looks like the following:

Polynomial Expressions and Equations

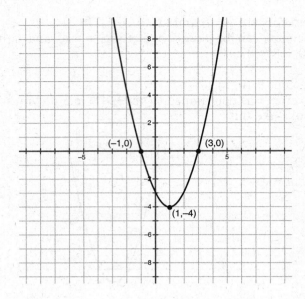

Graphing Quadratic Equations on the Graphing Calculator

There are two types of questions on the Regents Exam that involve graphing. For the questions in Parts II, III, and IV, you will have to produce the graph on graph paper. For the multiple-choice questions in Part I, though, you may just have to identify the graph from a list of choices. The graphing calculator can serve as a good way to check a graph that you produced by hand or to see what the graph looks like so you can select the correct graph from a set of choices.

For the TI-84:

Press the [Y=] button and input $x^2 - 5x + 6$ after the Y1=. Press [ZOOM] [6] to see the graph.

For the TI-Nspire:

From the home screen, press [B] to open the graphing Scratchpad. If the Entry Line is not visible, press [tab]. Input $x^2 - 5x + 6$ after f1(x)= and press [enter].

Focus and Directrix of a Parabola

Every parabola has an invisible line and invisible point associated with it. The invisible line is called the *directrix*, and the invisible point is called the *focus*. For a parabola that resembles a right-side-up "U," the focus is a bit above the vertex. The directrix is a horizontal line below the vertex. The distance between the focus and vertex is the same as the distance between the vertex and the directrix. For a parabola that resembles an upside-down "U," the focus is a bit below the vertex and the directrix is a bit above. Again, the distance between the focus and vertex is the same as the distance between the vertex and the directrix.

MATH FACTS

The distance between the vertex and the focus is the same as the distance between the vertex and the directrix. If the parabola is defined as $y = ax^2 + bx + c$ or as $y = a(x - h)^2 + k$, that distance is equal to $\dfrac{1}{4a}$ and is usually denoted by the variable *p*.

Polynomial Expressions and Equations

For the parabola $y = \frac{1}{4}x^2$, the value of p is $\frac{1}{4 \cdot \frac{1}{4}} = \frac{1}{1} = 1$. So the focus is a point located 1 unit above the vertex, at (0, 1). The directrix is a horizontal line located 1 unit below the vertex, at $y = -1$.

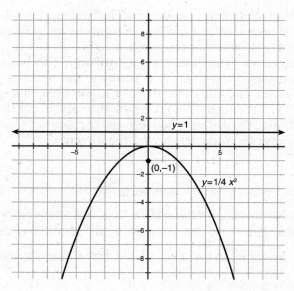

Math Facts

If the vertex of a parabola is at (h, k), the focus is at (h, $k + p$) and the directrix is at $y = k − p$, where $p = \dfrac{1}{4a}$.

Example 6

What are the coordinates of the focus and the equation of the directrix of the parabola defined by $y = \dfrac{1}{16}(x − 2)^2 − 5$?

Solution: Since this equation is in vertex form, the vertex is (−(−2), −5) = (2, −5).

Since $a = \dfrac{1}{16}$, $p = \dfrac{1}{4 \cdot \frac{1}{16}} = 4$. The focus is at (2, −5 + 4) = (2, −1), and the directrix is at $y = −5 − 4 = −9$.

Here is the graph of the parabola and its focus and directrix.

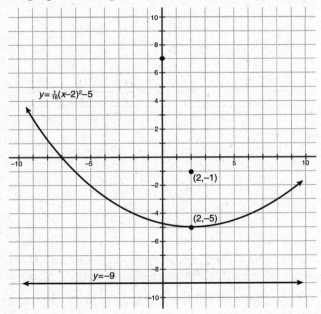

If the focus and directrix are known, it is possible to find the equation of the parabola related to that focus and directrix. The vertex is the midpoint of the vertical line segment connecting the focus to the directrix. The distance p is half the length of that vertical line segment, and the a-value for the quadratic equation equals $\dfrac{1}{4p}$.

Polynomial Expressions and Equations

If the focus is at (−1, 4) and the directrix is at $y = -2$, a vertical line segment connecting (−1, 4) to the line $y = -2$ would have length 6. The vertex would be the midpoint of that line segment, located at (−1, 1). Half the length of the segment is the value of p, which is 3 for this example. The formula relating a and p is $a = \dfrac{1}{4p} = \dfrac{1}{4 \cdot 3} = \dfrac{1}{12}$.

Since you have the vertex and the a-value, the equation in vertex form is $y = \dfrac{1}{12}(x+1)^2 + 1$.

Example 7

What is the equation in standard form of a parabola that has its focus at (0, 7) and a directrix of $y = 1$?

Solution: First sketch the focus and directrix on graph paper. Then draw the vertical segment joining the focus and the directrix.

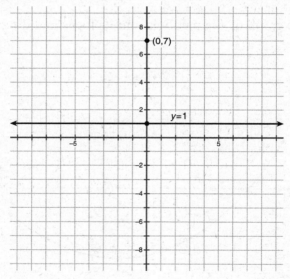

The value of p is half the length of the segment, and the vertex is the midpoint of the segment. The segment is length 6, so p is 3 and $a = \dfrac{1}{4p} = \dfrac{1}{12}$. The midpoint of the segment is at (0, 4). In vertex form, the equation is $y = \dfrac{1}{12}(x-0)^2 + 4$. In standard form, it is $y = \dfrac{1}{12}x^2 + 4$.

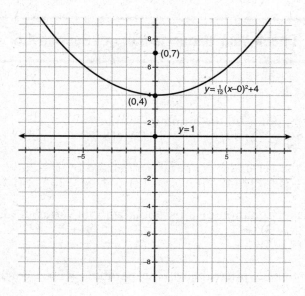

MATH FACTS

For any point on a parabola, the distance from that point to the parabola's focus is the same as the distance from that point to the parabola's directrix. In the graph below, the congruent line segments are marked.

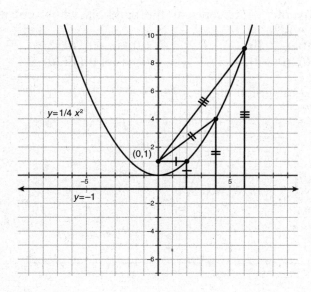

Polynomial Expressions and Equations

Check Your Understanding of Section 1.5

A. *Multiple-Choice*

1. Which is a point on the graph of $y = 2x^2$?
   (1) (3, 9)   (2) (9, 3)   (3) (18, 3)   (4) (3, 18)

2. What is the vertex of the parabola defined by $y = 2(x + 4)^2 - 5$?
   (1) (−4, −5)   (2) (4, −5)   (3) (−4, 5)   (4) (4, 5)

3. Which is an equation of a parabola with its vertex at (3, −1)?
   (1) $y = (x - 3)^2 + 1$
   (2) $y = (x - 3)^2 - 1$
   (3) $y = (x + 3)^2 + 1$
   (4) $y = (x + 3)^2 - 1$

4. What are the x-intercepts of the parabola whose equation is $y = (x - 1)^2 - 9$?
   (1) (4, 0), (−2, 0)
   (2) (−4, 0), (2, 0)
   (3) (3, 0), (−3, 0)
   (4) (0, 4), (0, −2)

5. What is the vertex of the parabola whose equation is $y = x^2 + 4x - 12$?
   (1) (−2, −16)
   (2) (−2, −24)
   (3) (2, 0)
   (4) (2, −16)

6. What are the x-intercepts of the parabola whose equation is $y = x^2 - 4x - 12$?
   (1) (6, 0), (−2, 0)
   (2) (0, 6), (0, −2)
   (3) (0, −6), (0, 2)
   (4) (−6, 0), (2, 0)

7. Which is the graph of $y = x^2 + 2x - 3$?

   (1)

36

(2)

(3)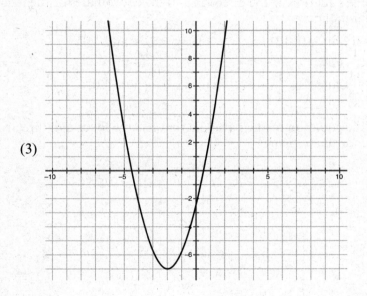

Polynomial Expressions and Equations

(4)
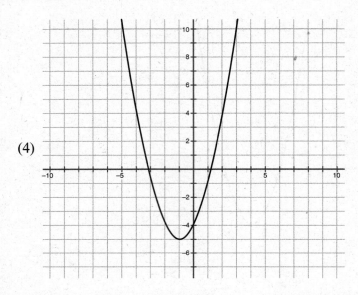

8. What is the equation of the directrix and the coordinates of the focus of the parabola $y = 3x^2$?

(1) $y = -12$ and $(0, 12)$

(2) $y = -\dfrac{1}{4}$ and $\left(0, \dfrac{1}{4}\right)$

(3) $y = -\dfrac{1}{12}$ and $\left(0, \dfrac{1}{12}\right)$

(4) $y = \dfrac{1}{12}$ and $\left(0, -\dfrac{1}{12}\right)$

9. What is the equation of the parabola that has the focus and directrix on this coordinate plane?

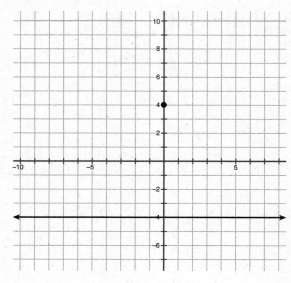

38

(1) $y = 12x^2$ (3) $y = \dfrac{1}{12}x^2$

(2) $y = x^2$ (4) $y = -\dfrac{1}{12}x^2$

10. Which is the graph of $y = -\dfrac{1}{4}(x+3)^2 + 5$?

(1)

(2)
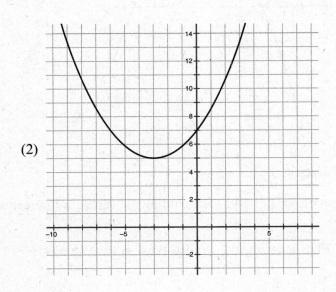

Polynomial Expressions and Equations

(3)

(4)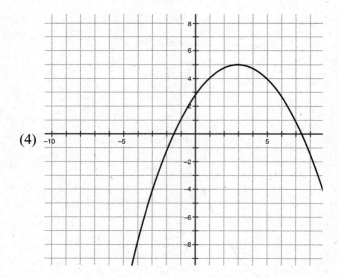

B. *Show how you arrived at your answers.*

1. Sketch the graph of $y = (x - 3)^2 - 1$, including the vertex, x-intercepts, and y-intercept.

2. $y = x^2 + bx + 7$ has a vertex at $(-4, -9)$. What is the value of b?

3. Convert $y = x^2 - 10x + 23$ into vertex form.

4. What is the equation of the parabola with its focus at (4, 3) and its directrix at $y = 1$?

5. If $m\overline{PD} = 8$, what is the length of $\overline{PF}$?

1.6 POLYNOMIAL GRAPHS

Graphs of cubic and other higher-degree polynomials can be sketched by knowing a few key points on the graph and having an understanding of the general shape of curves based on higher-degree polynomials. The x-intercepts and y-intercept are important as well as having an understanding of the *end behavior* at both ends of the curve.

Polynomial Expressions and Equations

Graphing Basic Cubic Equations

The graph of a quadratic function resembles a "U" (whether it is right-side up or upside-down). However, the graph of a cubic (3rd-degree) equation resembles a capital "N." The "N" can be either right-side up or upside-down.

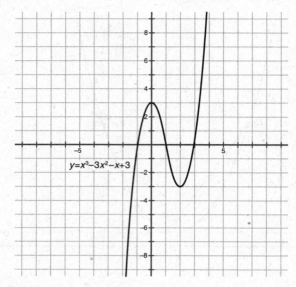

It is useful if the cubic function you want to graph is already in factored form, such as $f(x) = (x + 1)(x - 1)(x - 3)$. If you were to multiply this out, the coefficient of the x^3-term would be +1, so the a-value is positive. The graph resembles a right-side-up "N."

The x-intercepts can be found by substituting 0 for y to get the equation $0 = (x + 1)(x - 1)(x - 3)$.

This equation can be solved by finding the x-values that would make any of the factors equal to zero.

$x + 1 = 0$ or $x - 1 = 0$ or $x - 3 = 0$
$x = -1$ or $x = +1$ or $x = +3$

So the x-intercepts are $(-1, 0)$, $(1, 0)$, and $(3, 0)$.

The y-intercept can be found by substituting $x = 0$ into the equation and solving for y.

$$y = (0 + 1)(0 - 1)(0 - 3) = +3$$

So the y-intercept is $(0, 3)$.

Plot the 4 points, and join them with a curve like a right-side-up "N."

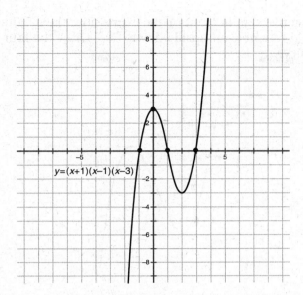

This equation can also be graphed with the graphing calculator.

Example 1

Sketch a graph of the function $f(x) = -\frac{1}{4}(x-2)(x-1)(x+4)$, including x-intercepts and y-intercept.

Solution: The x-intercepts can be found by using the equations $x - 2 = 0$, $x - 1 = 0$, and $x + 4 = 0$. The x-intercepts are $(2, 0)$, $(1, 0)$, and $(-4, 0)$. Since $y = -\frac{1}{4}(0-2)(0-1)(0+4) = -2$, the y-intercept is $(0, -2)$. Because

Polynomial Expressions and Equations

there is a negative sign in front of the expression, the *a*-value is negative so this should resemble an upside-down "N." Draw the four points, and sketch the curve.

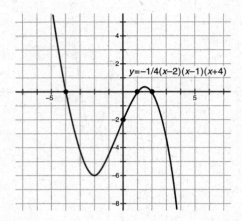

Graphing Polynomial Functions That Have Double Roots

A factored polynomial like $f(x) = (x - 2)(x - 2)(x + 3)$ has roots at 2, 2, and −3. Since there are two 2s, the graph of this function has only two *x*-intercepts: (2, 0) and (−3, 0). When there are two of the same root, it is called a *double root*. Graphing a polynomial that has double roots requires a special technique.

MATH FACTS

At a double root, the graph of the function "bounces off" the *x*-axis!

To graph $f(x) = (x - 2)(x - 2)(x + 3)$, first notice that the *a*-value is positive. So the graph will resemble a right-side-up "N."

The *x*-intercepts are at (2, 0) and (−3, 0). The *y*-coordinate of the *y*-intercept is at $y = (0 - 2)(0 - 2)(0 + 3) = 12$. So the *y*-intercept is (0, 12). Graph these three points.

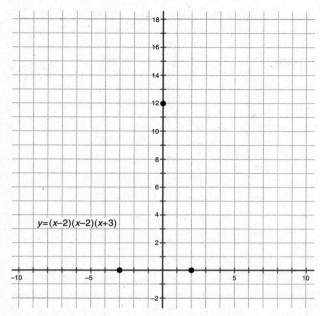

Since the graph should look like a right-side-up "N," start from the bottom left of the paper and draw toward the first x-intercept, $(-3, 0)$.

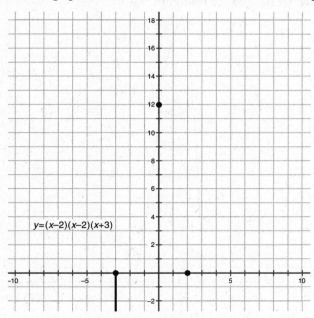

Then continue through the *y*-intercept, (0, 6).

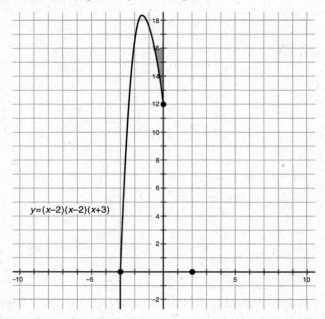

Then move to the next *x*-intercept, (2, 0).

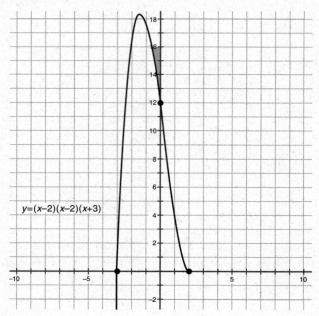

Since this is the double root, don't go through the x-axis. Instead, bounce off the axis and go back up to the right.

The graphing calculator agrees!

Example 2

Sketch the graph of $f(x) = \frac{1}{6}(x+4)(x^2 - 6x + 9)$.

Solution: This is not in factored form, but the $x^2 - 6x + 9$ can be factored to $(x - 3)^2$ or $(x - 3)(x - 3)$.

To graph $f(x) = \frac{1}{6}(x+4)(x-3)(x-3)$, the x-intercepts are at $(-4, 0)$ and $(3, 0)$. The y-intercept is at $(0, 6)$. The graph will resemble a right-side-up "N." It will bounce off the x-axis at the double root at the x-intercept $(3, 0)$.

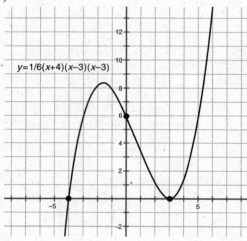

Example 3

If a, b, and c are all positive integers, which could be the graph of $f(x) = a(x - b)(x^2 - c^2)$?

(1)

(2)

(3)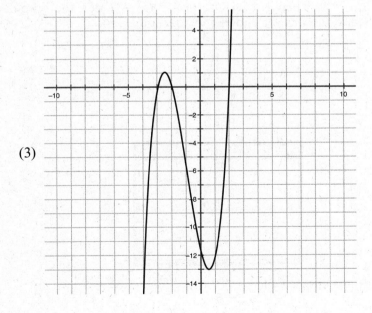

Polynomial Expressions and Equations

(4)
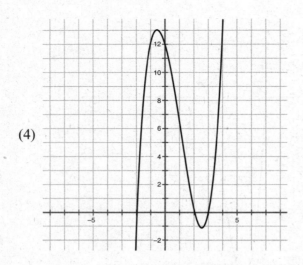

Solution: The $x^2 - c^2$ can be factored into $(x - c)(x + c)$, so the function becomes $f(x) = a(x - b)(x - c)(x + c)$. Since a is positive, the graph will resemble a right-side-up "N." The x-intercepts will be at $(b, 0)$, $(c, 0)$, and $(-c, 0)$.

An alternative is to make up three positive integers for a, b, and c, like $a = 1$, $b = 3$, $c = 2$. Then graph $y = 1(x - 3)(x^2 - 2^2)$ on the graphing calculator. See which answer choice most resembles it.

The solution is choice (4).

Check Your Understanding of Section 1.6

A. Multiple-Choice

1. Which could be the equation for this graph?

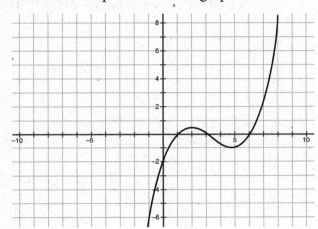

(1) $y = \dfrac{1}{9}(x-1)(x-3)(x-6)$ (3) $y = \dfrac{1}{9}(x-1)(x+3)(x-6)$

(2) $y = \dfrac{1}{9}(x-1)(x+3)(x+6)$ (4) $y = \dfrac{1}{9}(x+1)(x-3)(x-6)$

2. Which could be the equation for this graph?

(1) $y = \dfrac{1}{8}(x-2)^2(x-4)$ (3) $y = \dfrac{1}{16}(x-2)(x-4)^2$

(2) $y = -\dfrac{1}{4}(x-2)(x-4)$ (4) $y = -\dfrac{1}{8}(x-2)^2(x-4)$

Polynomial Expressions and Equations

3. What are the x-intercepts and y-intercept of $y = (x - 2)(x + 4)(x + 5)$?
   (1) $(-2, 0), (4, 0), (5, 0), (0, -40)$
   (2) $(-2, 0), (4, 0), (5, 0), (0, 40)$
   (3) $(2, 0), (-4, 0), (-5, 0), (0, -40)$
   (4) $(2, 0), (-4, 0), (-5, 0), (0, 40)$

4. If a and b are both positive numbers, which could be the graph of $y = (x + a)^2(x - b)$?

(1)

(2)

(3)

(4)

5. Which equation could this be the graph of?

(1) $y = x - 2$
(2) $y = (x - 2)^2$
(3) $y = (x - 2)^3$
(4) $y = (x - 2)^4$

6. If this is the graph of $y = a(x - 2)(x - 1)(x + 3)$, what is the value of a?

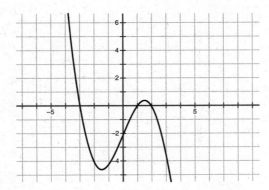

(1) $\dfrac{1}{3}$ (2) $-\dfrac{1}{6}$ (3) $\dfrac{1}{6}$ (4) $-\dfrac{1}{3}$

Polynomial Expressions and Equations

7. If a and b are positive integers, which could be the graph of $y = (x - b)(x^2 + 2ax + a^2)$?

(1)

(2)

(3)

(4)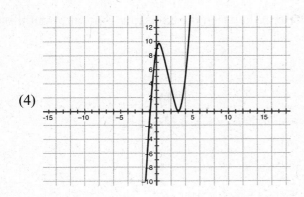

8. What are the x-intercepts of $y = x^3 - 2x^2 - 9x + 18$?
   (1) (2, 0), (3, 0), (4, 0)
   (2) (2, 0), (3, 0), (-3, 0)
   (3) (-2, 0), (3, 0), (-3, 0)
   (4) (-2, 0), (-9, 0), (18, 0)

9. What are the x-intercepts of the graph of $y = x^3 + 2x^2 - 8x$?
   (1) (2, 0), (-4, 0)
   (2) (0, 0), (-2, 0), (4, 0)
   (3) (0, 0), (2, 0), (8, 0)
   (4) (0, 0), (2, 0), (-4, 0)

10. What is the least number of x-intercepts an equation of the form $y = ax^3 + bx^2 + cx + d$ can have?
    (1) 0   (2) 1   (3) 2   (4) 3

B. *Show how you arrived at your answers.*

1. Sketch the graph of $y = \dfrac{1}{6}(x-2)(x+2)(x+3)$. Include all x- and y-intercepts.

2. What is the equation of this graph?

3. A graph of a cubic equation has a double root at $x = 2$ and a single root at $x = -1$. What could be the equation?

Polynomial Expressions and Equations

4. What is the main difference between the equations that represent these two graphs?

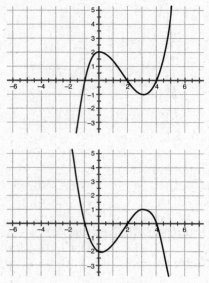

5. One of the x-intercepts of the cubic equation $y = x^3 + 4x^2 - 7x - 10$ is (2, 0). What are the other two x-intercepts?

Chapter Two
RATIONAL EXPRESSIONS AND EQUATIONS

2.1 ARITHMETIC WITH RATIONAL EXPRESSIONS

=== KEY IDEAS ===

A *rational expression* is a fraction that has a polynomial expression in the denominator. It often also has a polynomial expression in the numerator. An example of a rational expression is $\dfrac{x+3}{x^2+6x+8}$. Just like fractions involving integers, rational expressions can be simplified, reduced, multiplied, divided, added, and subtracted.

Reducing Rational Expressions

A *rational number* is a fraction, like $\dfrac{6}{10}$, that has an integer in both the numerator and the denominator. When both the numerator and denominator of a rational number are multiplied or divided by the same number, the result is a rational number that is equivalent to the original rational number.

$$\frac{6}{10} = \frac{2 \cdot 6}{2 \cdot 20} = \frac{12}{20}$$

$$\frac{6}{10} = \frac{6 \div 2}{10 \div 2} = \frac{3}{5}$$

$\dfrac{6}{10}, \dfrac{12}{20},$ and $\dfrac{3}{5}$ all represent the same number, 0.6.

Reducing a rational number is when the numerator and denominator are both divided by the same common factor. Factoring the numerator and denominator of a rational number makes it easier to reduce the fraction to *lowest terms*.

$$\frac{6}{10} = \frac{\cancel{2} \cdot 3}{\cancel{2} \cdot 5} = \frac{3}{5}$$

In the example above, we sometimes say that the 2s "canceled out."

Rational Expressions and Equations

MATH FACTS

Although dividing or multiplying the numerator and denominator by the same number results in an equivalent rational number, adding the same number to or subtracting the same number from the numerator and denominator does not produce an equivalent rational number. An example of this is $\dfrac{6}{10} \neq \dfrac{6-2}{10-2} = \dfrac{4}{8} = \dfrac{1}{2}$. For this reason, we sometimes say that we can cancel only numbers that are multiplied by both the numerator and the denominator, not numbers that are added to or subtracted from both the numerator and the denominator.

The value of a rational expression depends on the chosen value of x. The expression $\dfrac{5}{x+2}$ is equal to $\dfrac{5}{3}$ when $x = 1$. When $x = 2$, the value of the expression is $\dfrac{5}{4}$. When $x = -2$, the expression becomes $\dfrac{5}{0}$, which is undefined. Two rational expression are said to be equivalent if they evaluate to the same number for each x-value for which they are both defined.

One way to create a rational expression equivalent to another rational expression is to multiply the numerator and denominator by the same expression. When the numerator and denominator of the rational expression $\dfrac{5}{x+2}$ are multiplied by $x + 3$, the expression becomes $\dfrac{5(x+3)}{(x+2)(x+3)} = \dfrac{5x+15}{x^2+5x+6}$. This new expression is equivalent to the original one for every value of x except -3 since the new expression is undefined for $x = -3$. In contrast, the original expression is equal to -5 when $x = -3$.

Another way to create an equivalent rational expression is to reduce by finding a common factor in the numerator and denominator. Then cancel out those factors by dividing both the numerator and the denominator by that factor. This process is simplest when the numerator and denominator are already factored completely.

Example 1

Simplify the rational expression $\dfrac{3(x+5)}{(x+2)(x+5)}$.

58

Solution: The $(x + 5)$ factors can be canceled out, leaving the expression $\dfrac{3}{x+2}$.

If the expressions in the numerator and denominator of the rational expression are not already factored, factor them and cancel out equal factors.

Example 2

Simplify the rational expression $\dfrac{2x+8}{x^2+5x+4}$.

Solution: The numerator factors into $2(x+4)$. The denominator factors into $(x+4)(x+1)$. The rational expression then reduces to

$$\frac{2(x+4)}{(x+4)(x+1)} = \frac{2}{x+1}.$$

MATH FACTS

The expression $b - a$ can be factored into $-1(a - b)$. This can be useful if in the numerator there is an expression like $x - 2$ and in the denominator there is an expression like $2 - x$. The expression in the denominator can be factored into $-1(x - 2)$. Then the $(x - 2)$ factors can be canceled out.

Example 3

Simplify the expression $\dfrac{x^2-9}{9-3x}$.

Solution: The numerator factors into $(x - 3)(x + 3)$, and the denominator factors into $3(3 - x)$. So the expression is $\dfrac{(x-3)(x+3)}{3(3-x)}$. Since there is an $x - 3$ in the numerator and a $3 - x$ in the denominator, rewrite the $3 - x$ as $-1(x - 3)$. The expression is now $\dfrac{(x-3)(x+3)}{-3(x-3)} = \dfrac{x+3}{-3}$.

Multiplying Rational Expressions

One way to multiply rational numbers is to multiply the numerators and the denominators of the factors and then reduce the solution.

$$\frac{22}{10} \cdot \frac{35}{21} = \frac{770}{210}$$

Rational Expressions and Equations

Reducing the fraction $\dfrac{770}{210}$ to lowest terms, however, can be complicated.

A shortcut is to factor each of the numbers in the two fractions, combine these factors into one numerator and one denominator, and then cancel any common factors.

$$\frac{22}{10} \cdot \frac{35}{21} = \frac{2 \cdot 11}{2 \cdot 5} \cdot \frac{7 \cdot 5}{7 \cdot 3} = \frac{\cancel{2} \cdot 11 \cdot \cancel{7} \cdot \cancel{5}}{\cancel{2} \cdot \cancel{5} \cdot \cancel{7} \cdot 3} = \frac{11}{3}$$

This same process can be applied to multiplying two rational expressions. By first factoring all the numerators and denominators, the factors to be canceled out are much more obvious.

$$\frac{11x+33}{x^2-x-12} \cdot \frac{x^2-16}{3x+12}$$

After all numerators and denominators have been factored using the different factoring patterns described in Chapter 1, the result is the following.

$$\frac{11\cancel{(x+3)}\,\cancel{(x-4)}\,\cancel{(x+4)}}{\cancel{(x-4)}\,\cancel{(x+3)}\,3\cancel{(x+4)}} = \frac{11}{3}$$

Dividing Rational Expressions

Just like dividing rational numbers, rational expressions are divided by multiplying the first fraction by the reciprocal of the second fraction.

$$\frac{5}{(x+3)(x-2)} \div \frac{3(x+2)}{(x-2)(x+2)}$$

$$\frac{5}{(x+3)(x-2)} \cdot \frac{(x-2)(x+2)}{3(x+2)}$$

$$\frac{5\,\cancel{(x-2)}\,\cancel{(x+2)}}{(x+3)\,\cancel{(x-2)}\,3\cancel{(x+2)}}$$

$$\frac{5}{3(x+3)}$$

Adding Rational Expressions

Like adding rational numbers, adding rational expressions requires finding a lowest common denominator. One of four situations typically occurs when adding rational numbers or expressions.

Situation 1: The expressions already have a common denominator.

Since $\dfrac{2}{7}$ and $\dfrac{3}{7}$ have a common denominator of 7, $\dfrac{2}{7}+\dfrac{3}{7}=\dfrac{2+3}{7}=\dfrac{5}{7}$.

Similarly, the rational expressions $\dfrac{x+2}{x+3}$ and $\dfrac{x+5}{x+3}$ have a common denominator of $x+3$

$$\dfrac{x+2}{x+3}+\dfrac{x+5}{x+3}=\dfrac{(x+2)+(x+5)}{x+3}=\dfrac{x+2+x+5}{x+3}=\dfrac{2x+7}{x+3}$$

Situation 2: One denominator is a multiple of the other denominator.

For the fractions $\dfrac{2}{7}$ and $\dfrac{3}{14}$, the denominator 14 is a multiple of the denominator 7. The 7 needs to be multiplied by 2 to be equal to 14. Both the numerator and the denominator of the fraction $\dfrac{2}{7}$ must be multiplied by 2 to become the equivalent fraction $\dfrac{4}{14}$, which can then be added to $\dfrac{3}{14}$.

$$\dfrac{2}{7}+\dfrac{3}{14}$$

$$\dfrac{2\cdot2}{2\cdot7}+\dfrac{3}{14}$$

$$\dfrac{4}{14}+\dfrac{3}{14}=\dfrac{4+3}{14}=\dfrac{7}{14}=\dfrac{1}{2}$$

This situation can happen with rational expressions too.

$$\dfrac{2}{x+3}+\dfrac{3}{2x+6}$$

Factor the second denominator into $2(x+3)$. This is equal to 2 times the other denominator.

$$\dfrac{2\cdot2}{2(x+3)}+\dfrac{3}{2(x+3)}$$

$$\dfrac{4}{2(x+3)}+\dfrac{3}{2(x+3)}=\dfrac{7}{2(x+3)}=\dfrac{7}{2x+6}$$

61

Rational Expressions and Equations

Situation 3: The denominators have no common factor.

If the denominators have no common factor, the lowest common denominator is the product of the two denominators.

This happens when adding two rational numbers like $\frac{2}{3} + \frac{3}{5}$. Since 3 and 5 have no common factor, the least common denominator is $3 \cdot 5 = 15$. Both denominators must become 15, and the numerators must change accordingly.

$$\frac{2}{3} + \frac{4}{5}$$

The numerator and the denominator of the first fraction get multiplied by the denominator of the second fraction. The numerator and the denominator of the second fraction get multiplied by the denominator of the first fraction.

$$\frac{5 \cdot 2}{5 \cdot 3} + \frac{3 \cdot 4}{3 \cdot 5} = \frac{10}{15} + \frac{12}{15} = \frac{22}{15}$$

With rational expressions, this situation frequently happens.

$$\frac{2}{x+1} + \frac{4}{x+3}$$

Since the expressions $(x + 1)$ and $(x + 3)$ have no common factor, the lowest common denominator is $(x + 1)(x + 3)$.

$$\frac{(x+3)2}{(x+3)(x+1)} + \frac{(x+1)4}{(x+1)(x+3)}$$

$$\frac{2x+6}{(x+1)(x+3)} + \frac{4x+4}{(x+1)(x+3)}$$

$$\frac{(2x+6)+(4x+4)}{(x+1)(x+3)} = \frac{2x+6+4x+4}{(x+1)(x+3)} = \frac{6x+10}{(x+1)(x+3)}$$

Situation 4: The two denominators share a common factor, but the larger denominator is not a multiple of the smaller one.

This situation is by far the most complicated one when adding rational numbers and rational expressions. This is seen with rational numbers when adding fractions like $\frac{7}{10}$ and $\frac{5}{6}$. The number 10 is not a multiple of 6, so this is not Situation 2. The two numbers have a common factor, 2, so this is not Situation 3. By factoring the two numbers $10 = 2 \cdot 5$ and $6 = 2 \cdot 3$, the least common multiple must have a factor of 2, 3, and 5, which makes it 30. Notice that if the two original denominators were multiplied, that would make a product of 60. This number

62

could serve as a common denominator, but it is not the least common denominator.

$$\frac{7}{10} + \frac{5}{6}$$

$$\frac{3 \cdot 7}{3 \cdot 10} + \frac{5 \cdot 5}{5 \cdot 6}$$

$$\frac{21}{30} + \frac{25}{30} = \frac{46}{30}$$

This situation often happens with rational expressions.

$$\frac{7}{x^2 - 9} + \frac{5}{x^2 + 7x + 12}$$

Factor the denominators.

$$\frac{7}{(x-3)(x+3)} + \frac{5}{(x+3)(x+4)}$$

The common factor must have an $(x - 3)$, an $(x + 3)$, and an $(x + 4)$ in it. So the smallest factor is $(x - 3)(x + 3)(x + 4)$.

The first denominator needs an $(x + 4)$ to become $(x - 3)(x + 3)(x + 4)$. The second denominator needs an $(x - 3)$ to become $(x - 3)(x + 3)(x + 4)$.

$$\frac{(x+4)7}{(x+4)(x-3)(x+3)} + \frac{(x-3)5}{(x-3)(x+3)(x+4)}$$

$$\frac{7x+28}{(x+4)(x-3)(x+3)} + \frac{5x-15}{(x+4)(x-3)(x+3)}$$

$$\frac{(7x+28)+(5x-15)}{(x+4)(x-3)(x+3)}$$

$$\frac{7x+28+5x-15}{(x+4)(x-3)(x+3)}$$

$$\frac{12x+13}{(x+4)(x-3)(x+3)}$$

Example 4

Simplify $\dfrac{3}{x^2 + x - 20} + \dfrac{2}{x^2 - 16}$.

Rational Expressions and Equations

Solution:

$$\frac{3}{(x+5)(x-4)} + \frac{2}{(x+4)(x-4)}$$

The common denominator is $(x + 5)(x - 4)(x + 4)$.

$$\frac{(x+4)3}{(x+5)(x-4)(x+4)} + \frac{(x+5)2}{(x+5)(x-4)(x+4)}$$

$$\frac{3x+12}{(x+5)(x-4)(x+4)} + \frac{2x+10}{(x+5)(x-4)(x+4)}$$

$$\frac{(3x+12)+(2x+10)}{(x+5)(x-4)(x+4)}$$

$$\frac{3x+12+2x+10}{(x+5)(x-4)(x+4)}$$

$$\frac{5x+22}{(x+5)(x-4)(x+4)}$$

Subtracting Rational Expressions

Subtracting rational expressions is nearly the same as adding them. An extra complication, which happens frequently on the Regents exam, is you must be careful distributing the negative sign through the parentheses of the second expression.

Even with two expressions that already have the same denominator, distributing the negative sign is an important issue.

Example 5

Simplify $\dfrac{3x+4}{x+2} - \dfrac{x-2}{x+2}$.

(1) $\dfrac{2x}{x+2}$

(2) $\dfrac{2x+2}{x+2}$

(3) $\dfrac{2x+4}{x+2}$

(4) $\dfrac{2x+6}{x+2}$

64

Solution: Choice (4) is correct.

$$\frac{(3x+4)-(x-2)}{x+2}$$

$$\frac{3x+4-x+2}{x+2}$$

$$\frac{2x+6}{x+2}$$

The most common incorrect answer is choice (2). This is what you would get if you distributed the negative sign through the second expression incorrectly as $3x + 4 - x - 2$. This error is one of the most common errors in all of high school math, so be careful while doing subtraction problems that involve polynomials!

Check Your Understanding of Section 2.1

A. Multiple-Choice

1. What is $\dfrac{x^2-4}{x^2+x-6}$ reduced to simplest terms?

   (1) $\dfrac{x+2}{x+3}$ (2) $\dfrac{-1}{x-2}$ (3) $\dfrac{x-2}{x-3}$ (4) $\dfrac{-4}{x-6}$

2. What is $\dfrac{2x+6}{x^2-2x-15}$ reduced to simplest terms?

   (1) $\dfrac{6}{x^2-15}$ (2) $\dfrac{2x}{x^2-15}$ (3) $\dfrac{2}{x-5}$ (4) $\dfrac{6}{2x-3}$

3. What is $\dfrac{x^2+5x+6}{x^2-9} \cdot \dfrac{5x-10}{x^2-4}$ reduced to simplest terms?

   (1) $\dfrac{5}{x-3}$ (2) $\dfrac{5}{x+2}$ (3) $\dfrac{5}{x-2}$ (4) $\dfrac{5}{x+4}$

4. What is $\dfrac{12-3x}{5x-5} \cdot \dfrac{x^2-1}{x^2-3x-4}$ reduced to simplest terms?

   (1) $\dfrac{3}{5}$ (2) $\dfrac{12}{5}$ (3) $-\dfrac{3}{5}$ (4) $-\dfrac{12}{5}$

Rational Expressions and Equations

5. What is $\dfrac{4x+16}{x^2+5x+6} \div \dfrac{x^2-16}{5x+15}$ reduced to simplest terms?

(1) $\dfrac{20}{x+2}$

(2) $\dfrac{20}{x-4}$

(3) $\dfrac{20}{(x+2)(x+3)}$

(4) $\dfrac{20}{(x+2)(x-4)}$

6. When $\dfrac{x+2}{x-3} \cdot \dfrac{x+4}{x+5}$ is multiplied, which of the following does it have the same answer as?

(1) $\dfrac{x+2}{x-3} \div \dfrac{x+4}{x+5}$

(2) $\dfrac{x+2}{x-3} + \dfrac{x+4}{x+5}$

(3) $\dfrac{x+2}{x-3} - \dfrac{x+4}{x+5}$

(4) $\dfrac{x+2}{x-3} \div \dfrac{x+5}{x+4}$

7. What is $\dfrac{1}{x} + \dfrac{1}{x+2}$?

(1) $\dfrac{2}{2x+2}$

(2) $\dfrac{2x+2}{x(x+2)}$

(3) $\dfrac{1}{x(x+2)}$

(4) $\dfrac{2}{x(x+2)}$

8. What is $\dfrac{4}{(x+2)(x+3)} + \dfrac{3}{(x+2)(x-2)}$?

(1) $\dfrac{7x+1}{(x+2)(x-2)(x+3)}$

(2) $\dfrac{7}{(x+2)(x-2)(x+3)}$

(3) $\dfrac{7x+3}{(x+2)(x-2)(x+3)}$

(4) $\dfrac{7}{(x+2)^2(x-2)(x+3)}$

9. What is $\dfrac{5}{x-1} - \dfrac{3}{x+2}$?

(1) $\dfrac{2x+7}{(x-1)(x+2)}$

(2) $\dfrac{2x+13}{(x-1)(x+2)}$

(3) $\dfrac{2}{-3}$

(4) $\dfrac{2x+3}{(x-1)(x+2)}$

10. What is $\dfrac{x}{x^2+7x+12} - \dfrac{3}{x^2-9}$?

(1) $\dfrac{x^2-6x+12}{(x+3)(x-3)(x+4)}$

(3) $\dfrac{x^2-6x-12}{(x+3)(x-3)(x+4)}$

(2) $\dfrac{x-3}{7x+21}$

(4) $\dfrac{x-3}{7x+3}$

B. *Show how you arrived at your answers.*

1. Ethan says that $\dfrac{2x+5}{2}$ can be reduced to $x + 5$. Braylon says this is not correct. Who is right and why?

2. James notices the following pattern:

$$\frac{1}{2} - \frac{1}{3} = \frac{1}{6} = \frac{1}{2\cdot 3}$$

$$\frac{1}{3} - \frac{1}{4} = \frac{1}{12} = \frac{1}{3\cdot 4}$$

$$\frac{1}{4} - \frac{1}{5} = \frac{1}{20} = \frac{1}{4\cdot 5}$$

He has a theory that, in general, $\dfrac{1}{x} - \dfrac{1}{x+1} = \dfrac{1}{x(x+1)}$. Prove that James is correct about his theory?

3. Simplify $\left(\dfrac{1}{x+h} - \dfrac{1}{x}\right) \div \left(\dfrac{x+h-x}{1}\right)$.

4. Talia simplified $\dfrac{5}{x+2} - \dfrac{2}{x-3}$ by this process:

$$\frac{5(x-3)-2(x+2)}{(x+2)(x-3)} = \frac{5x-15-2x+4}{(x+2)(x-3)} = \frac{3x-11}{(x+2)(x-3)}$$

There was an error in Talia's calculation. What was the error?

5. Fully simplify $\dfrac{x}{x+1} + \dfrac{x+1}{x+2}$.

6. The rational expression $\dfrac{x^3+12x^2+47x+60}{x^2-9}$ can be reduced. What is it in fully reduced form?

Rational Expressions and Equations

2.2 SOLVING RATIONAL EQUATIONS

KEY IDEAS

A *rational equation* is an equation that contains at least one rational expression. Rational equations often require finding a common denominator for all the terms involved. The process for solving a rational equation often produces extra solutions that need to be rejected.

Simple Rational Equations

The simplest rational equations are ones that already have a common denominator.

$$\frac{2x+5}{(x+2)(x-3)} = \frac{13}{(x+2)(x-3)}$$

Since these have the same denominator already, the denominators can be ignored to create the equation with just the numerators.

$$2x + 5 = 13$$
$$-5 = -5$$
$$2x = 8$$
$$\frac{2x}{2} = \frac{8}{2}$$
$$x = 4$$

With rational equations, it is necessary to check your answer in case the "ignoring" step somehow caused an incorrect answer to creep in. Simply plug your answer into the original problem.

$$\frac{2 \cdot 4 + 5}{(4+2)(4-3)} \stackrel{?}{=} \frac{13}{(4+2)(4-3)}$$

$$\frac{13}{6 \cdot 1} \stackrel{?}{=} \frac{13}{6 \cdot 1}$$

$$\frac{13}{6} \stackrel{\checkmark}{=} \frac{13}{6}$$

Solving Rational Equations with Cross Multiplication

Some rational equations have just one term on each side but the terms have different denominators.

$$\frac{2}{x+1} = \frac{4}{x+4}$$

The quickest way to solve an equation of this form is by *cross multiplication*. To cross multiply, make a new equation that has the product of one numerator with the other denominator on each side of the equal sign.

$$2(x+4) = 4(x+1)$$

$$
\begin{array}{rcl}
2x+8 &=& 4x+4 \\
-2x &=& -2x \\
\hline
8 &=& 2x+4 \\
-4 &=& -4 \\
4 &=& 2x \\
\frac{4}{2} &=& \frac{2x}{2} \\
2 &=& x
\end{array}
$$

To check if this is correct, substitute $x = 2$ into the original equation to see if the left side and the right side evaluate to the same number.

$$\frac{2}{2+1} \overset{?}{=} \frac{4}{2+4}$$

$$\frac{2}{3} \overset{\checkmark}{=} \frac{4}{6}$$

=| **MATH FACTS** |=

Cross multiplication is a shortcut for making both sides of the equal sign have a common denominator. You then ignore the denominators and make an equation out of the numerators. If $\dfrac{a}{b} = \dfrac{c}{d}$, use cross multiplication to get the common denominator bd: $\dfrac{ad}{bd} = \dfrac{bc}{bd}$ or $ad = bc$.

Solving Rational Equations with Multiple Terms

If the rational equation has more than one term on either side of the equal sign (normally the left side), those terms need to be combined. Then the resulting equation can be solved using the methods mentioned earlier in this section.

Rational Expressions and Equations

$$\frac{x-1}{x+2} + \frac{x}{x+1} = \frac{6x+5}{x^2+3x+2}$$

$$\frac{x-1}{x+2} + \frac{x}{x+1} = \frac{6x+5}{(x+2)(x+1)}$$

$$\frac{(x+1)(x-1)}{(x+1)(x+2)} + \frac{(x+2)x}{(x+2)(x+1)} = \frac{6x+5}{(x+2)(x+1)}$$

$$\frac{x^2-1}{(x+1)(x+2)} + \frac{x^2+2x}{(x+2)(x+1)} = \frac{6x+5}{(x+2)(x+1)}$$

$$\frac{2x^2+2x-1}{(x+2)(x+1)} = \frac{6x+5}{(x+2)(x+1)}$$

$$2x^2+2x-1 = 6x+5$$
$$\underline{-6x-5 = -6x-5}$$
$$2x^2-4x-6 = 0$$
$$2(x^2-2x-3) = 0$$
$$2(x-3)(x+1) = 0$$

$$x-3=0 \quad \text{or} \quad x+1=0$$
$$x=3 \quad \text{or} \quad x=-1$$

The process of ignoring the denominator can result in fake answers. If you substitute $x = 3$ into the original equation, it becomes $\frac{2}{5} + \frac{3}{4} = \frac{23}{20}$, which is true. However, if you substitute $x = -1$ into the original equation, the denominator of the second term becomes 0, so the expression is undefined. The $x = -1$ solution must be rejected, leaving only the solution $x = 3$.

A question like this where a solution must be rejected is common on the Algebra II Regents.

Example 1

Solve for all values of x that satisfy the equation $\frac{x}{x+1} + \frac{x-1}{x+4} = \frac{8x+5}{(x+1)(x+4)}$.

Solution: The common denominator for the two rational expressions on the left side of the equal sign is $(x + 1)(x + 4)$.

70

$$\frac{x}{x+1} + \frac{x-1}{x+4} = \frac{8x+5}{x^2+5x+4}$$

$$\frac{(x+4)x}{(x+4)(x+1)} + \frac{(x+1)(x-1)}{(x+1)(x+4)} = \frac{8x+5}{(x+1)(x+4)}$$

$$\frac{x^2+4x}{(x+4)(x+1)} + \frac{x^2-1}{(x+1)(x+4)} = \frac{8x+5}{(x+1)(x+4)}$$

$$\frac{(x^2+4x)+(x^2-1)}{(x+4)(x+1)} = \frac{8x+5}{(x+1)(x+4)}$$

$$\frac{x^2+4x+x^2-1}{(x+4)(x+1)} = \frac{8x+5}{(x+1)(x+4)}$$

$$\frac{2x^2+4x-1}{(x+4)(x+1)} = \frac{8x+5}{(x+1)(x+4)}$$

Since the two rational expressions have the same denominator, that denominator can be ignored.

$$2x^2 + 4x - 1 = 8x + 5$$
$$\underline{-8x - 5 = -8x - 5}$$
$$2x^2 - 4x - 6 = 0$$
$$2(x^2 - 2x - 3) = 0$$
$$2(x-3)(x+1) = 0$$

$$x - 3 = 0 \quad \text{or} \quad x + 1 = 0$$
$$x = 3 \quad \text{or} \quad x = -1$$

Although $x = 3$ does make the original equation true, $\frac{4}{5} + \frac{3}{8} = \frac{37}{40}$, the $x = -1$ makes two of the rational expressions undefined. So $x = -1$ has to be rejected. The only solution is $x = 3$.

Rate Word Problems

Some real-world word problems involving rates can be solved using rational equations.

Rational Expressions and Equations

Example 2

Arianna can complete a coding project in 3 hours. Elijah can complete the same coding project in 2 hours. How long will it take them if they work together?

Solution: Arianna's rate is $\dfrac{1}{3}$ project per hour. Elijah's rate is $\dfrac{1}{2}$ project per hour. Their rate together is $\dfrac{1}{x}$ project per hour. Write an equation to solve the problem.

$$\frac{1}{2} + \frac{1}{3} = \frac{1}{x}$$
$$\frac{3x}{6x} + \frac{2x}{6x} = \frac{6}{6x}$$
$$3x + 2x = 6$$
$$5x = 6$$
$$x = \frac{6}{5} = 1\frac{1}{5} = 1.2$$

Working together would take them 1.2 hours.

Example 3

Two cars are driving from Washington, D.C. to New York, a distance of 300 miles. The first car went 10 miles per hour faster than the second car and got to New York 1 hour sooner. How long did it take each car to make the trip?

Solution: If x is the first car's time, then $\dfrac{300}{x}$ is the car's speed. Since the second car took one more hour, $x + 1$ is the second car's time and $\dfrac{300}{x+1}$ is the second car's speed. Since the first car's speed is 10 mph more than the second car's speed, the equation is $\dfrac{300}{x+1} + 10 = \dfrac{300}{x}$.

The solution is $x = 5$. So the first car took 5 hours, and the second car took $x + 1 = 6$ hours.

Check Your Understanding of Section 2.2

Chapter Two   **RATIONAL EXPRESSIONS AND EQUATIONS**

A. Multiple-Choice

1. Solve for x.

$$\frac{3x+5}{x-2} = \frac{x+13}{x-2}$$

(1) 2   (2) 3   (3) 4   (4) 5

2. Solve for x.

$$\frac{5x-3}{2x^2-3x} = \frac{2x+12}{2x^2-3x}$$

(1) 2   (2) 3   (3) 4   (4) 5

3. Solve for x.

$$\frac{x+1}{x+2} = \frac{x+4}{x+6}$$

(1) 0   (2) 1   (3) 2   (4) 3

4. Solve for x.

$$\frac{x+1}{x+2} = \frac{2x+2}{x+7}$$

(1) $x = 1$ (3) $x = -1$ or $x = 3$
(2) $x = 3$ (4) No solutions

5. Solve for x.

$$\frac{3}{x+2} + \frac{2}{x-3} = \frac{15}{(x+2)(x-3)}$$

(1) 2   (2) 3   (3) 4   (4) 5

6. Solve for x.

$$\frac{1}{x-4} + \frac{3}{x-1} = \frac{7}{x^2-5x+4}$$

(1) 4   (2) 5   (3) 6   (4) 7

73

Rational Expressions and Equations

7. Solve for x.

$$\frac{x}{x-3} + \frac{2}{x+1} = \frac{x+2}{x^2-2x-3}$$

(1) $x = 2$ or $x = -4$ (3) $x = 2$
(2) $x = -2$ or $x = 4$ (4) $x = -4$

8. Solve for x.

$$\frac{3}{x-4} - \frac{2}{x+5} = \frac{13}{(x-4)(x+5)}$$

(1) -10 (2) -6 (3) 10 (4) 6

9. Solve for x.

$$\frac{6}{x+4} - \frac{4}{x+2} = \frac{12}{x^2+6x+8}$$

(1) -6 (2) 6 (3) -8 (4) 8

10. Solve for x.

$$\frac{x+2}{x-4} - \frac{3}{x+1} = \frac{23}{x^2-3x-4}$$

(1) $x = 3$ or $x = -3$ (3) $x = -3$
(2) $x = 3$ (4) $x = 4$ or $x = -2$

B. *Show how you arrived at your answers.*

1. A group of people contribute equal amounts of money to get a $24 gift for a friend. If two more people contributed and they all paid equal amounts, they would each pay $1 less. How many people were there originally?

2. Leah solves $\dfrac{x}{(x-2)(x+2)} - \dfrac{3}{(x+2)(x-4)} = \dfrac{-4}{(x-2)(x+2)(x-4)}$ and gets solutions of $x = 5$ and $x = 2$. Paxton says that just $x = 5$ is the solution. Who is correct?

3. Kevin and Noelle drive 240 miles to Boston to watch a Yankees vs. Red Sox game. On the way home, they drive 10 mph slower and get home 48 minutes later than they would have had they driven the same speed as before. How fast did they drive from New York to Boston?

74

4. What is the solution to $\dfrac{1}{x+2} + \dfrac{2}{x+3} + \dfrac{5}{x+4} = \dfrac{178}{(x+2)(x+3)(x+4)}$?

5. Solve for x.

$$\dfrac{x+3}{x-1} - \dfrac{x-1}{x+2} = \dfrac{26}{(x-1)(x+2)}$$

2.3 GRAPHING RATIONAL FUNCTIONS

Just like polynomial, exponential, and radical functions, rational functions can be graphed. The graphs of rational functions usually include horizontal and/or vertical *asymptotes*. Asymptotes are like invisible lines that curves get closer and closer to. The graph of a rational function can help solve certain rational equations.

Vertical and Horizontal Asymptotes

Below is the graph of the rational function $R(x) = \dfrac{1}{x-2}$.

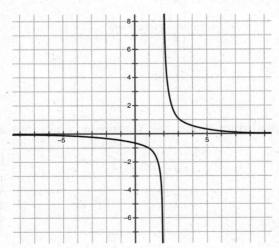

A characteristic of a graph of a rational function is that it often has one or more *vertical asymptotes*. A vertical asymptote is a vertical line that the graph never touches or crosses. In this graph, the vertical asymptote is the vertical line at $x = 2$. If you evaluate $R(2) = \dfrac{1}{2-2} = \dfrac{1}{0}$, it is undefined. For numbers very close to 2, such as 2.000001, the function produces a very high number. For $x = 2.000001$, $R(2.000001) = \dfrac{1}{2.000001-2} = \dfrac{1}{0.000001} = 1,000,000$.

Some graphs of rational functions also have a *horizontal asymptote*. Like the vertical asymptote, it is a horizontal line that, usually for very high and very low x-values, the curve gets closer and closer to without touching or crossing. In the graph for $R(x) = \dfrac{1}{x-2}$, there is a horizontal asymptote at $y = 0$, the x-axis.

The asymptotes have an effect on the domain and range of the function. For $R(x) = \dfrac{1}{x-2}$, the domain is everything except $x = 2$ and the range is everything except $y = 0$.

When graphing rational functions on the graphing calculator, be sure to use parentheses carefully or to use the fraction template.

Graph the more complicated function $R(x) = \dfrac{x-5}{(x+1)(x+4)}$ using a graphing calculator.

For the TI-84:

If you are in classic mode, enter the function into the Y= screen. Make sure to use proper parentheses.

If you are in mathprint mode, the Y= screen is more intuitive.

For the TI-Nspire:

After the f1(x) in the entry line of the graphing scratchpad, press the [template] button and select the fraction template. Enter $x - 5$ into the numerator and $(x + 1)(x + 4)$ into the denominator.

Checking to See if Two Rational Expressions Are Equivalent Using the Graphing Calculator

For a multiple-choice question that asks you to simplify a rational expression, the graph of the original expression can be compared to the graphs of the answer choices.

Example

Simplify $\dfrac{3x+4}{x+2} - \dfrac{x-2}{x+2}$.

(1) $\dfrac{2x}{x+2}$

(2) $\dfrac{2x+2}{x+2}$

(3) $\dfrac{2x+4}{(x+2)}$

(4) $\dfrac{2x+6}{x+2}$

Rational Expressions and Equations

Solution: This example was done earlier in the chapter with algebra but can also be done with the graphing calculator. By creating the graph of the original expression and the graph of the four answer choices, only choice (4) has the same graph as the original. Just look at the TI-84 screen shots below.

Graphs of choices:

Solving Rational Equations with the Graphing Calculator

The equation $\dfrac{2}{x+1} = \dfrac{4}{x+4}$, which was solved with algebra earlier in this chapter, can also be solved with the intersect feature of the graphing calculator. To solve a rational equation with the graphing calculator, graph both the expression on the left side of the equal sign and the expression on the right side of the equal sign. The x-coordinate of the intersection point is the solution to the equation.

Since the two expressions intersect at a point with an *x*-coordinate of 2, the solution is $x = 2$.

Check Your Understanding of Section 2.3

A. Multiple-Choice

1. What are the asymptotes of the graph of $R(x) = \dfrac{3}{x+4}$?

   (1) $x = 4, \ y = 0$ (3) $x = 0, \ y = 4$
   (2) $x = -4, \ y = 0$ (4) $x = 0, \ y = -4$

2. Which of the following could be the equation for this graph?

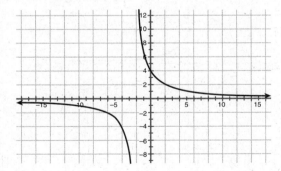

   (1) $R(x) = \dfrac{12}{x+3}$ (3) $R(x) = \dfrac{x-3}{12}$

   (2) $R(x) = \dfrac{x+3}{12}$ (4) $R(x) = \dfrac{12}{x-3}$

Rational Expressions and Equations

3. Which of the following is the graph of $y = \dfrac{8}{x+2}$?

(1)

(2)

(3)

(4)

4. Use a graphing calculator to determine which equation has the same graph as $y = \dfrac{2}{x+1} + \dfrac{3}{x}$.

(1) $y = \dfrac{5x+3}{x(x+1)}$

(3) $y = \dfrac{5x+3}{2x+1}$

(2) $y = \dfrac{5}{2x+1}$

(4) $y = \dfrac{5}{x(x+1)}$

5. Which of the following has the same graph as $y = \dfrac{x^2+5x+6}{x^2-4}$?

(1) $y = \dfrac{x+2}{x-2}$

(3) $y = 1 - \dfrac{5}{4}x + 6$

(2) $y = \dfrac{x+3}{x-2}$

(4) $y = \dfrac{x-3}{x+2}$

Chapter Two   RATIONAL EXPRESSIONS AND EQUATIONS

81

Chapter Three

EXPONENTIAL AND LOGARITHMIC EXPRESSIONS AND EQUATIONS

3.1 PROPERTIES OF EXPONENTS

Key Ideas

An expression like 2^5 is called an *exponential expression*. It has a *base*, which in this case is the number 2, and an *exponent*, which in this case is the number 5. For positive integer exponents, the expression can be evaluated by multiplying the base by itself the number of times of the exponent.

The expression 2^5 is equal to 32 because $2 \cdot 2 \cdot 2 \cdot 2 \cdot 2 = 32$. This is very different from $2 \cdot 5 = 10$. So $2^5 \ne 2 \cdot 5$.

Multiplying Exponential Expressions

When two exponential expressions that have the same base are multiplied, the product can be written as an exponential expression that also has that base but whose exponent is the sum (not the product) of the exponents. If the bases are different, there is no simple way to multiply the expressions.

Example 1

Simplify $2^3 \cdot 2^5$.

   (1) 2^{15} (2) 4^{15} (3) 2^8 (4) 4^8

   Solution: Choice (3) is correct: 2^8. To see why the rule is to keep the base and add, rather than multiply, the two exponents, simplify $2^3 \cdot 2^5$ the long way.

$$2^3 \cdot 2^5 = (2 \cdot 2 \cdot 2) \cdot (2 \cdot 2 \cdot 2 \cdot 2 \cdot 2) = 2 \cdot 2 \cdot 2 \cdot 2 \cdot 2 \cdot 2 \cdot 2 \cdot 2 = 2^8$$

Dividing Exponential Expressions

When two exponential expressions that have the same base are divided, the product can be written as an exponential expression that also has that base but whose exponent is the difference of the exponents. If the bases are different, there is no simple way to divide the expressions.

Example 2

Simplify $2^{12} \div 2^4$.

    (1) 1^3 (2) 1^8 (3) 2^3 (4) 2^8

Solution: Choice (4) is correct: 2^8. To see why the rule is to keep the base and subtract the exponent, write the expression as a fraction and reduce.

$$\frac{(2^{12})}{2^4} = \frac{2 \cdot 2 \cdot 2 \cdot 2 \cdot 2 \cdot 2 \cdot 2 \cdot 2 \cdot \cancel{2} \cdot \cancel{2} \cdot \cancel{2} \cdot \cancel{2}}{\cancel{2} \cdot \cancel{2} \cdot \cancel{2} \cdot \cancel{2}} = 2 \cdot 2 \cdot 2 \cdot 2 \cdot 2 \cdot 2 \cdot 2 \cdot 2 = 2^{}$$

Raising a Power to a Power

To raise an exponential expression to a power, keep the same base as in the original expression. Multiply the two exponents to find the new exponent.

Example 3

Simplify $(2^3)^2$.

Solution: Keep the base and multiply the two exponents. The answer is 2^6. To see why, expand.

$$(2^3)^2 = (2^3)(2^3) = 2^{3+3} = 2^{3 \cdot 2} = 2^6$$

Example 4

Which expression is equivalent to 1.07^x?

    (1) 1.07^{12x} (2) $(1.07^{12})^x$ (3) 1.12^{7x} (4) $\left(1.07^{\frac{1}{12}}\right)^{12x}$

Solution: Choice (4) is correct. To raise an expression that has an exponent in it to a power, multiply the two exponents. In this case $\frac{1}{12} \cdot 12x = 1x = x$. So choice (4) is equivalent to 1.07^x.

The power rule for exponents can also be used in reverse. If there is an expression like 2^{3x}, it can be converted to $(2^3)^x = 8^x$. The expressions 2^{3x} and 8^x are equivalent.

Exponential and Logarithmic Expressions and Equations

When the exponent is not a positive integer, things get a bit more complicated. For instance 2^0 is *not* equal to 0, as you might expect. Instead, 2^0 is equal to 1. Negative integer exponents may also seem unusual, even impossible. However, 2^{-3} is not undefined or even -8. Test it on a calculator. The answer is 0.125!

These unusual situations arise from the rule for dividing exponential expressions. $2^5 \div 2^5 = 2^{5-5} = 2^0$. However, any number divided by itself is equal to 1, so $2^0 = 1$.

$2^1 \div 2^4 = 2^{1-4} = 2^{-3}$. However, $\dfrac{2^1}{2^4}$ reduces to $\dfrac{1}{2^3} = \dfrac{1}{8} = 0.125$.

==== **MATH FACTS** ====

There are three main properties of exponents.

$$x^a \cdot x^b = x^{a+b}$$

$$\frac{x^a}{x^b} = x^{a-b}$$

$$\left(x^a\right)^b = x^{ab}$$

==== **MATH FACTS** ====

Any number (besides 0) raised to the 0 power is equal to 1. Any number (besides 0) raised to a negative power is equal to the reciprocal of that number raised to the positive version of that power.

$$b^0 = 1$$

$$b^{-n} = \frac{1}{b^n}$$

Example 5

What is the value of 3^{-4} rounded to the nearest thousandth?

Solution:

$$3^{-4} = \frac{1}{3^4} = \frac{1}{81} \approx 0.012$$

84

Example 6

Write an expression equivalent to $\dfrac{x^{-2}}{y^{-3}}$ that uses just positive exponents.

Solution: Remember that $x^{-2} = \dfrac{1}{x^2}$ and $y^{-3} = \dfrac{1}{y^3}$.

$$\frac{x^{-2}}{y^{-3}} = \frac{\dfrac{1}{x^2}}{\dfrac{1}{y^3}} = \frac{1}{x^2} \cdot \frac{y^3}{1} = \frac{y^3}{x^2}$$

Notice how the x^{-2} moved to the denominator as x^2 while the y^{-3} moved to the numerator as a y^3. In general, when there is a negative exponent, the expression containing that exponent can move from numerator to denominator or from denominator to numerator with the negative exponent changing into a positive exponent.

To raise a number to a fractional power, like $4^{\frac{3}{2}}$, first notice that $4^1 = 4$ and $4^2 = 16$. Since the exponent $\dfrac{3}{2}$ is between 1 and 2, the answer should be something between 4 and 16. However, it is not halfway between 4 and 16, which is 10 as you might expect. Instead, if you test it on a calculator, you get 8!

═══════════════ **MATH FACTS** ═══════════════

To raise a base to a fractional power $\dfrac{n}{d}$, take the dth root of the number and raise it to the nth power.

$$b^{\frac{n}{d}} = \left(\sqrt[d]{b}\right)^n$$

Example 7

What is the value of $8^{\frac{1}{3}}$?

Solution:

$$8^{\frac{1}{3}} = \left(\sqrt[3]{8}\right)^1 = 2^1 = 2$$

Notice that when the numerator of the exponent is 1, the solution just becomes the dth root of the base where d is the denominator. This is because the multiplication rule for exponents says that

Exponential and Logarithmic Expressions and Equations

$8^{\frac{1}{3}} \cdot 8^{\frac{1}{3}} \cdot 8^{\frac{1}{3}} = 8^{\frac{1}{3} + \frac{1}{3} + \frac{1}{3}} = 8^1 = 8$, while the definition of cube root says that

$\sqrt[3]{8} \cdot \sqrt[3]{8} \cdot \sqrt[3]{8} = 8$. So $8^{\frac{1}{3}} = \sqrt[3]{8}$.

Example 8

What is the value of $8^{\frac{5}{3}}$?

   Solution:

$$8^{\frac{5}{3}} = \left(\sqrt[3]{8}\right)^5 = 2^5 = 32$$

The Distributive Property for Exponents

If an expression in parentheses is the product of numbers and/or variables, the entire expression in the parentheses can be raised to a power by raising each of the factors to that power and multiplying them together. For example, $(2 \cdot 5)^3$ can be calculated by solving $2^3 \cdot 5^3 = 8 \cdot 125 = 1,000$. This is also the solution if you simplified inside the parentheses first: $10^3 = 1,000$. This property is needed when variables are involved.

Example 9

Simplify $(3x^2y^3)^2$.

   Solution:

$$3^2(x^2)^2(y^3)^2 = 9x^{2 \cdot 2}y^{3 \cdot 2} = 9x^4y^6$$

Example 10

Simplify the expression $\dfrac{\sqrt[3]{x^5}}{(x^2)^{\frac{1}{3}}}$

   Solution: First express the numerator and the denominator as x raised to a power represented by a single number.

$$\sqrt[3]{x^5} = x^{\frac{5}{3}}$$

$$(x^2)^{\frac{1}{3}} = x^{2 \cdot \frac{1}{3}} = x^{\frac{2}{3}}$$

$$\frac{\sqrt[3]{x^5}}{(x^2)^{\frac{1}{3}}} = \frac{x^{\frac{5}{3}}}{x^{\frac{2}{3}}} = x^{\frac{5}{3} - \frac{2}{3}} = x^1$$

Check Your Understanding of Section 3.1

A. Multiple-Choice

1. Simplify $3^4 \cdot 3^5$.
   (1) 3^{20} (2) 3^9 (3) 9^9 (4) 9^{20}

2. Simplify $(5^2)^3$.
   (1) 5^8 (2) 5^6 (3) 5^5 (4) 5^9

3. What is a^{-4}?
   (1) $-(a^4)$ (2) $a^{\frac{1}{4}}$ (3) $\dfrac{1}{a^4}$ (4) $\sqrt[4]{a}$

4. What is $a^{\frac{3}{4}}$?
   (1) $\sqrt[3]{a^4}$ (2) $a^{-\frac{3}{4}}$ (3) $a^{-\frac{4}{3}}$ (4) $\sqrt[4]{a^3}$

5. Simplify $(2x^3 y^4)^3$.
   (1) $8x^{27}y^{64}$ (2) $6x^9 y^{12}$ (3) $8x^9 y^{12}$ (4) $6x^{27}y^{64}$

6. What is $8^{-\frac{2}{3}}$?
   (1) $\dfrac{1}{4}$ (2) $-\dfrac{1}{4}$ (3) 4 (4) -4

7. What is 5^0?
   (1) 0 (2) 1 (3) 5 (4) Undefined

8. What is $\dfrac{8x^4 y^2}{2x^2 y^5}$?
   (1) $\dfrac{6x^2}{y^3}$ (2) $\dfrac{6x^2}{y^2}$ (3) $\dfrac{4x^2}{y^3}$ (4) $\dfrac{4x^3}{y^2}$

9. What is $7^{\frac{1}{2}} \cdot 7^{\frac{3}{2}}$?
   (1) 47 (2) 48 (3) 49 (4) 50

10. What is $11^{-4} \cdot 11^4$?
    (1) 0 (2) 1 (3) 11 (4) 121

B. *Show how you arrived at your answers.*

1. Ashlynn says that $5^3 \cdot 5^4 = 5^7$. Colin says that it is equal to 5^{12}. Who is right and why?

2. In 5th grade, Charles learned that $5.2 \times 10^{-4} = 0.00052$. Show how the properties of negative exponents justify this answer.

3. If $2^{10} = 1{,}024$, how can you quickly calculate 2^{11} if your calculator does not have an exponent key?

4. If $9^1 = 9$ and $9^2 = 81$, what is the value of $9^{1.5}$? (Hint: change 1.5 into an improper fraction.)

5. What is the value of $8^{-2} \cdot 25^{\frac{3}{2}} \cdot 27^{-\frac{2}{3}}$?

3.2 SOLVING EXPONENTIAL EQUATIONS BY GUESS AND CHECK OR BY GRAPHING

KEY IDEAS

An *exponential equation* is one in which the variable is an exponent. An example is the equation $2^x = 32$. Some exponential equations have integer solutions, some have rational solutions (fractions), and some have irrational solutions. One way to solve exponential equations is through guess and check. Another way is to use the intersect feature of a graphing calculator.

Solving Exponential Equations with Guess and Check

To solve $2^x = 32$ with guess and check, first think of a number that might be reasonable. Suppose you guess that $x = 10$. To check, calculate 2^{10}, which is not 32 but, 1,024. So 10 was too big of a guess. Since the answer to this question is an integer, after enough guessing and checking, you will come to the answer $x = 5$ since $2^5 = 32$.

Example 1

Solve $3^x = 729$ by guess and check.

    Solution: Since $3^6 = 729$, the solution is $x = 6$.

    When a positive integer is raised to a negative power, the result is a number between 0 and 1. In an exponential equation where the number after the equals sign is between 0 and 1, the exponent is negative.

Example 2

Which value of x makes $5^x = \dfrac{1}{125}$?

    (1) -2 (2) -3 (3) $\dfrac{1}{2}$ (4) $\dfrac{1}{3}$

    Solution: Since $\dfrac{1}{125}$ is between 0 and 1, the answer must be negative.

Since $5^{-2} = \dfrac{1}{5^2} = \dfrac{1}{25}$, the denominator is too small. Since $5^{-3} = \dfrac{1}{5^3} = \dfrac{1}{125}$,

the answer is $x = -3$, which is choice (2).

    When the solution to an exponential equation is not an integer, guess and check can be used to find an approximate answer.

Example 3

To the nearest tenth, what is the solution to the equation $2^x = 42$?

    Solution: Since $2^5 = 32$, which is too small, and since $2^6 = 64$, which is too big, the answer is between 5 and 6. Test the numbers 5.1, 5.2, 5.3, and so on until you find a number that is close to 42.

$$2^{5.1} \approx 34$$
$$2^{5.2} \approx 37$$
$$2^{5.3} \approx 39$$
$$2^{5.4} \approx 42$$

Sometimes it is possible to get an exact solution even when the solution is not an integer. In those cases, the exponent is a rational number (a fraction).

Exponential and Logarithmic Expressions and Equations

Example 4

What is the exact solution to $4^x = 32$?

Solution: Since $4^2 = 16$ and $4^3 = 64$, the answer is between 2 and 3. Since $32 = 2^5$ and $2 = \sqrt{4} = 4^{\frac{1}{2}}, 32 = \left(4^{\frac{1}{2}}\right)^5 = 4^{\frac{1}{2} \cdot 5} = 4^{\frac{5}{2}}$. So the solution is $x = \dfrac{5}{2}$.

Exponential equations with exponents on both sides of the equal sign are easiest to solve when the two bases are the same.

Example 5

What value of x makes $3^x = 3^7$?

Solution: Since the bases are the same, the exponents must be equal. So $x = 7$.

Example 6

What value of x makes $3^{2x+3} = 3^7$?

Solution: Since the bases are the same, the exponents must be equal. This means that $2x + 3 = 7$. This can be solved with algebra: $x = 2$.

If the bases are different, they can sometimes be converted into the same base and then solved by equating the exponents.

Example 7

What value of x makes $3^{2x-4} = 9^4$?

Solution: Since 9 can be written as 3^2, the right side of the equation becomes $(3^2)^4$. This is equivalent to 3^8 because powers can be raised to powers by multiplying the exponents.

The equation is now $3^{2x-4} = 3^8$.

$$2x - 4 = 8$$
$$x = 6$$

Graphs of Two-Variable Exponential Equations

An equation of the form $y = a \cdot b^x + c$ where a, b, and c are numbers is a *two-variable exponential equation*. Like all two-variable equations, the solution set is a set of ordered pairs that make the equation true.

90

For example, the equation $y = 2^x$ has the ordered pairs (0, 1), (1, 2), (2, 4), and (3, 8) as four of the elements of its solution. Other points can be found by creating a chart where different integer values are chosen for x and the y-value is then calculated.

x	2^x	y
3	2^3	8
2	2^2	4
1	2^1	2
0	2^0	1
−1	$2^{-1} = \dfrac{1}{2^1}$	$\dfrac{1}{2}$
−2	$2^{-2} = \dfrac{1}{2^2}$	$\dfrac{1}{4}$
−3	$2^{-3} = \dfrac{1}{2^3}$	$\dfrac{1}{8}$

When the seven ordered pairs from the chart are graphed, it looks like this. The domain of this graph is all real numbers.

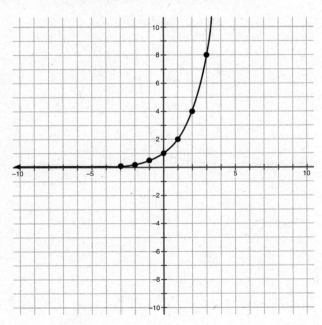

Exponential and Logarithmic Expressions and Equations

For a two-variable exponential equation with a *b*-value greater than 1, the shape of the graph has this shape, like a playground slide going up to the right.

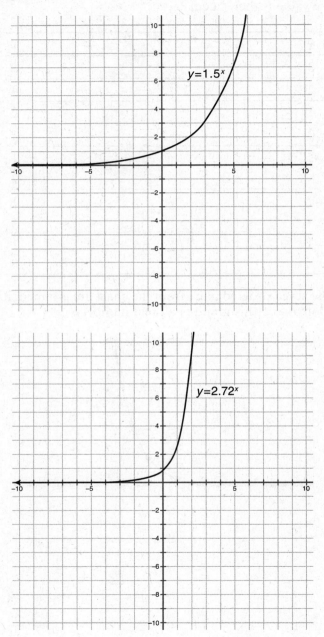

For a two-variable exponential equation with a b value between 0 and 1, the shape of the graph has this shape, like a playground slide going down to the right.

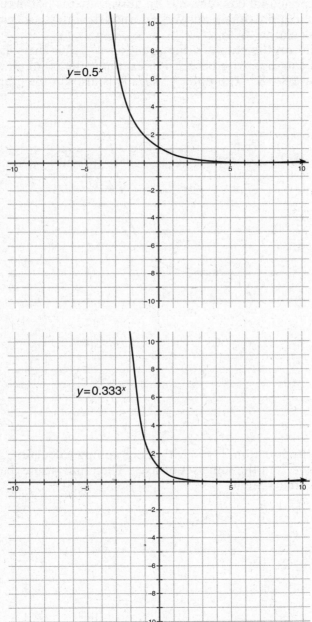

Solving Exponential Equations with a Graphing Calculator

When an exponential equation has an answer that is not an integer, one way to get an approximate answer is to use the intersect feature of the graphing calculator.

Each point on the graph of the equation $y = 2^x$ is the solution to an exponential equation with base 2. For example, the point (3, 8) is on the graph because $8 = 2^3$. The point (2.807, 7) is also on the graph because $7 \approx 2^{2.807}$.

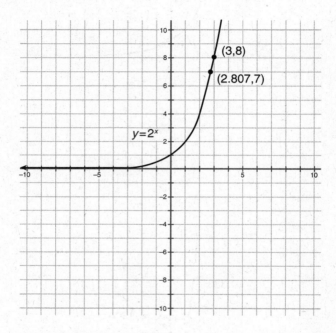

To solve the equation $2^x = 7$ with the graphing calculator, graph both $y = 2^x$ and $y = 7$ on the same set of axes. Then find where they intersect to determine the x-value.

For the TI-84:
Press [y=] and put 2^x after Y1 and 7 after Y2. Press [ZOOM] [6] to see the graphs. Then press [2nd], [TRACE], [5], [ENTER], [ENTER]. Move the cursor near the intersection and press [ENTER] again.

For the TI-Nspire:
From the home screen, press [B] to get to the graphing scratchpad. Then press [tab]. In the entry line, put 2^x after f1(x)= and press [enter] and put 7 after f2(x)= and press [enter]. Press [menu], [6], [4]. Move the cursor to the left of the intersection. Press [enter]. Then move the cursor to the right of the intersection and press [enter] again.

The approximate solution of $2^x = 7$ is shown to be about 2.81.

Example 8

Use the graphing calculator to find, to the nearest hundredth, the solution to the equation $3^x = 5$.

Solution: Using either the TI-84 or the TI-Nspire, the intersection of the graphs of $y = 3^x$ and $y = 5$ is at (1.46, 5).

Exponential and Logarithmic Expressions and Equations

Check Your Understanding of Section 3.2

A. Multiple-Choice

1. Solve $4^x = 256$.
   (1) 4   (2) 8   (3) 16   (4) 64

2. Solve $6^x = \dfrac{1}{36}$.
   (1) 2   (2) −2   (3) −3   (4) −6

3. Which value is nearest to x if $3^x = 40$?
   (1) 13.3   (2) 3   (3) 3.2   (4) 3.4

4. Solve $8^x = 16$.
   (1) $\dfrac{4}{3}$ (2) 2   (3) $\dfrac{3}{4}$ (4) $\dfrac{3}{2}$

5. Solve $7^x = 1$.
   (1) −1   (2) 1   (3) 0   (4) $\dfrac{1}{7}$

6. Solve $9^x = \dfrac{1}{81}$.
   (1) $\dfrac{1}{2}$ (2) −2   (3) 2   (4) $-\dfrac{1}{2}$

7. Solve $4^{2x+3} = 4^{5x-9}$.
   (1) 1   (2) 2   (3) 3   (4) 4

8. Solve $8^{x+4} = 2^{x+4}$.
   (1) 2   (2) −2   (3) 1   (4) −4

9. Which of the following is the graph of $y = 3^x$?

(1)

(2)

(3)

(4)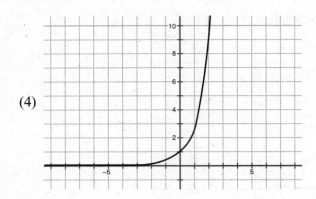

10. The graph of which equation is shown?

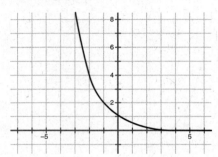

(1) $y = 2^x$ (2) $y = 3^x$ (3) $y = 2^{-x}$ (4) $y = 3^{-x}$

B. *Show how you arrived at your answers.*

1. Below is a graph of $y = 20$ and $y = 3^x$. Based on the graph, what is an approximate solution to the equation $3^x = 20$?

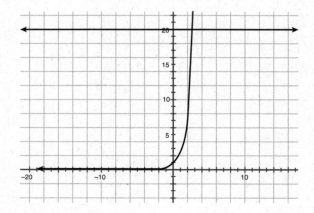

2. Charles says this is the graph of $y = \left(\dfrac{1}{2}\right)^x$. Hope says that it is the graph of $y = 2^{-x}$. Who is correct?

3. Solve for x if $\left(\dfrac{1}{4}\right)^{x-3} = \left(\dfrac{1}{8}\right)^{2x-6}$.

4. $4^x = 100$ is between which two integers? Explain your reasoning?

5. Bianca is solving the equation $3 \cdot 2^x = 36$ as follows:

$$3 \cdot 2^x = 36$$
$$6^x = 36$$
$$x = 2$$

Find the error in Bianca's reasoning.

3.3 LOGARITHMS

=== KEY IDEAS ===

An alternative to using graphs to solve an exponential equation is to use a *logarithm*. The logarithm feature of a calculator can be used to solve exponential equations quickly and accurately.

Using the Definition of Logarithm to Solve Equations

A *logarithm* is the number that a number needs to be raised to in order to get another number. Logarithms are written in the form $\log_b a$, where

Exponential and Logarithmic Expressions and Equations

b is known as the base of the logarithm. To evaluate the expression $\log_2 32$, ask yourself, "To what power must 2 be raised to get 32?" Since $2^5 = 32$, the value of $\log_2 32$ is 5.

Example 1

What is the value of $\log_5 125$?

Solution: This is the same as asking $5^x = 125$. Since $5^3 = 125$, $\log_5 125 = 3$.

Example 2

What is the value of $\log_4 \dfrac{1}{64}$?

Solution: When a positive number is raised to a negative power, the result is a number less than 1. In this case since $4^{-3} = \dfrac{1}{4^3} = \dfrac{1}{64}$, the answer is -3.

Example 3

What is the value of $\log_7 1$?

Solution: Since $7^0 = 1$, $\log_7 1$ must be equal to 0.

MATH FACTS

If the log has no base written, it is assumed to be base 10. So log 100 is the same as the expression $\log_{10} 100$, which is equal to 2 because $10^2 = 100$.

Converting Log Equations into Equivalent Exponential Equations

Any log equation can be converted into an equivalent exponential equation. The equation $\log_b a = c$ can be rearranged to the equation $b^c = a$.

Example 4

Rewrite the equation $\log_5 9 = x$ as an equation that does not involve logarithms.

Solution: $5^x = 9$

100

Example 5

Rewrite the equation $3^x = 17$ as an equation that involves logarithms.

Solution: $\log_3 17 = x$

Calculating Logarithms with Base 10 or Base e with a Calculator

All scientific and graphing calculators have two buttons for logarithms. The log button is for log base 10, and the ln button is for log base *e*. For other bases, the two calculators used in this book have a built-in function for calculating logarithms.

The log button is used to evaluate expressions like $\log_{10} 50$. The expression $\log_{10} 50$ can also be written with the shorthand log 50. Since $10^1 = 10$ and $10^2 = 100$, we should expect $\log_{10} 50$ to be something between 1 and 2. By using the log key on the TI-84 or by pressing [ctrl] [10^x] on the TI-Nspire, we see that $\log_{10} 50 \approx 1.69897$.

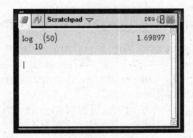

The number *e*, like the number π, is an irrational number. It equals approximately 2.72, although a better approximation is 2.71828182845904523536. The number *e* serves an important role in exponential equations used in real-world modeling.

The ln button is used to evaluate expressions like $\log_e 50$. The expression $\log_e 50$ can also be written with the shorthand ln 50. Since $e^3 = 2.72^3 = 7.39$ and $e^4 = 2.72^4 = 54.6$, we should expect $\log_e 50$ to be something between 3 and 4. By using the ln key on the graphing calculator, we see that $\log_e 50 \approx 3.9120$.

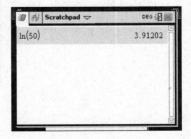

=== MATH FACTS ===

Using log x is a shorthand way of writing $\log_{10} x$. Using ln x is a shorthand way of writing $\log_e x$. The number e is an important mathematical constant that is equal to approximately 2.72. It is **not** a variable.

Example 6

Evaluate $\log_{10}$ 750 to the nearest hundredth.

Solution: Typing log 750 into the calculator gives approximately 2.88.

Example 7

Evaluate $\log_e$ 750 to the nearest hundredth.

Solution: Typing ln 750 into the calculator gives approximately 6.62.

Calculating Logarithms with Bases Other than 10 or e

The expression $\log_5 50$ is somewhere between 2 and 3 because $5^2 = 25 < 50$ while $5^3 = 125 > 50$. What is the answer exactly? Since it is an irrational number, getting a decimal approximation to the nearest thousandth is generally all we need.

The TI-84 and the TI-Nspire both have functions that calculate these logarithms.

For the TI-84:
Press [MATH] and then scroll to menu option A for logBASE. Enter 5 as the small number after the log and 50 as the large number.

For the TI-Nspire:
Press the log key by pressing [ctrl] and then [10^x]. Fill in the number 5 as the base and 50 as the number.

 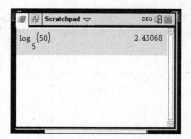

So $\log_5 50 \approx 2.43068$.

If your calculator does not have the log base function, another way to calculate $\log_b a$ is by calculating $\dfrac{\log a}{\log b}$. For $\log_5 50$, this would be $\dfrac{\log 50}{\log 5}$. If you use this method, make sure the numerator and the denominator both use the same type of logarithm (log or ln) but not both. So $\dfrac{\log 50}{\log 5} = \dfrac{\ln 50}{\ln 5}$. However, you cannot use $\dfrac{\log 50}{\ln 5}$ or $\dfrac{\ln 50}{\log 5}$.

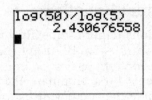

Example 8

Estimate $\log_4 200$ to the nearest hundredth.

Solution: Use the log base function on your calculator: 3.82

Example 9

Estimate $\log_2 0.35$ to the nearest hundredth.

Solution: Use the log base function on your calculator: −1.51

Solving Exponential Equations with a Calculator

Section 3.2 showed you how to solve exponential equations like $2^x = 7$ by graphing $y = 2^x$ and $y = 7$ and then finding the x-coordinate of the intersection point. Now you can solve the same equation by converting the equation into $\log_2 7 = x$ and using the calculator to evaluate $\log_2 7$.

$$2^x = 7$$
$$x = \log_2 7$$
$$x \approx 2.8074$$

Exponential and Logarithmic Expressions and Equations

Example 10

Estimate the solution to the equation $4^x = 450$. Round to the nearest hundredth.

Solution:

$$4^x = 450$$

$$x = \log_4 450$$

$$x \approx 4.41$$

Multistep Exponential Equations

Some exponential equations require some algebra steps to convert them into equations in the form $b^x = a$. By using the addition, subtraction, multiplication, and division properties of equality, this can often be accomplished in two steps. First eliminate the constant by adding or subtracting. Then eliminate the coefficient by multiplying or dividing.

For example, the equation $2 \cdot 3^x + 5 = 50$ could be solved in this way.

$$2 \cdot 3^x + 5 = 50$$
$$\underline{-5 = -5}$$
$$\frac{2 \cdot 3^x}{2} = \frac{45}{2}$$
$$3^x = 22.5$$
$$x = \log_3 22.5$$
$$x \approx 2.834$$

Sometimes the exponent is not just an x but, instead, is a more complicated expression. When this happens, there is more algebra to do after the log step to isolate the x.

For example, the equation $7^{2x-3} = 19$ could be solved in this way.

$$7^{2x-3} = 19$$
$$2x - 3 = \log_7 19$$
$$2x - 3 \approx 1.513$$
$$\underline{+3 = +3}$$
$$2x \approx 4.513$$
$$x \approx 2.257$$

104

Example 11

Solve for x rounded to the nearest hundredth in the following equation $3 \cdot 5^{5x-2} + 2 = 200$.

Solution:

$$3 \cdot 5^{5x-3} + 2 = 200$$
$$\underline{\phantom{3 \cdot 5^{5x-3}} -2 = -2}$$
$$\frac{3 \cdot 5^{5x-3}}{3} = \frac{198}{3}$$
$$5^{5x-3} = 66$$
$$5x - 3 = \log_5 66$$
$$5x - 3 \approx 2.603$$
$$\underline{ +3 = +3}$$
$$\frac{5x}{5} \approx \frac{5.603}{5}$$
$$x \approx 1.12$$

Leaving the Solution to an Exponential Equation in Unsimplified Form

Sometimes in a multiple-choice question about an exponential equation, the answer choices are not simply numbers. Instead, they are more involved mathematical expressions involving logarithms. These can be solved without using the calculator log functions.

Example 12

Which expression is a solution to the equation $3e^{5x} = 300$?

(1) $\dfrac{\ln 100}{5}$ (2) $\dfrac{\ln 300}{5}$ (3) $\dfrac{\ln 5}{300}$ (4) $\dfrac{\ln 5}{100}$

Exponential and Logarithmic Expressions and Equations

Solution: Choice (1) is correct.

$$\frac{3e^{5x}}{3} = \frac{300}{3}$$

$$e^{5x} = 100$$

$$5x = \log_e 100$$

$$\frac{5x}{5} = \frac{\log_e 100}{5}$$

$$x = \frac{\log_e 100}{5} = \frac{\ln 100}{5}$$

It is also possible to answer this question by calculating the solution to the original equation, which is $x \approx 0.921$. Then check each of the answer choices to see which one is also approximately equal to 0.921.

The Graph of a Two-Variable Logarithmic Equation

The graph of the function $y = \log_2 x$ is closely related to the graph of the function $y = 2^x$. The ordered pairs (1, 2), (2, 4), (3, 8), (0, 1), and $\left(-1, \frac{1}{2}\right)$ are solutions to $y = 2^x$. The ordered pairs (2, 1), (4, 2), (8, 3), (1, 0), and $\left(\frac{1}{2}, -1\right)$ are solutions to $y = \log_2 x$. Notice that for each ordered pair that satisfies $y = 2^x$, there is a "partner" ordered pair that satisfies $y = \log_2 x$ in which the x-coordinate and y-coordinate have been "swapped." When both equations are graphed on the same set of axes, the graph of $y = \log_2 x$ is the reflection of $y = 2^x$ over the line $y = x$.

106

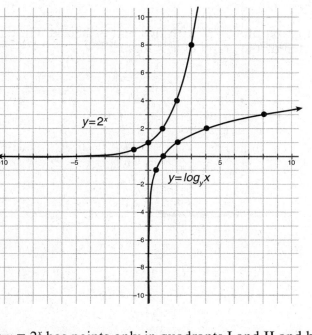

Note that $y = 2^x$ has points only in quadrants I and II and has a horizontal asymptote at $y = 0$. Also note that $y = \log_2 x$ has points only in quadrants I and IV and has a vertical asymptote at $x = 0$. The domain of $y = \log_2 x$ is $x > 0$.

Logarithmic graphs can also be created on the graphing calculator.

Example 13

Below is a graph of $y = 3^x$ with five points labeled. On the same axes, make a sketch of the graph of $y = \log_3 x$.

Solution:

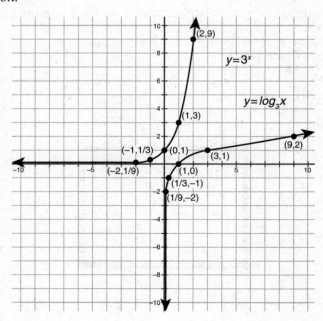

Check Your Understanding of Section 3.3

A. Multiple-Choice

1. What is $\log_2 32$?
   (1) 3   (2) 4   (3) 5   (4) 16

2. What is $\log_3 \dfrac{1}{81}$?
   (1) $\dfrac{1}{4}$ (2) $-\dfrac{1}{4}$ (3) 4   (4) -4

3. The equation $m^x = n$ is equivalent to which of the following?
   (1) $\log_n m = x$ (3) $\log_m n = x$
   (2) $\log_m x = n$ (4) $\log_n x = m$

4. $\log_{10} 500$ is between which two integers?
   (1) 2 and 3   (3) 4 and 5
   (2) 3 and 4   (4) 5 and 6

5. Rounded to the nearest tenth, what is the solution to $10^x = 350$?
   (1) 2.3   (2) 2.4   (3) 2.5   (4) 2.6

6. $\log_5 35$ is equal to which of the following?
   (1) $\dfrac{\log 5}{\log 35}$ (2) $\log 7$ (3) $\log 35$ (4) $\dfrac{\log 35}{\log 5}$

7. What is the solution to x to the nearest tenth in $3 \cdot 4^x = 270$?
   (1) 3.0   (2) 3.1   (3) 3.2   (4) 3.3

8. The number e is between which of the following?
   (1) 2 and 3   (3) 4 and 5
   (2) 3 and 4   (4) 5 and 6

9. Which equation has this graph?

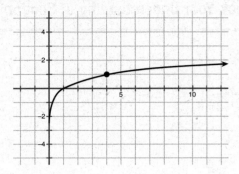

(1) $y = \log_4 x$ (3) $y = 4^x$

(2) $y = \log_{\frac{1}{4}} x$ (4) $y = \left(\dfrac{1}{4}\right)^x$

10. If the equation of the graph under the $y = x$ line is $y = \log_{\frac{3}{2}} x$, what is the equation of the graph above the $y = x$ line?

(1) $y = x^{\frac{3}{2}}$ (2) $y = \left(\dfrac{3}{2}\right)^x$ (3) $y = \dfrac{3}{2}x^2$ (4) $y = \left(\dfrac{2}{3}\right)^x$

B. Show how you arrived at your answers.

1. Which is greater: $\log_3 1{,}000$ or $\log_4 1{,}000$? Explain your reasoning.

2. If $x = \log_7 7^{123456789}$, what is the value of x?

3. Solve $3 \cdot 5^x + 7 = 262$ for x. Round to the nearest hundredth.

4. Logan solves $7^x = 263$ by calculating $x = \dfrac{\log 263}{\log 7}$. Calvin solves the same equation by calculating $x = \dfrac{\ln 263}{\ln 7}$. Who is correct and why?

5. Make a sketch of $y = \log_2 x$. Include all seven points that have coordinates where one (or both) of the coordinates are integers.

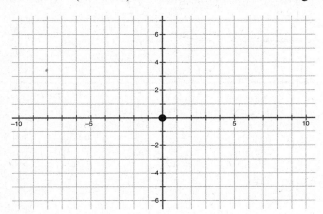

3.4 TRANSFORMED GRAPHS OF EXPONENTIAL AND LOGARITHMIC FUNCTIONS

KEY IDEAS

By knowing the graph of the basic exponential and logarithmic functions like $f(x) = 2^x$ or $g(x) = \log_2 x$, it is possible to create or identify the graphs of more complicated functions involving exponential or logarithmic expressions. Doing so requires you use transformations like horizontal and vertical shifts and also horizontal and vertical stretches and squeezes.

Graphs with Vertical Shifts

The graphs of $y = 2^x$ and $y = 2^x + 3$ are closely related. The graph of $y = 2^x + 3$ is what you get when each point on the graph of $y = 2^x$ is shifted up by 3 units. The horizontal asymptote of $y = 2^x + 3$ is at $y = 3$ since the asymptote of $y = 2^x$, which is $y = 0$, is also shifted up by 3 units.

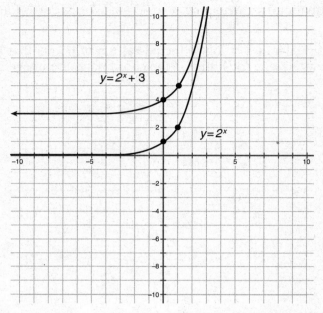

The graphs of $y = \log_2 x$ and $y = \log_2 x + 3$ are related in the same way. The graph of the second equation is the same as the graph of the first but shifted up by 3 units.

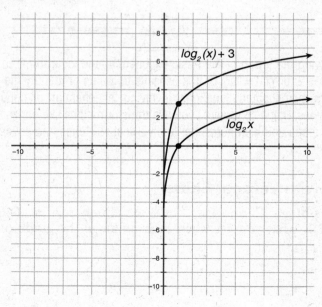

| MATH FACTS |

In general, the graph of $y = f(x) + a$ is a vertical shift of a units of the graph of $y = f(x)$. If a is positive, the shift is up. If a is negative, the shift is down.

Example 1

If $f(x) = 3^x$, make a sketch of the graphs of $y = f(x)$ and $y = f(x) - 2 = 3^x - 2$ on the same set of axes.

Solution: The graph of $y = f(x) - 2$ is a vertical shift down by 2 units of the graph of $y = f(x)$.

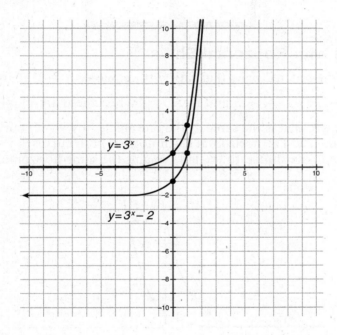

Exponential and Logarithmic Expressions and Equations

Graphs can also be made using the graphing calculator.

Graphs with Horizontal Shifts

Some of the ordered pairs that satisfy the equation $y = 2^{x+3}$ are $(-2, 2)$, $(-1, 4)$, $(-3, 1)$, $\left(-4, \dfrac{1}{2}\right)$, and $\left(-5, \dfrac{1}{4}\right)$. When these are graphed, they create a curve that is the same as the one created by $y = 2^x$ but shifted 3 units to the left. The horizontal asymptote of both curves is $y = 0$ since a horizontal line shifted to the left remains the same horizontal line.

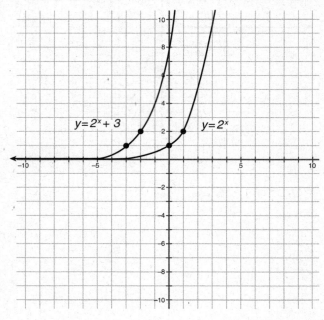

Some of the ordered pairs that satisfy the equation $y = \log_2(x+3)$ are $(-1, 1)$, $(1, 2)$, $(-2, 0)$, $\left(-\dfrac{5}{2}, -1\right)$, and $\left(-\dfrac{11}{4}, -2\right)$. When these are graphed, they create a curve that is the same as the one created by $y = \log_2 x$ but shifted 3 units to the left. The vertical asymptote of $y = \log_2(x+3)$ is at $x = -3$ but the vertical asymptote of $y = \log_2 x$ is at $x = 0$. Notice that the asymptote has been shifted to the left by 3 units.

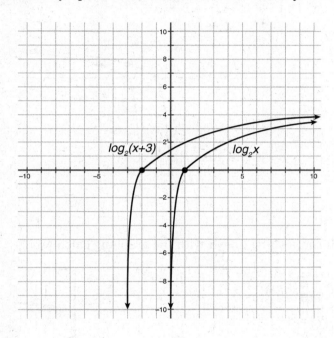

Example 2

If $g(x) = \log_3 x$, sketch on the same set of axes $y = g(x)$ and $y = g(x-2) = \log_3(x-2)$.

Solution: Shifting a curve left by -2 is equivalent to shifting it right by $+2$.

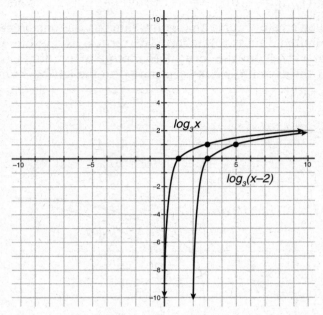

This graph can also be drawn with the graphing calculator.

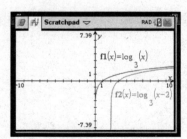

3.5 USING EXPONENTIAL OR LOGARITHMIC EQUATIONS IN REAL-WORLD SCENARIOS

KEY IDEAS

Many real-world scenarios can be modeled with exponential or logarithmic equations. These include exponential growth of populations or of money and also exponential decay of temperature or of radioactive material. If an equation for a model is provided, the equation can be used to answer questions about the scenario. If an equation is not provided, it is possible to create an equation that can then be used to answer questions.

Using an Exponential Equation that Is Already Provided

The population of Regentsland can be modeled by the equation $P = 20 \cdot 1.03^T$, where T is the number of years since 2000 and P is the population in millions. Using this formula, two types of questions can be answered about the model.

Type 1:
Determine the population for a given year.
To find the population in the year 2020, substitute $T = 20$ into the equation and solve for P.
$P = 20 \cdot 1.03^{20} \approx 36$ million people

Type 2:
Determine the year for a given population.
To find when the population of Regentsland will be 50 million people, substitute $P = 50$ into the equation and solve for T. This solution will involve logarithms.

$$50 = 20 \cdot 1.03^T$$
$$\frac{50}{20} = \frac{20 \cdot 1.03^T}{20}$$
$$2.5 = 1.03^T$$
$$T = \log_{1.03} 2.5 \approx 40 \text{ years} = 2040$$

Example 1

The temperature of a hot chocolate can be modeled by the equation $T = 130 \cdot 0.87^M + 70$, where T is the temperature in degrees Fahrenheit

Exponential and Logarithmic Expressions and Equations

and M is the number of minutes since the hot chocolate has been removed from the stove.

a) How hot will the hot chocolate be after 5 minutes?

b) How many minutes will it take for the hot chocolate to be at 86 degrees?

Solution:

a) Substitute 5 for M. $T = 130 \cdot 0.87^5 + 70 \approx 135$ degrees

b) Substitute 86 for M.

$$86 = 130 \cdot 0.87^T + 70$$

$$\frac{-70 = \qquad\qquad -70}{16 = 130 \cdot 0.87^T}$$

$$\frac{16}{130} = \frac{130 \cdot 0.87^T}{130}$$

$$0.1231 = 0.87^T$$

$$M = \log_{.87} 0.1231 \approx 15 \text{ minutes}$$

Example 2

A formula that relates the monthly payment, $\$M$, to pay off a loan of $\$P$ borrowed at an interest rate of r over a period of n months is

$$M = \frac{P\left(\dfrac{r}{12}\right)\left(1 + \dfrac{r}{12}\right)^n}{\left(1 + \dfrac{r}{12}\right)^n - 1}.$$

How many months will it take to pay off a $\$60{,}000$ student loan at 3% interest if the payment is $\$300$ a month? How much money will have been paid by the time the loan is paid off?

Solution: $P = 60{,}000$, $r = 0.03$, $M = 300$, and the unknown to solve for is n.

$$300 = \frac{60{,}000\left(\dfrac{0.03}{12}\right)\left(1 + \dfrac{0.03}{12}\right)^n}{\left(1 + \dfrac{0.03}{12}\right)^n - 1}$$

$$\frac{300}{1} = \frac{150 \cdot 1.0025^n}{1.0025^n - 1}$$

118

To continue solving this question, cross multiply to create the following equation.

$$300 \cdot 1.0025^n - 300 = 150 \cdot 1.0025^n$$

$$\frac{+300 = \qquad\qquad\qquad +300}{300 \cdot 1.0025^n = 150 \cdot 1.0025^n + 300}$$

The 1.0025^n terms need to be combined to make this into an exponential equation with just one term containing a variable.

$$300 \cdot 1.0025^n = 150 \cdot 1.0025^n + 300$$

$$-150 \cdot 1.0025^n = -150 \cdot 1.0025^n$$

$$150 \cdot 1.0025^n = 300$$

$$\frac{150 \cdot 1.0025^n}{150} = \frac{300}{150}$$

$$1.0025^n = 2$$

$$\log_{1.0025} 2 = n$$

$$277 \approx n$$

It will take approximately 277 months to pay off the loan. The total amount paid in that time will be $300 \cdot 277 \approx \$83,000$.

Creating Exponential Equations for Real-World Scenarios

Many real-world scenarios can be modeled with equations of the form $y = a \cdot (1 + r)^x$, where a and r are replaced with constants. The a-value is the initial value for y, while the r-value is the *growth rate*. When r is positive, the model has exponential growth. When r is negative, the model has exponential decay. The $(1 + r)$ is sometimes called b and is known as the *growth factor* when the equation is in the form of $y = a \cdot b^x$.

Example 3

A viral video has 300 views (V) the first day (D), and the number of views grows at a rate of 15% each day. Create an equation that models this scenario. Use your equation to determine what day the video will have 300,000 views.

Solution: Since 300 is the initial value and 0.15 is the growth rate, the equation is

$$V = 300 \cdot (1 + 0.15)^D$$

Exponential and Logarithmic Expressions and Equations

To find the day that there are 300,000 views, substitute 300,000 for the variable V and solve using logarithms.

$$300,000 = 300 \cdot 1.15^D$$
$$\frac{300,000}{300} = \frac{300 \cdot 1.15^D}{300}$$
$$1,000 = 1.15^D$$
$$D = \log_{1.15} 1,000$$
$$D = 49 \text{ days}$$

Example 4

A ball is dropped from the top of a 50-foot-tall building. After each bounce, the ball rises to 80% of the highest point of the last bounce. Create an equation that relates the height the ball rises to (H) to the number of bounces (B). Use this equation to determine when the ball will bounce to a height of 10 feet.

Solution: In this situation, the 80% is the growth factor, b, and not the growth rate, r. The equation is $H = 50 \cdot 0.8^B$.

To find how many bounces until the ball bounces to a height of 10 feet, substitute 10 for H in the equation and solve for B.

$$10 = 50 \cdot 0.8^B$$
$$\frac{10}{50} = \frac{50 \cdot 0.8^B}{50}$$
$$0.2 = 0.8^B$$
$$B = \log_{0.8} 0.2 = 7 \text{ bounces}$$

Creating Exponential Equations Based on Compound Interest

Banks generally offer *compound interest*. This means that you get interest on your original money as well as interest on your interest. Different types of interest are compounded annually, compounded monthly, and compounded continuously.

For interest compounded yearly, the formula is $A = P(1 + r)^t$ where P is the initial amount (the principal), r is the interest rate, t is the number of years the money has been in the bank, and A is the amount of money.

If \$1,000 is deposited into a bank that offers 3% interest compounded annually, the formula that relates A and P after t years is

$$A = 1,000 \cdot (1 + 0.03)^t = 1,000 \cdot 1.03^t$$

120

MATH FACTS

For a finite number of compoundings per year, n, the formula that relates the initial amount of money (P) to the amount of money it grows to (A) after a number of years (T) at an interest rate of r is $A = P\left(1 + \dfrac{r}{n}\right)^{nt}$.

Example 5

How long will it take for $1,000 to grow to $2,000 in a bank that offers 1.3% interest compounded monthly?

Solution: Since the interest is compounded monthly, the value of n is 12. The formula is

$$A = 1,000 \cdot \left(1 + \frac{0.013}{12}\right)^{12t} \approx 1,000 \cdot 1.00108^{12t}$$

Substitute 2,000 for A and solve for t.

$$2,000 \approx 1,000 \cdot 1.00108^{12t}$$

$$\frac{2,000}{1,000} \approx \frac{1000 \cdot 1.00108^{12t}}{1,000}$$

$$2 \approx 1.00108^{12t}$$

$$12t \approx \log_{1.00108} 2 \approx 642$$

$$\frac{12t}{12} \approx \frac{642}{12}$$

$$t \approx 54 \text{ years}$$

When the interest is *compounded continuously*, meaning that it is compounded every instant of every second, the formula that relates P and A after t years is $A = Pe^{rt}$. Since the base of the exponential part of the equation is e, the ln button of the calculator can be used to solve for t when A, P, and r are known.

Exponential and Logarithmic Expressions and Equations

Example 6

How long will it take for $700 to grow to $2,100 if a bank offers 4% interest compounded continuously?

Solution: The equation is $2{,}100 = 700e^{0.04t}$.

$$2{,}100 = 700e^{0.04t}$$

$$\frac{2{,}100}{700} = \frac{700e^{0.04t}}{700}$$

$$3 = e^{0.04t}$$

$$0.04t = \log_e 3 = \ln 3$$

$$0.04t \approx 1.0986$$

$$\frac{0.04t}{0.04} \approx \frac{1.0986}{0.04}$$

$$t \approx 27 \text{ years}$$

Creating Equivalent Exponential Expressions Related to Real-World Scenarios

Exponential equations that arise in real-world scenarios often have a coefficient in front of the exponent. For example, in the equation $A = p \cdot 1.00108^{12t}$, the exponent has a coefficient of 12. It is possible to create an equation that is equivalent to the original equation that has a different coefficient in the exponent. This process utilizes the properties of exponents that $(x^a)^b = x^{ab}$ and that $x^{ab} = (x^a)^b$.

For the example $A = P \cdot 1.00108^{12t}$, the 1.00108^{12t} can be changed to $(1.00108^{12})^t$. This can then be simplified to 1.013^t. So the equivalent equation is $A = P \cdot 1.013^t$.

This process can also work in reverse. If the equation for a scenario involving annual compound interest is $A = P \cdot 1.07^t$, it can be changed into the equivalent expression $A = P \cdot \left(1.07^{\frac{1}{12}}\right)^{12t} \approx P \cdot 1.0057^{12t}$ to make it look more like an equation that represents interest that is compounded monthly. This equivalent expression enables us to conclude that the monthly interest rate is approximately 0.57%.

122

Check Your Understanding of Section 3.5

A. Multiple-Choice

1. The population of West Algebra can be modeled by the equation $P = 30 \cdot 1.04^T$, where T is the number of years since 2000 and P is the population in millions. How many million people will there be in 2020?
   (1) 63.7   (2) 64.7   (3) 65.7   (4) 66.7

2. The population of Barrontopia can be modeled by the equation $P = 20 \cdot 1.03^T$, where T is the number of years since 2000 and P is the population in millions. In what year will the population be 33 million?
   (1) 2017   (2) 2018   (3) 2019   (4) 2020

3. The temperature of a cup of herbal tea can be modeled by the equation $T = 90 \cdot 0.7^M + 75$, where T is the temperature and M is the number of minutes since the tea was taken off the stove. How hot will the tea be 15 minutes after it is taken off the stove?
   (1) 75 degrees   (3) 79 degrees
   (2) 77 degrees   (4) 81 degrees

4. The temperature of a slice of pizza can be modeled by the equation $T = 70 \cdot 0.85^M + 80$, where T is the temperature and M is the number of minutes the pizza has been out of the oven. When will the pizza be 88 degrees?
   (1) 7 minutes   (3) 11 minutes
   (2) 9 minutes   (4) 13 minutes

5. The amount of money Aria has in the bank after T years is determined by the equation $A = 1{,}000 \cdot 1.0512^T$. After how many years will Aria have $2,000 in the bank?
   (1) 12.9   (2) 13.9   (3) 14.9   (4) 15.9

6. The amount of money Zachary has in the bank after T years is determined by the equation $A = 1{,}000e^{0.05T}$. After how many years will Zachary have $2,000 in the bank?
   (1) 11.9   (2) 12.9   (3) 13.9   (4) 14.9

Exponential and Logarithmic Expressions and Equations

7. Which equation relates the population (P) in millions to the time that has passed (T) if the growth rate is 4% per year and the starting population is 10 million people?
   (1) $P = 10(0.96)^T$
   (2) $P = 10(1.04)^T$
   (3) $P = 10 + 10(0.04)^T$
   (4) $P = 10(1.4)^T$

8. Which equation relates the amount of money (A) to the amount originally deposited (P) in a bank that offers 2% interest compounded continuously for T years?
   (1) $A = Pe^{0.2T}$
   (2) $A = P(1 + 0.02)^T$
   (3) $A = P(1 + 0.2)^T$
   (4) $A = Pe^{0.02T}$

9. Which equation is equivalent to the equation $y = 50(1.07)^{12x}$?
   (1) $y = 50(2.1)^x$
   (2) $y = 50(1.5)^{2x}$
   (3) $y = 50(1.3)^{6x}$
   (4) $y = 53.5^{12x}$

10. What is the growth rate for the exponential equation $y = 300(1.7)^x$?
    (1) 70%   (2) 7%   (3) 170%   (4) 17%

B. *Show how you arrived at your answers.*

1. The population of the Commonwealth of Common Core Land can be modeled by the equation $P = 8 \cdot 1.07^T$, where P is the population in millions and T is the number of years since 2000.
   a) To the nearest million, how big will the population be in 2030?
   b) To the nearest year, when will the population be 100 million?

2. The temperature of a pint of ice cream is related to the number of minutes the ice cream has been out of the freezer can be represented by the equation $T = -45 \cdot 0.9^M + 70$.
   a) What will the temperature of the ice cream be 12 minutes after it has been taken out of the freezer?
   b) How long will it take for the ice cream to reach 41 degrees?

3. Rodney puts $200 into a savings account that offers 12% interest compounded monthly. Create an equation that relates the amount of money in the bank (A) to the number of years it is in the bank (T).

4. Reagan puts $100 into First United North Bank, which offers 5% interest compounded annually. David puts $100 into First United South Bank, which offers 4.9% interest compounded continuously.

   Who will have more money in the bank after 2 years?

124

5. A sheet of paper is $\dfrac{1}{200}$ inches (or $\dfrac{1}{1,267,200}$ miles!) thick. If the paper is folded, its thickness doubles. After n folds, the thickness is $T = \dfrac{1}{1,267,200} \cdot 2^N$, where T is in miles.

How many folds will it take until the paper's thickness reaches the moon, which is 240,000 miles away?

Chapter Four
RADICAL EXPRESSIONS AND EQUATIONS

4.1 SIMPLIFYING RADICALS

Key Ideas

A *radical expression* is one that involves a $\sqrt{}$ sign (radial sign). Radical expressions are often involved in the solutions to polynomial equations. To work with radicals, you must know how to put them into simplified form and how to combine them.

Definition of Radicals

The square root of a number is the thing that must be multiplied by itself to get that number. The symbol for square root is $\sqrt{}$, also called *the radical sign*. An example is $\sqrt{25} = 5$ because $5^2 = 25$. Even though it is also true that $(-5)^2 = 25$, the symbol $\sqrt{}$ means just the positive number that when squared is equal to the number inside the radical sign. A number that is not a perfect square, like 7, still has a square root although that square root is an irrational number. $\sqrt{7}$ is between 2 and 3 since $2^2 = 4$ and $3^2 = 9$. More precisely, $\sqrt{7}$ is approximately 2.645751311. You can verify this by multiplying 2.645751311 by 2.645751311 to get approximately 7. $\sqrt{25} \cdot \sqrt{25} = 25$, $\sqrt{7} \cdot \sqrt{7} = 7$; and, in general, for a positive value represented by x, $\sqrt{x} \cdot \sqrt{x} = x$.

If there is a small number outside the radical sign, it no longer indicates a square root. If there is a small 3 outside the radical sign, it becomes a *cube root* sign and is equal to the number that must be cubed to become the number under the radical sign. An example is $\sqrt[3]{64} = 4$ because $4^3 = 64$. The small number outside the radical sign is called the *index*. When there is no index, it is implied to be a 2. So a radical sign with no index is called the *square root sign*.

Multiplying Radicals

Two radical expressions that have the same index can be multiplied by multiplying the numbers inside the radical sign. For example $\sqrt{4} \cdot \sqrt{9} = \sqrt{4 \cdot 9} = \sqrt{36} = 6$. This is easily verified for this example since $2 \cdot 3 = 6$.

MATH FACTS

In general, $\sqrt{a} \cdot \sqrt{b} = \sqrt{ab}$. This rule also works in reverse for factoring radicals: $\sqrt{ab} = \sqrt{a} \cdot \sqrt{b}$.

Simplifying Square Roots

If the number inside a square root sign has a factor that is a perfect square, the radical can be *simplified*. The $\sqrt{50}$ can be simplified since one of the factors of 50 is 25, which is a perfect square.

$\sqrt{50} = \sqrt{25 \cdot 2} = \sqrt{25} \cdot \sqrt{2} = 5 \cdot \sqrt{2} = 5\sqrt{2}$. The 5 in this expression is not an index but a coefficient in front of the radical sign. The multiplication sign between the coefficient and the radical sign is not necessary.

Example 1

Simplify the expression $\sqrt{18}$.

Solution: Since $18 = 9 \cdot 2$, $\sqrt{18} = \sqrt{9} \cdot \sqrt{2} = 3\sqrt{2}$.

It is not always clear whether or not a large number has a factor that is a perfect square. By factoring the number into its *prime factors*, it is possible to group the matching factors into pairs and use the fact that $\sqrt{x} \cdot \sqrt{x} = x$.

$$\sqrt{2592} = \sqrt{2 \cdot 2 \cdot 2 \cdot 2 \cdot 2 \cdot 3 \cdot 3 \cdot 3 \cdot 3} = \sqrt{2 \cdot 2} \cdot \sqrt{2 \cdot 2} \cdot \sqrt{2} \cdot \sqrt{3 \cdot 3} \cdot \sqrt{3 \cdot 3}$$
$$= 2 \cdot 2 \cdot \sqrt{2} \cdot 3 \cdot 3 = 36\sqrt{2}$$

Radical expressions that involve variables can also be simplified with this approach.

Example 2

Simplify the expression $\sqrt{9x^5 y^8}$.

Solution: This can be factored into

$$\sqrt{3 \cdot 3 \cdot x \cdot x \cdot x \cdot x \cdot x \cdot y \cdot y \cdot y \cdot y \cdot y \cdot y \cdot y \cdot y}.$$

Pair up matching factors when possible.

$$\sqrt{3 \cdot 3} \cdot \sqrt{x \cdot x} \cdot \sqrt{x \cdot x} \cdot x \cdot \sqrt{y \cdot y} \cdot \sqrt{y \cdot y} \cdot \sqrt{y \cdot y} \cdot \sqrt{y \cdot y}$$
$$3 \cdot x \cdot x \cdot \sqrt{x} \cdot y \cdot y \cdot y \cdot y$$
$$3x^2 y^4 \sqrt{x}$$

Radical Expressions and Equations

Adding and Subtracting Radicals

Radicals can be added or subtracted only if they have the same index and the same number inside the radical sign. They are combined the same way that like terms are combined with polynomials. For example, $3\sqrt{2}+4\sqrt{2}=7\sqrt{2}$. Subtraction works the same way: $8\sqrt{3}-6\sqrt{3}=2\sqrt{3}$.

If two radical expressions have the same index but different numbers inside the radical sign, you cannot immediately add or subtract them. Sometimes after simplifying the expressions, they will have the same number inside the radical sign and can then be added or subtracted.

Example 3

Simplify the terms of $\sqrt{50}+\sqrt{18}$ and combine if possible.

Solution: After simplifying each term, it becomes $5\sqrt{2}+3\sqrt{2}=8\sqrt{2}$.

Check Your Understanding of Section 4.1

A. Multiple-Choice

1. What is $\sqrt{71}\cdot\sqrt{71}$?
   (1) $\sqrt{71}$ (2) 71 (3) 5,041 (4) 71^2

2. Which of the following is equivalent to $\sqrt{12}$?
   (1) $3\sqrt{2}$ (2) $2\sqrt{6}$ (3) $6\sqrt{2}$ (4) $2\sqrt{3}$

3. Which of the following is equivalent to $\sqrt{18x^9y^6}$?
   (1) $3x^4y^3\sqrt{2x}$ (3) $3x^3y^3$
   (2) $9x^3y^2\sqrt{y^2}$ (4) $3x^3y^2\sqrt{y^2}$

4. What is $2\sqrt{7}+6\sqrt{7}$?
   (1) $8\sqrt{14}$ (2) $8\sqrt{7}$ (3) $12\sqrt{7}$ (4) $12\sqrt{14}$

5. What is $\sqrt{12}+\sqrt{75}$?
   (1) $7\sqrt{3}$ (2) $\sqrt{87}$ (3) $7\sqrt{6}$ (4) $3\sqrt{7}$

6. Which of the following is equivalent to $\sqrt[3]{192}$?
   (1) $64\sqrt[3]{3}$ (2) $3\sqrt[3]{64}$ (3) $3\sqrt[3]{4}$ (4) $4\sqrt[3]{3}$

7. Which of the following is equivalent to $\sqrt[3]{27x^7y^{11}}$?
   (1) $3x^3y^5\sqrt[3]{3xy}$
   (2) $3x^2y^3\sqrt[3]{xy^2}$
   (3) $3x^2y^3\sqrt[3]{x^2y^2}$
   (4) $3x^2y^3\sqrt[3]{xy}$

8. What is $\sqrt[3]{71} \cdot \sqrt[3]{71} \cdot \sqrt[3]{71}$?
   (1) $\sqrt{71}$
   (2) $\sqrt[3]{71^2}$
   (3) 71
   (4) 357,911

9. Simplify $\dfrac{6+\sqrt{28}}{2}$.
   (1) $3+\sqrt{14}$
   (2) $3+\sqrt{7}$
   (3) $\sqrt{17}$
   (4) $\sqrt{10}$

10. What is $2\sqrt{3} \cdot 5\sqrt{3}$?
    (1) 30
    (2) $10\sqrt{3}$
    (3) $\sqrt{30}$
    (4) $7\sqrt{3}$

B. Show how you arrived at your answers.

1. Spencer simplifies $\sqrt{25+144}$ as $\sqrt{25}+\sqrt{144}=5+12=17$. Mia simplifies it as $\sqrt{169}=13$. Who is correct?

2. Daniel calculates $\sqrt{49}+\sqrt{576}=\sqrt{625}=25$. Is this correct? Why or why not?

3. In this right triangle, what is the length of the hypotenuse simplified in simplest terms?

Radical Expressions and Equations

4. What is $\dfrac{6}{\sqrt{3}} \cdot \dfrac{\sqrt{3}}{\sqrt{3}}$?

5. What is $(5+2\sqrt{3})(5-2\sqrt{3})$?

4.2 IMAGINARY AND COMPLEX NUMBERS

Key Ideas

An *imaginary* number is a number that, when multiplied by itself, results in a negative number. The most basic imaginary number is $\sqrt{-1}$ since $\left(\sqrt{-1}\right)^2 = \sqrt{-1} \cdot \sqrt{-1} = -1$. The number $\sqrt{-1}$ is abbreviated by the symbol i. Based on this definition, $i^2 = -1$. Though this is represented by a letter, i is not a variable. When imaginary numbers are combined with real numbers, the result is called a *complex number*. Complex numbers can be added, subtracted, and multiplied with rules similar to polynomial adding, subtracting, and multiplying.

Simplifying Imaginary Numbers

The square root of any negative number can be represented as an expression involving an i. To simplify $\sqrt{-9}$, split it into $\sqrt{9 \cdot (-1)} = \sqrt{9} \cdot \sqrt{-1} = 3i$. If the number inside the radical is not a perfect square, like $\sqrt{-7}$, the i will be written to the left of the radical sign, $\sqrt{-7} = \sqrt{-1 \cdot 7} = \sqrt{-1} \cdot \sqrt{7} = i\sqrt{7}$. If the number inside the radical has a factor that is a perfect square, like $\sqrt{-18}$, the i will be written between the coefficient and the radical: $\sqrt{-18} = \sqrt{9 \cdot (-1) \cdot 2} = \sqrt{9} \cdot \sqrt{-1} \cdot \sqrt{2} = 3i\sqrt{2}$.

Adding and Subtracting Imaginary Numbers

To add or subtract two imaginary numbers like $5i + 2i$, they are considered to be like terms. So they can be combined as you would do with variables, $5i + 2i = 7i$. For subtraction, it would be $5i - 2i = 3i$.

Multiplying Imaginary Numbers

Multiplying imaginary numbers like $5i \cdot 2i$ is similar to multiplying expressions like $5x \cdot 2x$. Since $5x \cdot 2x$ would be $10x^2$, $5i \cdot 2i$ is $10i^2$. Since $i^2 = -1$, $10i^2 = 10(-1) = -10$.

Example 1

Multiply $3i\sqrt{2} \cdot 5i\sqrt{3}$.

Solution: $3i\sqrt{2} \cdot 5i\sqrt{3} = 15i^2\sqrt{6} = -15\sqrt{6}$

Powers of *i*

i can be raised to other powers besides 2. When *i* is raised to an integer power, the only possible values are i, -1, $-i$, or 1.

Power		Solution
0	i^0	1
1	i^1	i
2	i^2	-1
3	$i^3 = i^2 \cdot i = (-1)i$	$-i$
4	$i^4 = i^3 \cdot i = (-i)i = -i^2 = -(-1)$	1

After the power of 3, the powers of *i* cycle between the four values: 1, i, -1, and $-i$. *i* raised to any multiple of 4 will be 1.

To raise *i* to a high power, like i^{47}, find the multiple of 4 smaller than 47 that is closest to it and rewrite as $i^{44} \cdot i^3 = 1 \cdot i^3 = i^3 = -i$.

The graphing calculator has a built-in *i*. On the TI-84, you access it by pressing [2ND] [.]. On the TI-Nspire, you press the [pi] button to access the constants menu. Calculating powers of *i* on the TI-84 outputs cryptic numbers. If you see something like $-3E-13$, this is shorthand for -0.0000000000003, which is approximately 0.

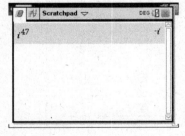

Complex Numbers

When a real number, like 3, is added to an imaginary number, like $2i$, together the sum is written as $3 + 2i$. A *complex number* is a number of the form $a + bi$ where a and b are real numbers.

Radical Expressions and Equations

Adding and Subtracting Complex Numbers

Complex numbers are added by adding the real parts and the imaginary parts separately and writing the answer in $a + bi$ form.

$$(3 + 2i) + (5 - 7i) = 8 - 5i$$

Subtracting complex numbers requires care as you distribute the -1 through the second expression.

$$(3 + 2i) - (5 - 7i) = 3 + 2i - 5 + 7i = -2 + 9i$$

The graphing calculator can also add or subtract complex numbers.

Multiplying Complex Numbers

Complex numbers can be multiplied the way binomials are multiplied by using the FOIL shortcut. Anytime an i^2 is encountered in a solution, it should be converted to a -1.

To multiply $(3 + 2i)(5 - 7i)$ with FOIL:

$$3 \cdot 5 + 3(-7i) + 2i(5) + 2i(-7i) = 15 - 21i + 10i - 14i^2$$
$$= 15 - 21i + 10i + 14 = 29 - 11i.$$

The graphing calculator can also multiply complex numbers.

Example 2

Multiply the numbers $3 - 4i$ and $2 + 5i$ and put the answer into $a + bi$ form.

Solution: $(3 - 4i)(2 + 5i) = 6 + 15i - 8i - 20i^2 = 6 + 15i - 8i + 20$
$= 26 + 7i$

Example 3

Write $(4 + 3yi)(3 - 2i) - (4 - 3yi)(3 - 2i)$ in $a + bi$ form, where y is a real number.

Solution: The short way to solve this is by first noticing that there is a $(3 - 2i)$ in each of the main expressions. This can be factored out to make the entire expression:

$$(3 - 2i)((4 + 3yi) - (4 - 3yi))$$
$$(3 - 2i)(4 + 3yi - 4 + 3yi)$$
$$(3 - 2i)(6yi)$$
$$18yi - 12yi^2$$
$$12y + 18yi$$

Complex Solutions to Quadratic Equations

If in the process of using the quadratic formula a negative number appears inside the radical, the quadratic equation is said to have two complex solutions.

Example 4

Use the quadratic equation to solve $x^2 - 4x + 13 = 0$ with $a = 1$, $b = -4$, and $c = 13$.

Solution: $x = \dfrac{-(-4) \pm \sqrt{(-4)^2 - 4(1)(13)}}{2(1)} = \dfrac{4 \pm \sqrt{16 - 52}}{2} = \dfrac{4 \pm \sqrt{-36}}{2}$

Since there is a negative number inside the radical sign, the solutions will be complex numbers.

$$x = \frac{4 \pm 6i}{2} = 2 \pm 3i$$

There are two complex solutions: $x = 2 + 3i$ or $x = 2 - 3i$.

Graphing Complex Numbers on the Complex Plane

A number like $2 + 3i$ cannot be graphed on the standard number line. Instead, complex numbers are graphed on something called *the complex plane*. The complex plane is like a number line for complex numbers. The

Radical Expressions and Equations

complex plane resembles the two-axis coordinate plane. Instead of the axes representing the *x*-coordinate and the *y*-coordinate, the axes represent the real part of the complex number and the imaginary part of the complex number.

Below is the complex plane. The number $2 + 3i$ is plotted as a single point in the position (2, 3).

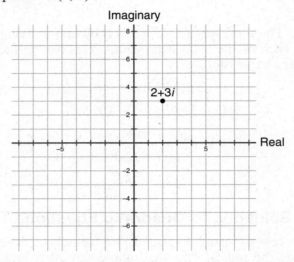

Real numbers like 2 can be thought of as $2 + 0i$ and are graphed on the real axis of the complex plane. Imaginary numbers like $2i$ can be thought of as $0 + 2i$ and are graphed on the imaginary axis of the complex plane.

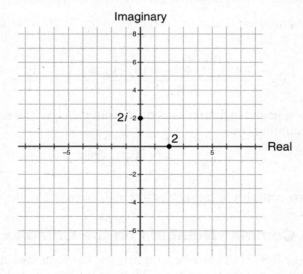

═══ **MATH FACTS** ═══

The absolute value of a complex number is the distance on the complex plane between that number and the number $0 + 0i$. The absolute value of a complex number $a + bi$ can be calculated with the formula $|a + bi| = \sqrt{a^2 + b^2}$.

Example 5

What is $|3 - 4i|$?

    (1) 5
    (2) −5
    (3) 7
    (4) −7

Solution: $|3 - 4i| = \sqrt{3^2 + (-4)^2} = \sqrt{9 + 16} = \sqrt{25} = 5$ so choice (1) is the answer.

Check Your Understanding of Section 4.2

A. Multiple-Choice

1. What is $\sqrt{-25}$?
    (1) $5i$ (2) 5   (3) $-5i$ (4) -5

2. What is $7i - 2i$?
    (1) 5   (2) $5i^2$ (3) -5 (4) $5i$

3. What is $-7i \cdot 3i$?
    (1) 21   (2) -21 (3) $21i$ (4) $-21i$

4. What is i^{66}?
    (1) -1 (2) 1   (3) $-i$ (4) i

5. What is $(3 + 5i) + (2 - 7i)$?
    (1) $5 + 2i$ (3) $-5 - 2i$
    (2) $-5 + 2i$ (4) $5 - 2i$

Radical Expressions and Equations

6. What is $(3 + 5i) - (2 - 7i)$?
   (1) $1 - 2i$
   (2) $1 + 12i$
   (3) $-1 - 2i$
   (4) $-1 - 12i$

7. What is $(3 + 5i)(2 - 7i)$?
   (1) $-29 - 11i$
   (2) $41 - 11i$
   (3) $29 - 11i$
   (4) $-41 - 11i$

8. What is $(3 + 5i)(3 - 5i)$?
   (1) 34
   (2) $-16 - 30i$
   (3) -16
   (4) $34 - 30i$

9. Solve the quadratic equation $x^2 - 6x + 34 = 0$.
   (1) $3 \pm 5i$
   (2) $-3 \pm 5i$
   (3) $13, -7$
   (4) $-13, 7$

10. What is $|12 - 5i|$?
    (1) $\sqrt{119}$
    (2) 12
    (3) 13
    (4) 7

B. *Show how you arrived at your answers.*

1. Plot the point $-6 + 8i$ on the complex plane. What is the absolute value of $-6 + 8i$?

2. Find the 4 answers to the equation $x^4 + 13x^2 + 36 = 0$.

3. What is $\left(\sqrt{2} + i\sqrt{2}\right)^2$?

4. What is $\left(1 + \frac{1}{2}i\right)^2 + \left(1 + \frac{1}{2}i\right)$?

5. A complex number whose absolute value is greater than 2 is not in something called the Mandelbrot Set. Show that $1.9 + 0.8i$ is not in the Mandelbrot Set.

4.3 SOLVING RADICAL EQUATIONS

KEY IDEAS

A *radical equation* is one that involves the square root, or cube root, or other kind of root of a variable. If a radical equation also involves the same variable raised to the first power, the equation often has multiple solutions. One step in a radical equation is usually to, at some point, square, or cube, both sides of the equation.

One-Step Radical Equations

When you square the square root of a perfect square, the result is the number that was originally under the radical sign. So $\left(\sqrt{9}\right)^2 = 3^2 = 9$. This is also true for numbers that are not perfect squares like $\left(\sqrt{7}\right)^2 = 7$. In general, we can say $\left(\sqrt{x}\right)^2 = x$.

The equation $\sqrt{x} = 9$ can be solved by squaring the expression on both sides of the equal sign.

$$\sqrt{x} = 9$$
$$\left(\sqrt{x}\right)^2 = 9^2$$
$$x = 81$$

Two-Step Radical Equations

Before squaring both sides of a radical equation, the radical term must be isolated. In the equation $\sqrt{x} + 3 = 7$, the $\sqrt{x}$ must be isolated by subtracting 3 from both sides of the equation. Then the equation can be completed by squaring both sides of the equation to eliminate the radical sign.

$$\sqrt{x} + 3 = 7$$
$$\underline{-3 = -3}$$
$$\sqrt{x} = 4$$
$$\left(\sqrt{x}\right)^2 = 4^2$$
$$x = 16$$

If the expression under the radical sign is not simply an x, there will be additional steps after the radical sign has been eliminated.

Radical Expressions and Equations

$$\sqrt{2x+1} = 5$$
$$\left(\sqrt{2x+1}\right)^2 = 5^2$$
$$2x+1 = 25$$

Now this equation must be solved using techniques from algebra.

$$2x+1 = 25$$
$$\underline{-1 = -1}$$
$$2x = 24$$
$$\frac{2x}{2} = \frac{24}{2}$$
$$x = 12$$

Example 1

Solve for x in the equation $2\sqrt{x+4} - 3 = 11$.

Solution: Two steps are needed to isolate the radical term.

$$2\sqrt{x+4} - 3 = 11$$
$$\underline{\phantom{2\sqrt{x+4}}+3 = +3}$$
$$2\sqrt{x+4} = 14$$
$$\frac{2\sqrt{x+4}}{2} = \frac{14}{2}$$
$$\sqrt{x+4} = 7$$

Now square both sides, and then isolate the x.

$$\left(\sqrt{x+4}\right)^2 = 7^2$$
$$x+4 = 49$$
$$\underline{-4 = -4}$$
$$x = 45$$

Radical Equations Involving a Linear Term

A more complicated type of radical equation is one like $\sqrt{x} + x = 6$. By testing different values, it can be seen that $x = 4$ is a number that makes the equation true since $\sqrt{4} + 4 = 2 + 4 = 6$. A different approach is needed if the equation requires an algebraic solution, if there is more than one solution, or if the solutions are irrational numbers.

138

As with the simpler equations, the first step is still to isolate the radical term. For this example, this means to subtract x from both sides of the equation.

$$\sqrt{x} + x = 6$$
$$\underline{-x = -x}$$
$$\sqrt{x} = 6 - x$$

Just like before, the next step is to eliminate the radical sign by squaring both sides of the equation. Squaring the right side, however, involves using polynomial multiplication (FOIL) described in Chapter 1.

$$\left(\sqrt{x}\right)^2 = \left(6 - x\right)^2$$
$$x = 36 - 12x + x^2$$

The problem has been turned into a quadratic equation that can be solved by eliminating all the terms from one side of the equation and then using one of the approaches from Chapter 1 to solve that equation.

$$x = 36 - 12x + x^2$$
$$\underline{-x = -x}$$
$$0 = x^2 - 13x + 36$$
$$0 = (x - 4)(x - 9)$$
$$x - 4 = 0 \quad \text{or } x - 9 = 0$$
$$x = 4 \qquad \text{or } x = 9$$

This process did get the solution $x = 4$, but it also seems to have found another solution. When you substitute $x = 9$ into the original equation, the solution does not work out properly.

$$\sqrt{9} + 9 = 6$$
$$3 + 9 = 6$$
$$12 \neq 6$$

The solution $x = 9$ is not true!

When this happens, the number is not an answer to the equation and we do not include it in the solution set. The clearest way to indicate this on a test is to cross out the $x = 9$ solution and write the word "reject."

The only solution to this equation is $x = 4$.

Radical Expressions and Equations

MATH FACTS

Squaring both sides of an equation can sometimes make equations that are not true into equations that are true. For example, it is not true that $-5 = 5$, but it is true that $(-5)^2 = 5^2$. This is why it is necessary to check any solutions that come from solving a radical equation that eventually became a quadratic equation.

Example 2

Solve for all values of x that satisfy the equation $\sqrt{x} - x = -6$

Solution:

$$\sqrt{x} - x = -6$$
$$\underline{+x = +x}$$
$$\sqrt{x} = -6 + x$$
$$(\sqrt{x})^2 = (-6 + x)^2$$
$$x = 36 - 12x + x^2$$
$$\underline{-x = -x}$$
$$0 = x^2 - 13x + 36$$
$$0 = (x - 4)(x - 9)$$
$$x - 4 = 0 \text{ or } x - 9 = 0$$
$$x = 4 \text{ or } x = 9$$

Each of these potential solutions needs to be substituted back into the original equation to check to see if they work.

Substitute $x = 4$: $\sqrt{4} - 4 = 2 - 4 = -2 \neq -6$. This solution does not work.

Substitute $x = 9$: $\sqrt{9} - 9 = 3 - 9 = -6 = -6$. This solution does work.

The solution is only $x = 9$.

Example 3

Find all solutions to the equation $\sqrt{x-2} + x = 8$.

Solution: Isolate the radical term, square both sides, and solve the resulting quadratic equation.

140

$$\sqrt{x-2}+x=8$$
$$\underline{\phantom{\sqrt{x-2}}-x=-x}$$
$$\sqrt{x-2}=8-x$$
$$\left(\sqrt{x-2}\right)^2=(8-x)^2$$
$$x-2=64-16x+x^2$$
$$\underline{-x+2=-x+2}$$
$$0=x^2-17x+66$$
$$0=(x-11)(x-6)$$
$$x-11=0\quad\text{or}\quad x-6=0$$
$$x=11\quad\text{or}\quad x=6$$

Of these two potential solutions, only $x = 6$ works since $\sqrt{11-2}+11=14$, not 8.

Example 4

What is the solution set for the equation $\sqrt{2x}+x=4$?

    (1) {2, 8}
    (2) {8}
    (3) {2}
    (4) {}

Solution: Choice (3) is the correct answer.

This question can be solved with algebra like the others and will eventually become a quadratic equation with two solutions, $x = 2$ and $x = 8$. However, the $x = 8$ solution will need to be rejected since it does not make the original equation true.

Since this is a multiple-choice question, the quickest and most accurate way to answer it would be to check the two numbers $x = 2$ and $x = 8$ to see that only $x = 2$ makes the original equation true.

Radical Equations Involving Two Radical Terms

Solving an equation that involves two radical terms like $\sqrt{x+1}+\sqrt{x-4}=5$ is a lengthy process. It is tempting to try to square both sides to eliminate all the radicals. However, this does not work since after squaring the left side with FOIL, there will be outer and inner terms to deal with.

To solve this sort of equation, first isolate one of the radical terms. Then square both sides very carefully. Continue by isolating the other radical term and squaring both sides again.

141

Radical Expressions and Equations

$$\sqrt{x+1}+\sqrt{x-4}=5$$
$$\underline{-\sqrt{x-4}=-\sqrt{x-4}}$$
$$\sqrt{x+1}=5-\sqrt{x-4}$$
$$\left(\sqrt{x+1}\right)^2=\left(5-\sqrt{x-4}\right)^2$$
$$x+1=25-10\sqrt{x-4}+x-4$$
$$\underline{-x=-x}$$
$$1=21-10\sqrt{x-4}$$
$$\underline{-21=-21}$$
$$-20=-10\sqrt{x-4}$$
$$\frac{-20}{-10}=\frac{-10\sqrt{x-4}}{-10}$$
$$2=\sqrt{x-4}$$
$$2^2=\left(\sqrt{x-4}\right)^2$$
$$4=x-4$$
$$\underline{+4=+4}$$
$$8=x$$

Check to see if the answer should be rejected.
$$\sqrt{8+1}+\sqrt{8-4}=\sqrt{9}+\sqrt{4}=3+2=5$$

The solution is $x = 8$.

Check Your Understanding of Section 4.3

A. Multiple-Choice

1. What is the solution to $\sqrt{x}+2=9$?
   (1) −49   (2) 49   (3) −7   (4) 7

2. What is the solution to $\sqrt{x+2}=9$?
   (1) −79   (2) 1   (3) 79   (4) −1

3. What is the solution to $3\sqrt{x-2}+5=17$?
   (1) −18   (2) 18   (3) −16   (4) 16

142

4. What value(s) of x make the equation $\sqrt{x} + x = 20$ true?
   (1) 14   (2) 25   (3) 16   (4) 16, 25

5. What value(s) of x make the equation $\sqrt{x} - x = -20$ true?
   (1) 16   (2) 25   (3) 16, 25   (4) 36

6. Find all solutions to the equation $\sqrt{8x + 9} - x = 2$.
   (1) 5, −1   (2) −1   (3) 5   (4) −5, 1

7. Find all solutions to the equation $\sqrt{x + 3} + x = 9$.
   (1) 13, 6   (2) 13   (3) 6   (4) 4

8. Find all solutions to the equation $\sqrt{30x + 45} - 2x = 3$.
   (1) $-\dfrac{3}{2}$, 6   (2) $-\dfrac{3}{2}$ (3) 6   (4) $\dfrac{3}{2}$, −6

9. For what value(s) does $\sqrt{x} = x$?
   (1) 1   (2) 0   (3) 1, 0, −1   (4) 1, 0

10. Solve for x: $\sqrt{x - 3} + \sqrt{x + 9} = 6$.
   (1) 6   (2) 7   (3) 8   (4) 9

B. *Show how you arrived at your answers.*

1. Solve for x.

$$5\sqrt{x - 4} + 3 = 18$$

2. Solve for x.

$$\sqrt{2x - 8} + x = 4$$

3. The equation $x = 3$ has a solution set of {3}. If you square both sides of the equation, what is the solution set of the new equation?

4. Jace tried to solve the equation $\sqrt{x - 4} + \sqrt{x + 4} = 4$ by first squaring both sides. He got:

$$\left(\sqrt{x - 4} + \sqrt{x + 4}\right)^2 = 4^2$$

$$x - 4 + x + 4 = 16$$

$$2x = 16$$

$$x = 8$$

However, 8 is not a solution to the equation. What did Jace do wrong?

5. What value(s) of x satisfy this equation?
$$\sqrt{x+6} - \sqrt{4-x} = 2$$

4.4 GRAPHS OF RADICAL FUNCTIONS

KEY IDEAS

The graphs of equations involving a radical like $y = \sqrt{x}$ or $y = \sqrt[3]{x}$ have shapes that are related to the graphs of the polynomial graphs $y = x^2$ and $y = x^3$, respectively. Graphs of functions involving radicals can be transformed by shifting left, right, up, or down and through vertical stretches and horizontal squeezes.

The Graph of the Square Root Function

The graph of the function $f(x) = \sqrt{x}$ includes the points (0, 0), (1, 1), (4, 2), and (9, 3). Since the $\sqrt{}$ symbol evaluates only to the positive square root of a number, the y-coordinates of all the points (the *range*) are all numbers greater than or equal to 0. Though the concept of imaginary numbers was introduced in the last section for the purpose of graphing on the coordinate axes, there is no place to graph a point like $(-4, 2i)$. So the x-coordinates of the points on this graph (the *domain*) will also be all numbers greater than or equal to 0. For this reason, the entire graph will be in quadrant I.

Below is the graph of the function $f(x) = \sqrt{x}$.

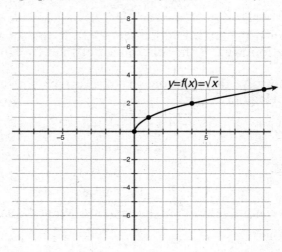

This curve is actually a half of a parabola reflected over the line $y = x$. Unlike the graphs of the polynomial functions introduced in Chapter 1, this graph's domain is not all real numbers but is $x \geq 0$.

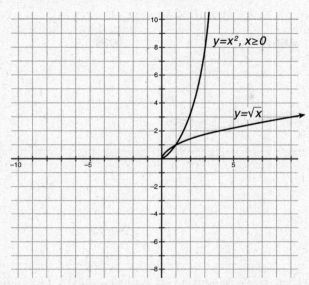

Transformations of the Square Root Function

The function $g(x) = \sqrt{x+2}$ has a graph very similar to the graph of $f(x) = \sqrt{x}$. In fact, $g(x)$ can be written as $f(x + 2)$. The graph of $f(x + 2)$ is the result of shifting the graph of $f(x)$ to the *left* by 2 units.

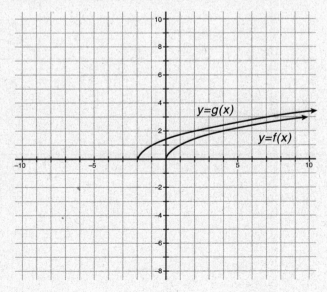

Radical Expressions and Equations

The four most common transformations of the function $f(x)$ are shown in the following table.

$g(x) = \sqrt{x+2} = f(x+2)$ shifted to the *left* 2 units	$g(x) = 2\sqrt{x} = 2f(x)$ vertical *stretch* by a factor of 2 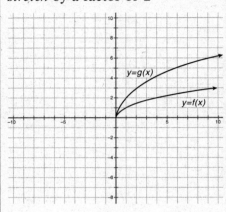
$g(x) = \sqrt{x-2} = f(x-2)$ shifted to the *right* 2 units	$g(x) = \sqrt{2x} = f(2x)$ horizontal *squeeze* by a factor of 2 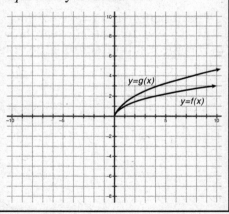

Graphically Solving Radical Equations

In Section 4.3, radical equations like $\sqrt{x} - x = -6$ were solved algebraically. Radical equations can also be solved with the intersect feature of the graphing calculator.

To get the most useful graph to analyze, first isolate the radical term by adding x to both sides.

$$\sqrt{x} - x = -6$$
$$+x = +x$$
$$\sqrt{x} = x - 6$$

On the graphing calculator graph on the same set of axes $y = \sqrt{x}$ and $y = x - 6$. Using the intersect feature of the graphing calculator, the two curves intersect at (9, 3). The x-intercept is the solution to the equation $x = 9$.

Check Your Understanding of Section 4.4

A. *Multiple-Choice*

1. Which graph is a sketch of $y = \sqrt{x} + 2$?

(1)

Radical Expressions and Equations

(2)

(3)

(4)

2. Which graph is a sketch of $y = \sqrt{x+2}$?

(1)

(2)

(3)

Radical Expressions and Equations

(4)

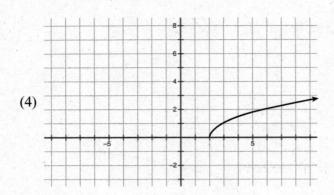

3. Which graph is a sketch of $y = \sqrt{x} - 2$?

(1)

(2)

150

(3)

(4)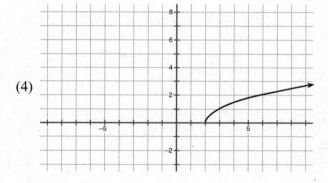

4. Which graph is a sketch of $y = \sqrt{x-2}$?

(1)

Radical Expressions and Equations

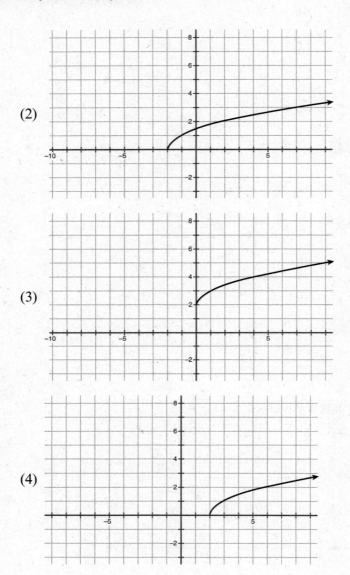

5. Which equation produces the following graph?

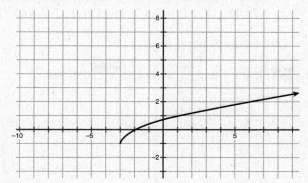

(1) $y = \sqrt{x+3} + 1$
(2) $y = \sqrt{x+3} - 1$
(3) $y = \sqrt{x-3} - 1$
(4) $y = \sqrt{x-3} + 1$

6. Which equation produces the following graph?

(1) $y = \sqrt{x-3} + 1$
(2) $y = \sqrt{x+3} + 1$
(3) $y = \sqrt{x-3} - 1$
(4) $y = \sqrt{x+3} - 1$

7. Solve the equation by graphing. $\sqrt{x+3} = -x+9$
   (1) 6   (2) 7   (3) 8   (4) 9

8. Solve the equation by graphing. $\sqrt{8x+9} - x = 2$
   (1) (5, −1)   (2) (−5, 1)   (3) 5   (4) −1

9. Solve the equation by graphing. $\sqrt{x-3} + \sqrt{x+9} = 6$
   (1) 5   (2) 6   (3) 7   (4) 8

Radical Expressions and Equations

10. Which of the following is a point on the graph of $y = 3\sqrt{2x} + 5$?
    (1) (8, 17)   (2) (8, 18)   (3) (8, 19)   (4) (8, 20)

B. *Show how you arrived at your answers.*

1. Sketch the graph of $y = \sqrt{x+4}$ on the axes below.

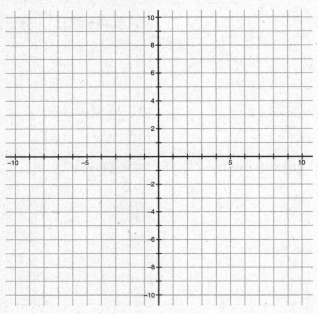

2. What is the equation for this graph?

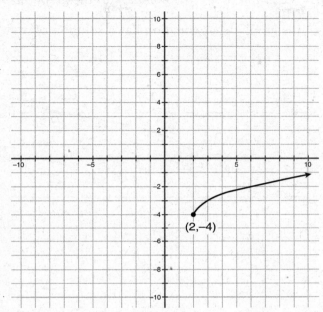

3. Ariana thinks that this graph is a vertical stretch by a factor of 2 of $y = \sqrt{x}$. Sienna thinks that it is a horizontal squeeze by a factor of 4. Who is correct?

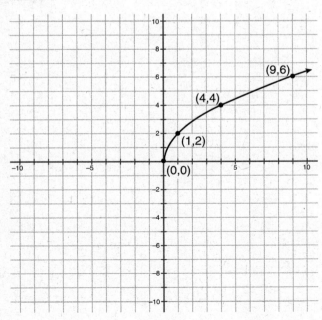

Radical Expressions and Equations

4. How does the graph of $y = \sqrt[3]{x}$ compare to the graph of $y = \sqrt{x}$?

5. Based on this graph, what equation has a solution of $x = 7$?

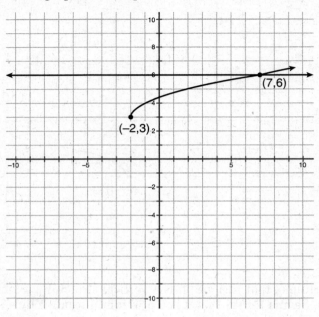

Chapter Five

TRIGONOMETRIC EXPRESSIONS AND EQUATIONS

5.1 UNIT CIRCLE TRIGONOMETRY

KEY IDEAS

The coordinates of the points on a circle can be described with the sine and the cosine of an angle. The ability to describe the location of points on a circle is a skill needed for various real-world applications related to physics.

Finding the Length of the Legs of a Right Triangle with a Unit Hypotenuse

Below is right triangle ABC with hypotenuse $AB = 1$ unit. Angle C is the right angle, and angle A has a measure of 40 degrees.

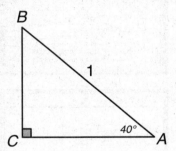

One way to think about sine 40° is that it is the length of the side opposite the 40° angle in a right triangle with a hypotenuse of length 1 unit. If your calculator is in degree mode and you enter sin(40), it should display approximately 0.6428. This means that in triangle ABC, side BC is approximately 0.6428 units long.

If your calculator is not in degree mode, set it to degree mode.

For the TI-84:

Press [MODE], move the cursor to DEGREE on the third line, and press [ENTER] to highlight it. Then press [2nd] [MODE] to get back to calculator mode.

For the TI-Nspire:

From the home screen, press [5] and then [2] for Document Settings. If the second field does not say "Degree," press [down] and [click] and change the setting to Degree. Then press [tab] until the "Make Default" button is highlighted. Press the [click] button.

If you have a recent operating system on the TI-Nspire, when you are working in the Scratchpad, you will see a small "deg" in the upper right hand corner.

Example 1

In the triangle below, what is the length of side *BC*?

Solution: The length of the side opposite the 72 degree angle in a right triangle with hypotenuse 1 is sine 72°, which is approximately 0.9511.

The cosine of an angle can be thought of as the length of the side adjacent to the angle in a right triangle that has a hypotenuse of length 1 unit.

In triangle *ABC* from above, cosine 40° will display the length of side *AC*, which is adjacent to angle *A*. (Side *AB* also seems to be adjacent to angle *A*. Since *AB* is already the hypotenuse, it can't be two things!) If you enter cos(40) on the calculator, it will display approximately 0.7660.

Trigonometric Expressions and Equations

MATH FACTS

In a right triangle with hypotenuse of length 1 unit, the length of a side opposite one of the acute angles is the sine of that angle. The length of a side adjacent to one of the acute angles is the cosine of that angle.

Example 2

If AB has length 1 and if angle B has a measure of 53°, find the lengths of sides AC and BC to the nearest hundredth.

Solution: Since AC is opposite to angle B, its length is sine 53°, which is approximately 0.80. Since BC is adjacent to angle B, its length is cosine 53°, which is approximately 0.60.

If the hypotenuse of a right triangle is known to be 1 and the length of either leg is known, it is possible to find one of the acute angles of the triangle with the $\sin^{-1}$ or the $\cos^{-1}$ function of the calculator.

In triangle ABC, hypotenuse AB has length 1 and leg BC has length 0.6293. Leg BC is opposite $\angle A$. So the measure of $\angle A$ can be found by

using the calculator to find which angle has a sine of 0.6293 by using the $\sin^{-1}$ function.
For the TI-84:
Press [2nd], [SIN] to access this feature.
For the TI-Nspire:
Press [trig] and [down] to access this feature.

Since $\sin^{-1}(0.6293) \approx 39°$ (make sure you are in degree mode), the measure of angle A is approximately 39°.

Example 3

In the triangle below, the length of AB is 1 and the length of AC is 0.5446. Rounded to the nearest degree, what is the measure of angle A?

Solution: Since AC is adjacent to angle A, the measure of angle A can be found by calculating $\cos^{-1}(0.5446)$. Rounded to the nearest degree, the angle is 57°.

Trigonometric Expressions and Equations

Even if the hypotenuse of the right triangle is not 1, cosine and sine can be used to find the lengths of the sides if one of the acute angles is known. In the triangle below, the hypotenuse is 5, angle A is 40 degrees, and the length of side BC is the unknown.

If the hypotenuse were 1, the length of BC would be $\sin 40° \approx 0.6428$. Since the hypotenuse is 5 times greater than 1, the side BC will be 5 times greater than 0.6428. Since $5(0.6428) \approx 3.2140$, this is the length of BC.

MATH FACTS

If the hypotenuse of the triangle has length r, the length of the side opposite angle A will be $r \cdot \sin \angle A$ and the length of the side adjacent to angle A will be $r \cdot \cos \angle A$.

To find an unknown acute angle when one of the legs is known and the hypotenuse is not 1, first divide the lengths of the known sides by the length of the hypotenuse. This changes the triangle into a similar triangle with hypotenuse 1. The angle can then be solved with $\cos^{-1}$ (if the adjacent side was known) or with $\sin^{-1}$ (if the opposite side was known).

Example 4

The length of AB is 8 and the length of BC is approximately 7.1904. What is the measure of $\angle A$ rounded to the nearest degree?

Solution: When the lengths of sides AB and BC are both divided by 8, which is the length of the hypotenuse, AB becomes 1 and BC becomes 0.8988. Since BC is the side opposite $\angle A$, the measure of $\angle A$ can be found with $\sin^{-1}$. Since $\sin^{-1}(0.8988)$ is approximately 64°, this is the measure of $\angle A$.

If in a right triangle only the two legs are known, the Pythagorean theorem can be used to find the hypotenuse first. Then either sine or cosine can be used to find the measure of the angle.

Example 5

The right triangle below has a vertex at (–3, 4). What is the measure of $\angle BOP$? What is the measure of angle AOP?

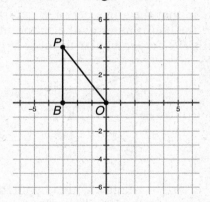

Solution: The length of OP can be calculated with the Pythagorean theorem $(-3)^2 + 4^2 = OP^2$. So $OP = 5$. Divide the three sides by 5, and triangle OPB becomes:

$$m\angle BOP = \sin^{-1}\left(\frac{4}{5}\right) \approx 53°$$

Since $\angle AOP$ is supplementary to $\angle BOP$, $m\angle AOP \approx 180° - 53° = 127°$.

Locating Points on the Unit Quarter Circle

The *unit quarter circle* is the part of a circle centered at (0, 0) with a radius of 1 and that is in the first quadrant.

If a radius is drawn so that the angle between radius *OP* and *OA* is 27°, it is possible to find the coordinates of the endpoint of the radius *P* by drawing line segment *PR* perpendicular to the *x*-axis.

Since triangle *ROP* is a right triangle with a hypotenuse of 1, the length of the sides can be determined with sine and cosine.
  $OR = \text{cosine } 27° \approx 0.89$
and
  $PR = \text{sine } 27° \approx 0.45$
  The coordinates of point *P* are (cosine 27°, sine 27°) ≈ (0.89, 0.45).

Example 6

In this unit quarter circle, angle *AOP* is 61°. What are the coordinates of point *P*?

  Solution: The coordinates of point *P* are (cosine 61°, sine 61°) ≈ (0.48, 0.87).

Locating Points on the Full Unit Circle

Even if ∠AOP is greater than 90°, the *x*-coordinate of point P will be the cosine of ∠AOP and the *y*-coordinate of point P will be the sine of ∠AOP.

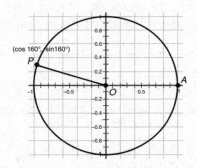

In the unit circle above, ∠AOP is 160°. The coordinates of point P are (cosine 160°, sine 160°). This can be approximated on the calculator as approximately (−0.94, 0.34).

This way of thinking about sine and cosine with a unit circle is related to, but different from, thinking about them with a right triangle.

If you reflect segment OP over the *y*-axis to become OP′, that acute ∠AOP′ would be 20°. The acute angle you get when you reflect point P into quadrant I is known as the *reference angle*.

If ∠AOP is greater than 360°, point P "wraps around" and will be in the same location as some other point that is less than 360°. For example, the point at 380° is at the same location as the point 380° − 360° = 20°. This is why the sine of 380° is the same as the sine of 20°.

Trigonometric Expressions and Equations

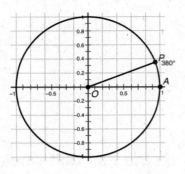

MATH FACTS

If circle O is a unit circle centered at (0, 0), point A is located at (1, 0), and point P is on the circle, the coordinates of point P are (cosine $\angle AOP$, sine $\angle AOP$). Depending on which quadrant P is in, sine, cosine, or both could be negative.

Example 7

If $\angle AOP$ is between 180° and 270°, what can be inferred about sine $\angle AOP$ and about cosine $\angle AOP$?

(1) They are both positive.
(2) Sine $\angle AOP$ is positive, but cosine $\angle AOP$ is negative.
(3) Sine $\angle AOP$ is negative, but cosine $\angle AOP$ is positive.
(4) They are both negative.

Solution: Choice (4) is correct. Every point in the quadrant III, including point P, has a negative x-coordinate and a negative y-coordinate. Since the x-coordinate is cosine $\angle AOP$ and since the x-coordinate is negative, cosine $\angle AOP$ must also be negative. Since the y-coordinate is sine $\angle AOP$ and since the y-coordinate is negative, sine $\angle AOP$ must also be negative.

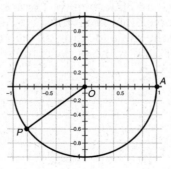

An alternate way to answer this question would be to choose some angle between 180° and 270°, such as 205°. Then enter cos(205) and sin(205) into the calculator (be sure you are in degree mode!) to see that they are both negative.

Example 8

Point P is on unit circle O with point A at $(1, 0)$. If the sine of $\angle AOP$ is negative and the cosine of $\angle AOP$ is positive, what is true about $\angle AOP$?

(1) It is between 0° and 90°.
(2) It is between 90° and 180°.
(3) It is between 180° and 270°.
(4) It is between 270° and 360°.

Solution: On the unit circle, the points that have a positive x-coordinate and a negative y-coordinate are in quadrant IV. The points on the unit circle in quadrant IV range from 270° to 360°, so the answer is choice (4).

Example 9

If sin $\angle AOP$ is negative, in what two quadrants can point P be in?

(1) I or II
(2) II or III
(3) III or IV
(4) I or IV

Solution: Since sine is related to the y-coordinate of point P and since points in quadrant III and IV have negative y-coordinates, the answer is choice (3).

If the coordinates of point P are known, it is possible to determine the measure of $\angle AOP$ with the graphing calculator. Calculating the angle's measure does differ a bit depending on what quadrant point P is in.

If point P is in quadrant I:

$\angle AOP = \sin^{-1}(|y\text{-coordinate}|)$ or $\cos^{-1}(|x\text{-coordinate}|)$

If point P is in quadrant II:

$\angle AOP = 180° - \sin^{-1}(|y\text{-coordinate}|)$ or $180° - \cos^{-1}(|x\text{-coordinate}|)$

If point P is in quadrant III:

$\angle AOP = 180° + \sin^{-1}(|y\text{-coordinate}|)$ or $180° + \cos^{-1}(|x\text{-coordinate}|)$

If point P is in quadrant IV:

$\angle AOP = 360° - \sin^{-1}(|y\text{-coordinate}|)$ or $360° - \cos^{-1}(|x\text{-coordinate}|)$

Trigonometric Expressions and Equations

Example 10

If point *P* has the coordinates (−0.8910, −0.4540), what is the measure of ∠*AOP* rounded to the nearest degree?

Solution: A sketch of the unit circle shows that the measure of the angle will be between 180° and 270°.

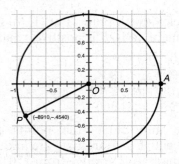

Since point *P* is in quadrant III, the angle can be calculated with either $180° + \sin^{-1}(0.4540)$ or $180° + \cos^{-1}(0.8910)$. Either way, the answer is 207°.

Notice that with the process described above, you should always put a positive number into the $\sin^{-1}$ or $\cos^{-1}$ function, even if the *x*-coordinate or *y*-coordinate is negative.

Example 11

If sin ∠*AOP* is 0.6 and cos ∠*AOP* is negative, what is the value of cos ∠*AOP*?

Solution: Since the sine is positive, point *P* must be in either quadrants I or II. Since cosine is negative, point *P* must be in either quadrants II or III. So point *P* must be in quadrant II since that is where both sine is positive and cosine is negative.

∠*AOP* can be calculated with the formula
$180° - \sin^{-1}(|y\text{-coordinate}|) = 180° - \sin^{-1}(0.6) \approx 143°$. Then use the calculator to find cosine $143° \approx -0.8$.

Example 12

If $\sin \angle AOP$ is $-\dfrac{5}{13}$ and if $\cos \angle AOP$ is positive, what is the value of $\cos \angle AOP$?

(1) $\dfrac{12}{13}$

(2) $-\dfrac{12}{13}$

(3) $\dfrac{5}{13}$

(4) $-\dfrac{5}{13}$

Solution: Since the sine is negative, point P must be in either quadrants III or IV. Since cosine is positive, point P must be in either quadrants I or IV. So point P must be in quadrant IV since that is where both sine is negative and cosine is positive.

$\angle AOP$ can be calculated with the formula

$360° - \sin^{-1}(|y\text{-coordinate}|) = 360° - \sin^{-1}\left(\dfrac{5}{13}\right) \approx 337°.$ Then use the

calculator to get cosine $337° \approx 0.9205$. Of the choices that are positive, $\dfrac{12}{13} \approx 0.9231$, which is very close to 0.9205. The answer is choice (1).

Trigonometric Expressions and Equations

Example 13

If the coordinates of A are $(-4, 0)$, of O are $(0, 0)$, and of P are $(-3, 4)$, what is the cosine of $\angle AOP$?

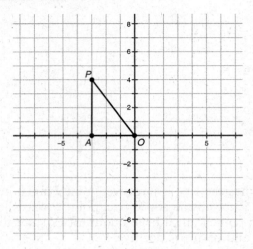

Solution: The sine of $\angle AOP$ is the x-coordinate of the point of intersection (Q) between the unit circle and the line OP. The lengths of the sides of triangle AOP are 3, 4, and 5. Since the hypotenuse OP has a length of 5, dividing all the sides by 5 will create a new triangle with a hypotenuse of 1. The lengths of the sides of similar triangle BOQ are $\frac{3}{5}$, $\frac{4}{5}$, and 1. Point Q has coordinates $\left(-\frac{3}{5}, \frac{4}{5}\right)$. So $\cos AOP = -\frac{3}{5}$.

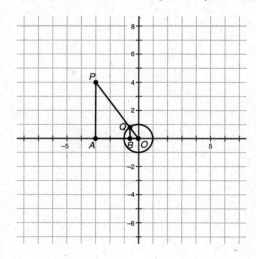

170

Check Your Understanding of Section 5.1

A. Multiple-Choice

1. In right triangle ABC, if $AC = \dfrac{5}{13}$ and $AB = 1$, what is the length of BC?

   (1) $\dfrac{5}{12}$ (2) $\dfrac{12}{13}$ (3) $\dfrac{13}{12}$ (4) $\dfrac{12}{5}$

2. In right triangle ABC, the measure of $\angle A$ is 25° and $AB = 1$. What is the length of AC?

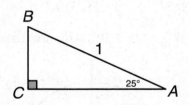

   (1) 0.9063   (2) 0.4226   (3) 0.2500   (4) 0.2778

Trigonometric Expressions and Equations

3. In right triangle DEF, the measure of $\angle D$ is 53° and $DE = 1$. What is the length of EF?

(1) 0.6018   (2) 0.5889   (3) 0.8000   (4) 0.7986

4. In right triangle EFG, if $GE = 0.7547$ and if $FG = 1$, what is the measure of $\angle G$?

(1) 31°   (2) 37°   (3) 41°   (4) 49°

5. In unit circle O, if $\angle AOB = 70°$, what are the coordinates of B?

(1) (0.94, 0.34)   (3) (0.2, 0.7)
(2) (0.34, 0.94)   (4) (0.4, 0.6)

6. In unit circle O, if $\angle AOB = 110°$, what are the coordinates of B?

(1) (0.94, −0.34)
(2) (−0.2, 0.7)
(3) (−0.41, 0.73)
(4) (−0.34, 0.94)

7. If B is on the unit circle, the x-coordinate of B is positive, and the y-coordinate of B is negative, in which quadrant is point B?

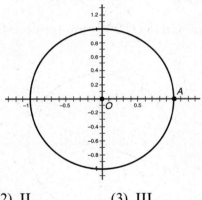

(1) I   (2) II   (3) III   (4) IV

Trigonometric Expressions and Equations

8. In unit circle O, the coordinates of point B are $(-0.91, -0.41)$. What is the measure of $\angle AOB$?

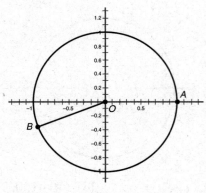

(1) 211°   (2) 237°   (3) 246°   (4) 204°

9. If $\sin \theta > 0°$ and $\cos \theta > 0$ and θ is in the standard position, in which quadrant is the terminal ray of θ?
   (1) I   (2) II   (3) III   (4) IV

10. Point B is on unit circle O. What are the coordinates of $\angle \theta$?

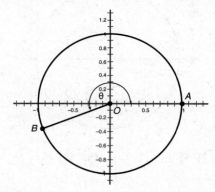

(1) $(-\cos \theta, -\sin \theta)$ (3) $(\cos \theta, -\sin \theta)$
(2) $(-\cos \theta, \sin \theta)$ (4) $(-\sin \theta, \cos \theta)$

B. *Show how you arrived at your answers.*

1. Right triangle *ABC* is similar to triangle *DEF*.

   a) What is the length of *AC*?
   b) What is the length of *DF*?

2. In right triangle *EFG* with hypotenuse $EG = 8$, what are the lengths of segments *EF* and *FG*?

3. What is the measure of $\angle AOB$?

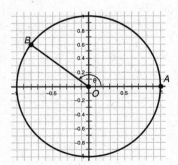

4. What are three angles between 90° and 360° that have a reference angle of 25°?

5. If $\sin \theta = -0.4848$ and $\theta < 270°$, what is the value of $\cos \theta$?

Trigonometric Expressions and Equations

5.2 TRIGONOMETRY EQUATIONS

KEY IDEAS

A trigonometry equation (often abbreviated as "trig equation") is one where the variable to be solved for is an angle. Trig equations often have multiple solutions, though it is also possible for them to have one or zero solutions.

The equation $\sin x = 0.5$ is an example of a trig equation that has two solutions between 0° and 360°. If you enter sine 30° into a calculator, it will say 0.5. If you, instead, enter sine 150° it will also say 0.5. So the solution set for this equation is {30°, 150°}.

Approximating Solutions to Trig Equations with the Unit Circle

The unit circle explains why there are generally two answers to trig equations. On the axes below is the unit circle and the horizontal line $y = 0.5$. The points on the circumference of the circle are 10 degrees apart from one another. As the horizontal line intersects the circle twice, there are two points on the circle that have a y-coordinate of 0.5, called P_1 and P_2. Also notice that angle BOP_2 is the same measure as angle AOP_1 because P_2 is a reflection over the y-axis of P_1. Angle AOP_1 is a 30° angle, so $\sin 30°$ is 0.5. Angle AOP_2 is a 150° angle, so sine 150° is also 0.5.

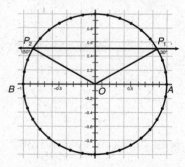

The equation $\sin x = -0.5$ also has two solutions. The line $y = -0.5$ intersects the unit circle at P_1 in quadrant III and at P_2 in quadrant IV. These points correspond to the angles 210° and 330°, respectively.

To use the unit circle to approximate trig equations involving cosine, a vertical line is needed instead.

Example 1

Use the unit circle below to approximate the two solutions (to the nearest 10°) to the equation $\cos x = 0.75$.

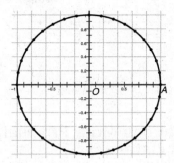

Solution: Points P_1 and P_2 both should have x-coordinates of 0.75 since the x-coordinate of a point on the unit circle is the cosine of the angle. Angle AOP_1 is approximately 40°, while angle AOP_2 is approximately 320°.

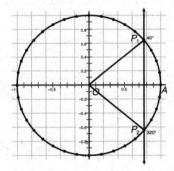

In cases where the vertical or horizontal line does not intersect the circle at all, the trig equation is said to have no solution. If the vertical or horizontal line is tangent to the circle, touching it at just one point, the trig equation has just one solution.

Trigonometric Expressions and Equations

Solving Trig Equations More Precisely with a Unit Circle and a Calculator

Finding the two solutions between 0° and 360° to a trig equation like $\sin x = -0.94$ requires four steps.

Step 1:

Sketch the proper vertical or horizontal line on the unit circle. For this example, it is a horizontal line at $y = -0.94$.

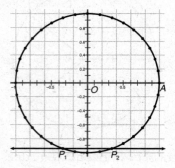

Step 2:

Identify the quadrants for any points of intersection between the unit circle and the line from step 1. For this example, point P_1 is in quadrant III and P_2 is in quadrant IV.

Step 3:

Determine the reference angle by taking the $\sin^{-1}$ or the $\cos^{-1}$ of the absolute value of the number after the equal sign. For this example, the reference angle is $\sin^{-1}(+0.94) \approx 70°$.

Step 4:

Depending on which quadrants the intersection points are in, use the following formulas.

- If point P is in quadrant I, one solution is x = reference angle.
- If point P is in quadrant II, one soution is $x = 180° -$ reference angle.
- If point P is in quadrant III, one solution is $x = 180° +$ reference angle.
- If point P is in quadrant IV, one solution is $x = 360° -$ reference angle.

For this example, since P_1 is in quadrant III and P_2 is in quadrant IV, the two solutions are $x \approx 180° + 70° = 250°$ and $x \approx 360° - 70° = 290°$.

Example 2

Use the unit circle below and the calculator to find the two solutions, to the nearest degree, between 0° and 360° to the trig equation $\cos x = -0.57$.

Solution: The sketch indicates that there are two points on the unit circle that have x-coordinates of -0.57. P_1 is in quadrant II, and P_2 is in quadrant III.

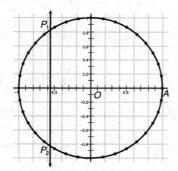

The reference angle is $\cos^{-1}(+0.57) \approx 55°$.

The two solutions are $x \approx 180° - 55° = 125°$ and $x \approx 180° + 55° = 235°$.

Check Your Understanding of Section 5.2

A. *Multiple-Choice*

For questions 1, 2, and 3, use this unit circle diagram.

1. Based on the unit circle diagram, find the solution(s) to the equation $\sin x = 0.8$.
   - (1) 35°, 325°
   - (2) 55°, 305°
   - (3) 55°
   - (4) 55°, 125°

2. Based on the unit circle diagram, find the solution(s) to the equation $\cos x = -0.6$.
   - (1) 55°
   - (2) 125°, 235°
   - (3) −55°
   - (4) 235°, 305°

3. Based on the unit circle diagram, find the solution(s) to the equation $\sin x = -1$.
   - (1) 270°
   - (2) 180°
   - (3) 90°
   - (4) 0°

4. Find the solution(s) to $\cos x = 0.5446$.
   - (1) 57° only
   - (2) 57° and 237°
   - (3) 57° and 123°
   - (4) 57° and 303°

5. Find the solution(s) to $\cos x = -0.5446$.
   - (1) 237° only
   - (2) 237° and 303°
   - (3) 57° and 303°
   - (4) 123° and 237°

6. If 0° ≤ x < 360°, what is the maximum number of solutions to the equation sin x = a where a is a real number?
   (1) 0 solutions
   (2) 1 solution
   (3) 2 solutions
   (4) 3 solutions

7. Find the solution(s) to cos x = −1 for 0° ≤ x < 360°.
   (1) 90°
   (2) 0°
   (3) 270°
   (4) 180°

8. At what value(s) of x does cos x = sin x?
   (1) 45°
   (2) 225°
   (3) 45° and 225°
   (4) 45° and 135°

9. If a° is one solution to sin x = k where k > 0, what is another solution?
   (1) 180° − a°
   (2) 180° + a°
   (3) 360° − a°
   (4) 360° + a°

10. What is the solution set for sin x = −1.5?
    (1) {215°}
    (2) {215°, 325°}
    (3) {325°}
    (4) {}

B. Show how you arrived at your answers.

1. On this unit circle, draw the two angles that have a sine of 0.7.

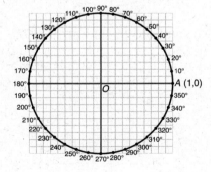

2. On this unit circle, draw the two angles that have a cosine of −0.4.

3. How can this unit circle be used to demonstrate that $\sin x = 2$ has no solutions?

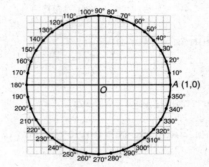

4. What are the two solutions to $\sin x = -0.8192$?

5. For which values of x, $0° \leq x < 360°$, is $\sin x > \cos x$?

5.3 RADIAN MEASURE

KEY IDEAS

Just as length can be measured in different units like feet or meters, angles can also be measured in different units. In addition to degrees, angles are sometimes measured in a unit called *radians*. One radian is equal to approximately 57.3°.

The circumference of the unit circle is $2 \cdot \pi \cdot 1 = 2\pi \approx 6.28$ units. In the diagram below, the edge of the unit circle is divided into 6 arcs each of length 1 unit and divided into 1 arc of length approximately 0.28 units. The 6 larger angles are each 1 radian. Since there are approximately 6.28 radians in a circle, the conversion factor is that 6.28 radians is approximately 360° making the approximate number of degrees in 1 radian $\frac{360°}{6.28} \approx 57.3°$.

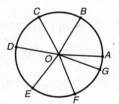

The *exact* number of radians in a circle is 2π. So the exact number of degrees in 1 radian is $\frac{360°}{2\pi} = \frac{180°}{\pi}$.

MATH FACTS

To convert radian measure to degrees, multiply the number of radians by $\frac{180°}{\pi}$.

Trigonometric Expressions and Equations

Example 1

Convert 3.49 radians to degrees, rounded to the nearest degree.

Solution:

$$3.49 \cdot \frac{180°}{\pi} \approx 200°$$

On the TI-84, access the π button by pressing [2ND] [^]. On the TI-Nspire, there is a [pi] button that opens a menu of 10 math symbols, including π.

Example 2

Convert $\frac{2\pi}{3}$ radians to degrees. Give an exact answer.

Solution:

$$\frac{2\pi}{3} \cdot \frac{180°}{\pi} = \frac{360°\pi}{3\pi} = 120°$$

MATH FACTS

To convert degrees to radians, divide by $\frac{180°}{\pi}$. Since dividing by a fraction can be accomplished by multiplying by the reciprocal of that fraction, it is instead quicker to multiply by $\frac{\pi}{180°}$.

Example 3

Convert 50° to radians, rounding to the nearest hundredth of a radian.

Solution:

$$50° \cdot \frac{\pi}{180°} \approx 0.87 \text{ radians}$$

Example 4

Convert 135° to radians. Give an exact answer.

Solution:

$$135° \cdot \frac{\pi}{180°} = \frac{135°\pi}{180°} = \frac{3\pi}{4} \text{ radians}$$

184

Some common radian to degree conversions are shown in the table.

Degrees	Radians
360°	2π
180°	π
90°	$\dfrac{\pi}{2}$
60°	$\dfrac{\pi}{3}$
45°	$\dfrac{\pi}{4}$
30°	$\dfrac{\pi}{6}$

Angles that are multiples of 30°, 45°, 60°, and 90° can be found by multiplying the radian equivalent of each angle by the appropriate factor. For example, since 270° = 3 · 90°, in radians $270° = 3 \cdot \dfrac{\pi}{2} = \dfrac{3\pi}{2}$.

Converting Radians and Degrees with a Calculator

The TI-84 and the TI-Nspire both have the ability to convert from degrees to radians or from radians to degrees.
For the TI-84:
   To convert radians to degrees, first set the calculator to degree mode. Then enter the number of radians to convert. Put this number in parentheses if it is something involving a fraction. Then press [2ND], [APPS] to get to the ANGLE menu. Press [3] for "r" to convert to radians and press [ENTER].

Trigonometric Expressions and Equations

To convert degrees to radians, first set the calculator to radian mode. Enter the number of degrees. Then go to the ANGLE menu. Select degrees by pressing [1] and then press [ENTER].

 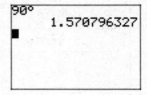

For the TI-Nspire:

To convert radians to degrees, enter the number of radians. Then press [ctrl], [catalog], and select the radian symbol. Then press [catalog], [1], [D], and select the DD function and press [enter].

 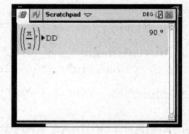

To convert degrees to radians, enter the number of degrees. Then press [ctrl], [catalog], and select the degree symbol. Then press [catalog], [1], [R], and select the Rad function and press [enter].

 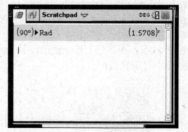

Converting Expressions in Terms of π to Radians

When using the calculator to convert degrees to radians, it will give a decimal approximation of the solution instead of an exact expression involving π. To rewrite as an expression involving π, divide the answer by π. The quotient will be the number that goes in front of the π in the exact answer.

To convert 150° to radians in terms of π on a calculator:

So 150° is equal to $\frac{5}{6}\pi$ radians.

Check Your Understanding of Section 5.3

A. Multiple-Choice

1. Convert 30° to radians.
   (1) $\frac{1}{6}$ (2) $\frac{\pi}{4}$ (3) $\frac{\pi}{6}$ (4) $\frac{\pi}{3}$

2. Convert 240° to radians.
   (1) $\frac{4}{3}$ (2) $\frac{5\pi}{6}$ (3) $\frac{4\pi}{3}$ (4) $\frac{5\pi}{4}$

3. Convert $\frac{\pi}{4}$ radians to degrees.
   (1) 90° (2) 60° (3) 30° (4) 45°

4. Convert $\frac{7\pi}{6}$ radians to degrees.
   (1) 200° (2) 205° (3) 210° (4) 215°

5. Convert 41° to radians, rounded to the nearest hundredth of a radian.
   (1) 0.66 (2) 0.68 (3) 0.70 (4) 0.72

6. Convert 308° to radians, rounded to the nearest hundredth of a radian.
   (1) 5.38 (2) 5.40 (3) 5.42 (4) 5.44

Trigonometric Expressions and Equations

7. Convert $\frac{1}{6}$ radians to degrees, rounded to the nearest tenth of a degree.
   (1) 30.0°   (2) 29.6°   (3) 19.6°   (4) 9.5°

8. Convert 2 radians to degrees, rounded to the nearest tenth of a degree.
   (1) 114.6°   (2) 94.6°   (3) 84.6°   (4) 360.0°

9. If circle O has a radius of 1 inch and $m\angle AOB$ is 1.3 radians, what is the length of arc $\widehat{AB}$?

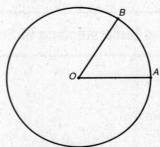

   (1) 1 inch   (2) 1.3 inches   (3) 1.6 inches   (4) 1.9 inches

10. In unit circle O, if the length of arc $\widehat{AB}$ is $\frac{\pi}{3}$, what are the coordinates of point B?

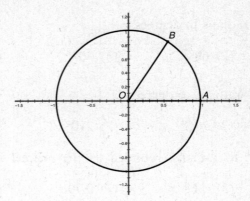

   (1) $\left(\frac{\pi}{6}, \frac{\pi}{3}\right)$
   (2) $\left(\frac{\sqrt{2}}{2}, \frac{\sqrt{2}}{2}\right)$
   (3) $\left(\frac{\sqrt{3}}{2}, \frac{1}{2}\right)$
   (4) $\left(\frac{1}{2}, \frac{\sqrt{3}}{2}\right)$

B. *Show how you arrived at your answers.*

1. Is 1 radian bigger or smaller than 60°? Explain your reasoning.

2. What is $\sin \frac{5\pi}{3}$ radians?

3. Arrange from smallest to biggest: 1 radian, 1 right angle, 1 degree.

4. In unit circle O, if $\angle AOB$ is an acute angle measured in radians, what is bigger: $\angle AOB$ or $\sin \angle AOB$?

5. Some calculators in addition to degrees and radians have a "gradians" mode. A gradian is a unit of angle measurement where 400 gradians are in a circle. How many gradians are in $\frac{\pi}{2}$ radians?

5.4 GRAPHS OF THE SINE AND COSINE FUNCTIONS

KEY IDEAS

When a graph is created in which the *x*-coordinate is the measure of an angle and the *y*-coordinate is either the sine or the cosine of that angle, the graph has the shape of a wave. This wave shape can be transformed in different ways to make waves that are larger, more compressed, or translated up or down.

Graphs of the Sine and Cosine Functions

Each point on the unit circle conveys three pieces of information: the angle AOP, the cosine of $\angle AOP$ (the *x*-coordinate of P), and the sine of $\angle AOP$ (the *y*-coordinate of P).

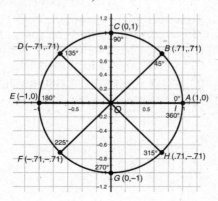

The information from this unit circle can be summarized in a chart.

Point	Angle	Cosine(x) (x-coordinate)	Sine(x) (y-coordinate)
A	0°	1	0
B	45°	0.71	0.71
C	90°	0	1
D	135°	−0.71	0.71
E	180°	−1	0
F	225°	−0.71	−0.71
G	270°	0	1
H	315°	0.71	−0.71
I	360° = 0°	1	0

To graph the sine function $f(x) = \sin(x)$, graph the nine points with the angle measure as the x-coordinate and the $\sin(x)$ as the y-coordinate. The final graph will resemble a backward "S" that is tilted on its side.

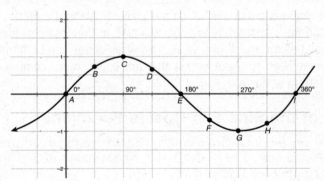

To graph the cosine function $g(x) = \cos(x)$, graph the nine points with the angle measure as the x-coordinate and the $\cos(x)$ as the y-coordinate. The final graph will resemble a backward "S" that is tilted on its side. In a cosine curve from 0° to 360°, the graph starts at a peak. In contrast in a sine curve, the graph starts at a point in the middle.

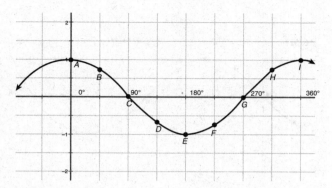

These graphs can also be created on a graphing calculator.
For the TI-84:
　Press [Y=] and enter sin(x) after the Y1=. Press [ZOOM], [7].

For the TI-Nspire:
　From the home screen, press [B] to get to the graph scratchpad. If the entry line is not visible, press [tab]. Enter sin(x) after f1(x) = and press [enter]. To get the sine, press the [trig] button. Press [menu], [4], [8] to get a better viewing window.

Since the window is set from −360 to +360, these calculator images show two cycles of the sine curve.
　If the angle is in radians, the graph will be the same as when the angle is in degrees. However, the scale on the x-axis will be different. The end of each cycle will be at 2π instead of at 360°. Instead of 0°, 90°, 180°, 270°, and 360°, the angles will be at 0 r, $\frac{\pi}{2}$ r, π r, $\frac{3\pi}{2}$ r, and 2π r. These values correspond to approximately 0 r, 1.57 r, 3.14 r, 4.71 r, and 6.28 r.

Trigonometric Expressions and Equations

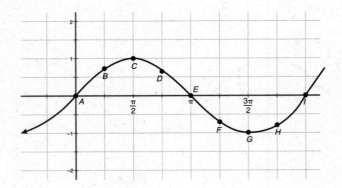

The Amplitude of Sine and Cosine Graphs

In a function like $g(x) = 3\sin(x)$, the number in front of the trig expression is known as the *amplitude*. This coefficient transforms the graph with a *vertical stretch*. The highest point on the graph is now (90°, 3) instead of (90°, 1). The lowest point on the graph is now (270°, −3) instead of (270°, −1).

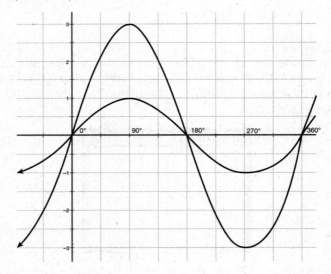

If the coefficient is negative, there is also a reflection of the curve over the x-axis. Here is the graph of $h(x) = -3\sin(x)$.

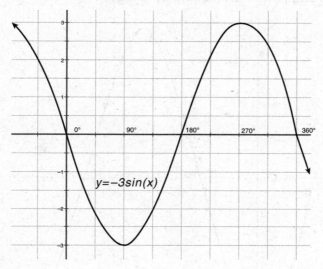

Example 1

Sketch the graph of $f(x) = \cos(x)$ and $g(x) = 2\cos(x)$ on the same set of axes from $0° \leq x \leq 360°$.

Solution:

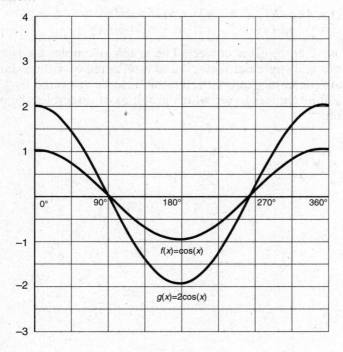

Trigonometric Expressions and Equations

Example 2

This is the graph of which equation?

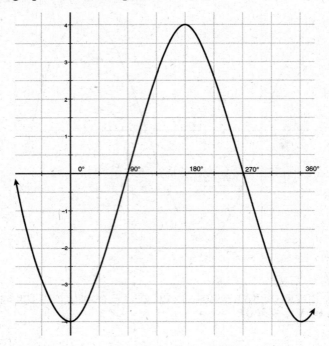

(1) $f(x) = 4\cos(x)$ (3) $f(x) = 4\sin(x)$
(2) $f(x) = -4\cos(x)$ (4) $f(x) = -4\sin(x)$

Solution: Choice (2) is correct. The graph resembles a cosine curve that has a vertical stretch of factor 4 and is reflected over the *y*-axis. Since this is a multiple-choice question, it could also be answered by graphing each of the four choices to see which graph most resembles the graph in the question.

Vertical Translations of Sine and Cosine Graphs

In a function like $g(x) = \sin(x) + 2$, the constant term affects the graph. The graph of $g(x) = \sin(x) + 2$ is like the graph of $f(x) = \sin(x)$ with each point translated up by 2 units. The invisible horizontal line through the middle of the curve, sometimes called the *axis* of the curve, is also translated up 2 units.

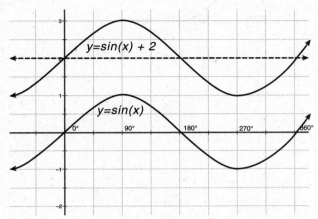

If the constant is negative, the graph will be shifted down.

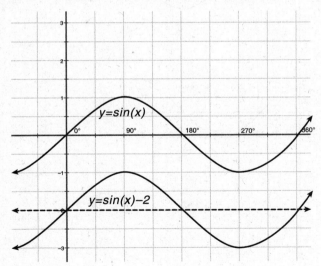

Example 3

Which of the following is the graph of $g(x) = \cos(x) - 1$?

(1)

(2)

(3)

(4)

Solution: Choice (4) is the answer. It is the graph of $f(x) = \cos(x)$ shifted down by 1 unit.

The Frequency of Sine and Cosine Graphs

The graph of the function $g(x) = \sin(2x)$ is like the graph of $f(x) = \sin(x)$ but with a *horizontal squeeze* by a factor of 2. The coefficient of the x, the 2 in this example, is known as the *frequency* since it is the number of complete waves that fit inside a 360° interval.

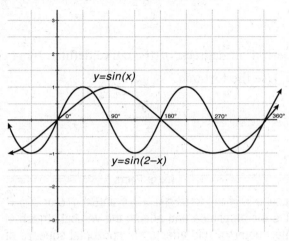

Example 4

Below is the graph of which function?

(1) $g(x) = 3\sin(x)$ (3) $g(x) = \sin(3x)$
(2) $g(x) = \sin(x) + 3$ (4) $g(x) = \cos(3x)$

Trigonometric Expressions and Equations

Solution: Since 3 cycles of the sine curve fit in a 360° interval, the coefficient of the *x* is 3. Choice (3) is correct.

The Period of Sine and Cosine Graphs

The *period* of a sine or cosine curve is the number of degrees (or radians) needed for the curve to complete one cycle. The period is related to the frequency (*B*) by the formula $P = \dfrac{360°}{B}$ for degrees or $P = \dfrac{2\pi}{B}$ for radians.

Example 5

What is the period, in degrees, of the curve defined by $g(x) = \cos(4x)$?

Solution: Use the formula.

$$P = \frac{360°}{4} = 90°$$

========== **MATH FACTS** ==========

When a function is of the form $g(x) = \pm A \sin(Bx) + C$ or $g(x) = \pm A \cos(Bx) + C$, the *A* is the amplitude and affects the vertical stretch of the curve. The *B* is the frequency and affects the horizontal squeeze of the curve. The *C* affects the vertical translation of the curve. The basic curves $g(x) = \sin(x)$ or $g(x) = \cos(x)$ have values $A = 1$, $B = 1$, and $C = 0$. A function can change one or more of the values to create wavelike curves that pass through different points

---

Graphs of Trig Functions with More than One Transformation

The graph of the function $g(x) = -3\sin(2x) + 1$ is like the graph of $f(x) = \sin(x)$ after four different transformations have been performed on it. The 3 causes a vertical stretch. The negative sign (−) in front of the 3 causes a reflection over the *x*-axis. The 2 causes a horizontal squeeze. The + 1 causes a vertical translation.

The graphing calculator will graph the transformed function.

Example 6

Which of the following is the graph of $g(x) = 2\cos(3x) + 2$?

Solution: Choice (2) is the answer. It has an amplitude of 2, a frequency of 3, and a vertical shift of 2. It is cosine instead of sine because it starts at a peak instead of in the middle of the curve.

Using the Graphing Calculator to Make a Table of Values

With a graphing calculator, you can quickly display a graph of a complicated trig function on the screen. However, if the question asks you to reproduce the graph accurately on graph paper, a table of values is useful. The table is best if it contains the entries that correspond to the five most important points on the trig graph.

Here are the steps for making a useful table of values for graphing the function $g(x) = -3\sin(2x) + 1$ on paper.

For the TI-84:

Press [Y=] and enter $3\sin(2x) + 1$ after Y1=. Press [2ND] [WINDOW] to open the table setup menu. Set "TblStart" to 0. Since the B-value is 2, set delta Tbl to (360/2)/4. (If you are in radian mode, do (2pi/2)/4 instead). The denominator of the fraction with 360 in the numerator should be the B-value. The independent and dependent settings should both be on "Auto." Press [2nd] [GRAPH] to see the table.

For the TI-Nspire:

In the graph scratchpad, enter $3\sin(2x) + 1$ after $f1(x) =$ in the entry line, and press [enter]. To see the table, hold down [ctrl] and press [T]. Press [menu], [2], [5] to edit the table settings. Make the table start value 0. Since the B-value is 2, set the table step value (360/2)/4 (or for radian mode (2pi/2)/4)) and press the OK button.

Finding the Equation of a Function from a Given Graph

There are five decisions to make when finding the equation of a function on which a trig graph was defined.

1) Is it a sine or a cosine function?
2) What is the vertical shift (the C-value)?
3) What is the amplitude (the A-value)?
4) What is the frequency (the B-value)?
5) Is the coefficient of A positive or negative?

These five questions can be answered in any order.

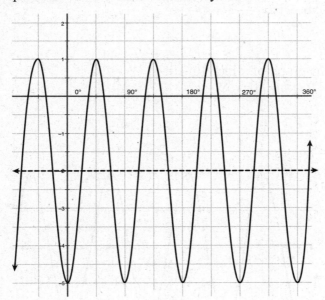

Step 1:

Since the y-intercept of this curve is not in the middle curve but is at one of the high or low points, this is a cosine curve.

Step 2:

Find the value of C, which is the vertical shift. Since the high point is at $y = +1$ and the low point is at $y = -5$, halfway between that is $y = -2$, which is the axis of the curve. So $C = -2$.

Trigonometric Expressions and Equations

Step 3:

Since the distance from the axis to one of the peaks or low points is 4 units, the amplitude is 4. This means $A = 4$.

Step 4:

Since the first cycle of this curve ends at the 90° point, the period is 90°. The formula that relates period (P) and frequency (B) is $P = \dfrac{360°}{B}$. So $90° = \dfrac{360°}{B}$ or $90°B = 360°$. So $B = 4$.

Step 5:

A cosine curve with a negative A-value starts from a low point and goes up to the right.
The equation is $y = -3\cos(4x) - 2$.

Example 7

Below is the graph of which function?

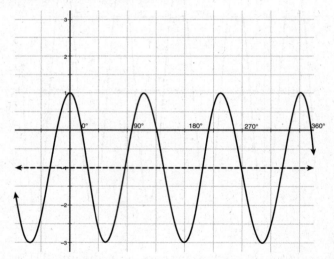

(1) $g(x) = 2\cos(3x) - 1$
(2) $g(x) = 3\cos(2x) - 1$
(3) $g(x) = 1\cos(3x) - 2$
(4) $g(x) = 2\cos(x) - 3$

Solution: Choice (1) is the correct answer. The amplitude (A) is 2. There are 3 (B) curves in the 360° interval. Since the axis is at $y = -1$, the curve has a vertical translation of -1 (C).

Example 8

The angles for this graph are in radians. What is the equation of the function on which this graph is based?

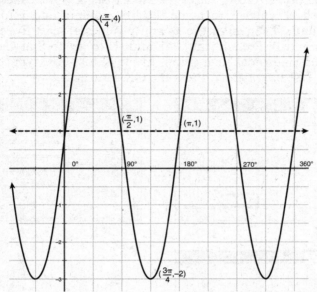

Solution: $f(x) = 3\sin(2x) + 1$. The C-value is 1 since the axis is at $y = 1$. The A-value is +3 since the high point is 3 units above the axis and the sine curve goes up first. To calculate the B-value, solve $P = \dfrac{2\pi}{B}$ with $P = \pi$. So $\pi = \dfrac{2\pi}{B}$, $B\pi = 2\pi$, $B = 2$.

When the angles are in radians, you can use the sinusoidal regression feature of a graphing calculator to find the equation for you.
For the TI-84:

Press [STAT], [1] and enter five points from the graph. The simplest points are the high and low points and also the points that are on the horizontal line that goes through the middle of the curve. The x-coordinates go into L1, and the y-coordinates go into L2. You can use the [2nd][^] to make a π, and it will turn it into a decimal in the list. Press [STAT], [RIGHT ARROW], and scroll down to [C] for SinReg. Then press [ENTER]. The calculator will tell you values of a, b, c, and d for the equation $y = a \cdot \sin(bx + c) + d$.

Trigonometric Expressions and Equations

The $C = -8.80942\text{E}-15$ is -0.00000000000000880942, which is the way the calculator is representing 0.

For the TI-Nspire:

From the home screen, select the Add Lists & Spreadsheet icon. On the spreadsheet, enter x to name the first column and y to name the second column. Then enter the x- and y-values at least five points into columns A and B. Press [menu], [4], and [1] to get to Stat Calculations. Press [C] for Sinusoidal Regression. In the X List box, type x.

In the y List box, type y. Press the OK button.

The equation is $y = 3\sin(2x) + 1$.

Example 9

If the angles are in radians, find the function on which the following graph is based.

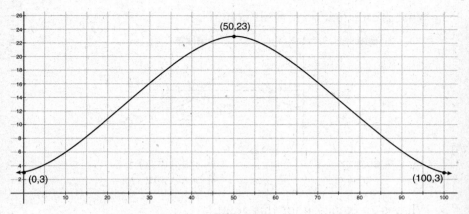

Solution: This is a cosine curve with a negative *A*-value. *A* is −10, and *C* is 13. The *B*-value can be calculated by $100 = \frac{2\pi}{B}$, $100B = 2\pi$, $B = \frac{2\pi}{100} = \frac{\pi}{50}$. So the equation is

$$f(x) = -10\cos\left(\frac{\pi}{50}x\right) + 13$$

Check Your Understanding of Section 5.4

A. Multiple-Choice

1. What is an equation for this graph?

(1) $y = \cos(x)$
(2) $y = -\cos(x)$
(3) $y = \sin(x)$
(4) $y = -\sin(x)$

2. What is an equation for this graph?

(1) $y = \cos(x)$
(2) $y = -\cos(x)$

(3) $y = \sin(x)$
(4) $y = -\sin(x)$

3. What is an equation for this graph?

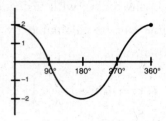

(1) $y = \cos(2x)$
(2) $y = 2\cos(x)$

(3) $y = \sin(2x)$
(4) $y = 2\sin(x)$

4. What is the graph of $y = -3\sin(x) + 1$?

(1)

(3)

(2)

(4)

5. What is the y-coordinate of one of the maximum points of $y = 2\cos(x) - 1$?
   (1) 3 (2) 2 (3) 1 (4) 0

6. Which of the following is the graph of $y = \sin(2x)$?

7. What is the period, in degrees, of the graph of $y = 4\sin\left(\dfrac{2}{3}x\right) + 1$?
   (1) 540° (2) 360° (3) 270° (4) 240°

8. What is the period, in radians, of the graph of $y = 4\cos\left(\dfrac{\pi}{10}x\right) - 3$?
   (1) 5 (2) 10 (3) 15 (4) 20

9. What could be the equation for the graph if the angles were measured in degrees?

   (1) $y = -2\sin(3x) + 2$
   (2) $y = 2\sin(3x) + 2$
   (3) $y = -3\sin(2x) + 2$
   (4) $y = 3\sin(2x) + 2$

10. What is an equation for this graph if the angle is measured in radians?

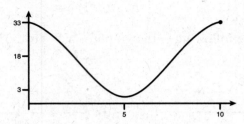

(1) $y = 15\sin\left(\dfrac{\pi}{5}x\right) + 18$ (3) $y = 15\cos(10x) + 18$

(2) $y = 18\cos(3x) + 15$ (4) $y = 15\cos\left(\dfrac{\pi}{5}x\right) + 18$

B. *Show how you arrived at your answers.*

1. Sketch one cycle of the graph of $y = 2\sin(x) - 1$.

2. Sketch one cycle of the graph of $y = -3\cos(2x) + 1$.

3. Create an equation for this graph with the angles in radians.

4. Create an equation for this graph with the angles in radians.

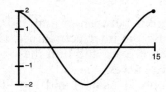

5. Sketch a graph of $y = -10\cos\left(\dfrac{\pi}{20}x\right) + 10$.

5.5 GRAPHICALLY SOLVING TRIG EQUATIONS

KEY IDEAS

Equations can be solved graphically by using the intersect feature of the graphing calculator. As long as the question doesn't specify that "only an algebraic solution will be accepted," using the graphical method is very quick and accurate.

Section 5.2 described a lengthy process for algebraically solving trig equations. These same problems can be solved very quickly with a graphing calculator.

To find the two solutions between $0°$ and $360°$ to $\sin x = -0.94$ using a graphing calculator, use the intersect feature. Graph each side of the equation on the same set of axes, and find the x-coordinate of the intersection points of the two curves.

For this example, graph $y = \sin(x)$ and $y = -0.94$ on the same axes. Make sure the calculator is set to degree mode and that the window ranges from a minimum x-value of 0 or less to a maximum x-value of 360 or more.

For the TI-84:

Press [2nd], [TRACE], [5], [ENTER], [ENTER] to select the sine curve and the horizontal line. Then move near one of the intersection points between 0 and 360. Press [ENTER] again. Do this again for the other intersection point.

For the TI-Nspire:

Press [menu], [6], [4]. Then move to a spot to the left of one of the intersection points between 0 and 360. Press [click]. Then move to a spot to the right of one of the intersection points. Press [click] again.

The process is the same if the angle is in radians, as is the case for the real-world scenarios in the next section. Just remember to change the x-values of the window so you can see from 0 to 2π.

Example

Find the two answers between 0 and 100 to the equation $18 = -10\cos\left(\dfrac{\pi}{50}x\right) + 13$ where the angle is in radians.

Solution: Set the calculator to radian mode. Then graph $y = 18$ and $y = -10\cos\left(\dfrac{\pi}{50}x\right) + 13$. The x-coordinates of the intersection points are approximately $x = 33$ and $x = 67$.

Trigonometric Expressions and Equations

Check Your Understanding of Section 5.5

A. Multiple-Choice

1. Graphically solve the equation $\sin(x) = 0.47$ for $0° \leq x < 360°$.
   (1) 28°
   (2) 28° and 332°
   (3) 28° and 208°
   (4) 28° and 152°

2. Graphically solve the equation $\cos(x) = -0.60$ for $0° \leq x < 360°$.
   (1) 127° and 233°
   (2) 53° and 127°
   (3) 53° and 307°
   (4) 127° and 307°

3. Graphically solve the equation $\sin(x) = 0.39$ for $0 \leq x < 2\pi$.
   (1) 0.4 and 2.7
   (2) 0.5 and 2.8
   (3) 0.6 and 2.9
   (4) 0.7 and 3.0

4. Graphically solve the equation $2\cos(3x) + 3 = 4$ for $0 \leq x < \dfrac{2\pi}{3}$.
   (1) 0.25 and 1.65
   (2) 0.30 and 1.70
   (3) 0.35 and 1.75
   (4) 0.40 and 1.80

5. Graphically solve the equation $15\cos\left(\dfrac{\pi}{5}x\right) + 17 = 30$, in radians, for $0 \leq x < 10$.
   (1) 0.6 and 9.4
   (2) 0.8 and 9.2
   (3) 1.0 and 9.0
   (4) 1.2 and 8.8

6. Graphically solve the equation $-30\cos\left(\dfrac{\pi}{6}x\right) + 70 = 80$, in radians, for $0 \leq x < 12$.
   (1) 3.4 and 8.6
   (2) 3.6 and 8.4
   (3) 3.8 and 8.2
   (4) 4.0 and 8.0

7. Graphically solve the equation $-40\sin\left(\frac{\pi}{15}x\right)+35=60$, in radians, for $0 \le x < 30$.
   (1) 18.2 and 26.8
   (2) 18.0 and 26.6
   (3) 17.8 and 26.4
   (4) 17.6 and 26.2

8. Graphically solve the equation $-30\cos\left(\frac{\pi}{25}x\right)+25=10$, in radians, for $0 \le x < 50$.
   (1) 8.1 and 41.9
   (2) 8.2 and 41.8
   (3) 8.3 and 41.7
   (4) 8.4 and 41.6

9. Graphically solve the equation $15\cos\left(\frac{\pi}{20}x\right)+10=3$, in radians, for $0 \le x < 40$.
   (1) 13 and 27
   (2) 12 and 28
   (3) 11 and 29
   (4) 10 and 30

10. How many solutions are there, for $0° \le x < 360°$, to the equation $\sin(2x) + 1 = \cos(x)$?
    (1) 0
    (2) 1
    (3) 2
    (4) 3

5.6 MODELING REAL-WORLD SCENARIOS WITH TRIG FUNCTIONS

KEY IDEAS

Many real-world scenarios can be modeled with trig functions. These include models based on spinning circles, bouncing springs, and weather patterns. Most things that go up, down, up, down in a repeating pattern can be modeled with a trig function involving sine or cosine. After creating a function to model a real-world scenario, that function can be used in an equation to calculate things about the model.

Trigonometric Expressions and Equations

The Ferris Wheel Problem

The most common, and easiest to follow, real-world scenario that can be modeled with a sine or cosine curve is the height above the ground of a car on a Ferris wheel.

The radius of this Ferris wheel is 10 feet, and the bottom of the Ferris wheel is 3 feet above the ground. The amount of time the Ferris wheel takes to make a complete revolution is 100 seconds.

When Evelyn gets on the Ferris wheel, she is exactly 3 feet above the ground. Then 50 seconds after starting, she will be 23 feet above the ground, at the peak of the Ferris wheel. Then 100 seconds after starting, she will be back to her starting position, again exactly 3 feet above the ground.

The graph of Evelyn's height above the ground will look like this.

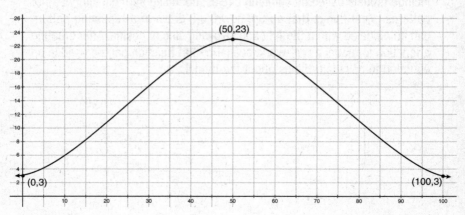

Using the methods from earlier in this chapter, the equation of this curve can be found to be $f(x) = -10\cos\left(\dfrac{\pi}{50}x\right) + 13$.

With this equation, two different kinds of questions can be answered.

1) How high above the ground will Evelyn be after a certain number of seconds?
2) After how many seconds will Evelyn be a certain number of feet above the ground?

To find how high above the ground Evelyn will be after 43 seconds, substitute 43 for x in the equation. $f(43) = -10\cos\left(\dfrac{\pi}{50} \cdot 43\right) + 13 \approx 22$ feet.

To find the two times between 0 and 100 when Evelyn will be 18 feet above the ground, solve the equation $18 = -10\cos\left(\dfrac{\pi}{50}x\right) + 13$. This can be done with algebra or with the graphing calculator. The graphing calculator approach is easiest.

By finding the x-coordinates of the intersection of the graphs of $y = 18$ and $y = -10\cos\left(\dfrac{\pi}{50}x\right) + 13$, you get $x \approx 33$ seconds and $x \approx 67$ seconds.

The Temperature Problem

Often the function for the scenario is already given and all that's left is solving equations with it. In this scenario, a table is provided that contains the temperature in New York over the course of two years.

Year	Temperature in NY (°F)										
2013	35.1	33.9	40.1	53.0	62.8	72.7	79.8	74.6	67.9	60.2	45.3	38.5
2014	28.6	31.6	37.7	52.3	64.0	72.5	76.1	74.5	69.7	59.6	45.3	40.5

A function that approximately models this data is $f(x) = -23\cos\left(\dfrac{\pi}{6}x\right) + 54$.

Trigonometric Expressions and Equations

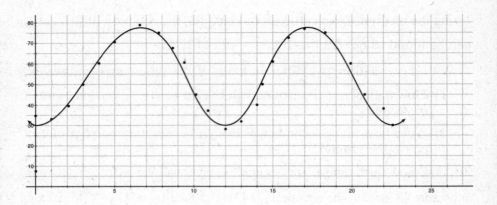

Use this function to answer the two types of questions.

Question 1: If the beginning of January is month $x = 0$, what is the temperature in the middle of March ($x = 2.5$)?

Substitute $x = 2.5$ into the function. $f(2.5) = -23\cos\left(\dfrac{\pi}{6} \cdot 2.5\right) + 54 \approx 48$ degrees

Question 2: At what two times is the temperature approximately 70 degrees?

Graph $y = 70$ and $y = -23\cos\left(\dfrac{\pi}{6}x\right) + 54$ on the graphing calculator. Find the x-coordinates of the two intersection points.

They intersect at $x = 4.5$ and $x = 7.5$. These x-values correspond to the middle of May and the middle of August.

Example

In New York on March 20 (which is the spring equinox), there are 12 hours of daylight and 12 hours of darkness. Approximately 91 days later on June 20 (which is the summer solstice), there are 15 hours of daylight and 9 hours of darkness. Approximately 91 days after that on September 22 (which is the fall equinox), there are 12 hours of daylight and 12 hours of darkness again. Approximately 91 days after that on December 21 (which is the winter solstice), there are 9 hours of daylight and 15 hours of darkness. Approximately 91 days after that, it is the spring equinox again.

Sketch the graph of a function $f(x)$ where x is the number of days that has passed since the spring equinox and $f(x)$ is the number of hours of daylight on that day. Find an approximate equation for this function.

Solution: The graph looks like this.

The equation is a sine curve with a positive A-value. The C-value is $+12$ since $y = 12$ is the axis of the sine curve. The A-value is $+3$ since the high point is 3 units above the axis and the low point is 3 units below the axis. The B-value can be solved with the equation $364 = \frac{2\pi}{B}$: $364B = 2\pi$, $B = \frac{2\pi}{364}$, $B \approx \frac{\pi}{182}$.

The function is approximately $f(x) = 3\sin\left(\frac{\pi}{182}x\right) + 12$.

Check Your Understanding of Section 5.6

B. *Show how you arrived at your answers.*

1. The height above ground of a Ferris wheel car can be modeled with the equation $h = -20\cos\left(\dfrac{\pi}{15}t\right) + 24$, where h is the height in feet and t is the time in seconds.
   a) How many seconds does it take for the Ferris wheel to make a complete revolution?
   b) What is the maximum height of the Ferris wheel?
   c) Sketch a graph that shows how the height of a Ferris wheel car relates to the amount of time since the beginning of the ride.

2. Matthew gets on a Ferris wheel at the bottom of the wheel, which is 5 feet above the ground. After 40 seconds, he is at the top of the wheel, 65 feet high.
   a) Sketch a graph that shows the relationship between time and height.

b) Create an equation that relates h and t.
c) Determine how high up Matthew will be after 55 seconds.
d) Determine the two times between 0 and 60 seconds when Matthew is 41 feet above the ground.

3. Below is the graph of the average temperature in Regentsville in each of the months where 0 = January, 1 = February, 2 = March, 3 = April, 4 = May, 5 = June, 6 = July, 7 = August, 8 = September, 9 = October, 10 = November, 11 = December, and 12 = next January.

a) Create an equation that relates temperature (y) and month (x).
b) Use that equation to find out the temperature in March ($x = 2$).
c) Use that equation to find the two times when the temperature is 70 degrees.

5.7 TRIGONOMETRY IDENTITIES

A *trig identity* is an equation in which the left side and the right side both contain trig expressions. Complicated trig expressions can sometimes be simplified by replacing parts of it with equivalent expressions and then simplifying the new expression.

MATH FACTS

The expression $(\sin x)^2$ is more commonly written as $\sin^2 x$. This is to prevent it from being confused with $\sin(x^2)$. The same applies to $(\cos x)^2$. It is more commonly written as $\cos^2 x$.

Trigonometric Expressions and Equations

There are five main trig identities.

1. $\sin^2 x + \cos^2 x = 1$ or

   $\sin^2 x = 1 - \cos^2 x$ or

   $\cos^2 x = 1 - \sin^2 x$

2. $\tan x = \dfrac{\sin x}{\cos x}$

3. $\csc x = \dfrac{1}{\sin x}$

4. $\sec x = \dfrac{1}{\cos x}$

5. $\cot x = \dfrac{1}{\tan x}$

Trig identities can be used to identify trig expressions that are equivalent to other trig expressions.

Rewrite All Expressions in Terms of Sine and Cosine

One way to simplify a trig expression is first to use the five basic identities described earlier to rewrite each expression in terms of sine and cosine. After this is done, you may be able to simplify further by reducing the fraction that remains.

The expression $\dfrac{\tan x}{\sec x}$ can be shown to be equivalent to sin x with the following steps:

$$\frac{\tan x}{\sec x}$$

$$\frac{\dfrac{\sin x}{\cos x}}{\dfrac{1}{\cos x}}$$

$$\frac{\sin x}{\cancel{\cos x}} \cdot \frac{\cancel{\cos x}}{1}$$

$$\sin x$$

222

Example 1

Show the steps to demonstrate that $\dfrac{\sec x}{\csc x}$ is equivalent to $\tan x$.

    Solution:

$$\frac{\sec x}{\csc x}$$

$$\frac{\dfrac{1}{\cos x}}{\dfrac{1}{\sin x}}$$

$$\frac{1}{\cos x} \cdot \frac{\sin x}{1}$$

$$\frac{\sin x}{\cos x}$$

$$\tan x$$

Example 2

Show that $\tan x \cdot \csc x = \sec x$.

    Solution:

$$\tan x \cdot \csc x$$

$$\frac{\sin x}{\cos x} \cdot \frac{1}{\sin x}$$

$$\frac{\cancel{\sin x}}{\cos x \cdot \cancel{\sin x}}$$

$$\frac{1}{\cos x}$$

$$\sec x$$

Trigonometric Expressions and Equations

Combining Fractions in Trig Identity Problems

Sometimes in a trig identity problem, fractions must be combined. In this case, a common denominator needs to be found. Often when the numerators are later combined, another trig identity helps further simplify the expression.

Example 3

Show that $\sin x \cdot \tan x + \cos x = \sec x$.

Solution:

$$\sin x \cdot \tan x + \cos x$$

$$\sin x \cdot \frac{\sin x}{\cos x} + \cos x$$

$$\frac{\sin^2 x}{\cos x} + \frac{\cos x}{1}$$

Here the common denominator for the two fractions is $\cos x$.

$$\frac{\sin^2 x}{\cos x} + \frac{\cos x \cdot \cos x}{\cos x}$$

$$\frac{\sin^2 x + \cos^2 x}{\cos x}$$

The first trig identity can be used to simplify the numerator.

$$\frac{1}{\cos x}$$

$$\sec x$$

More Complicated Trig Identities

A more complicated type of trig identity relies on the fact that an identity is the difference of perfect squares. For instance, $\cos^2 x = 1 - \sin^2 x$ can be factored into $(1 - \sin x)(1 + \sin x)$. Similarly, $\sin^2 x = 1 - \cos^2 x = (1 - \cos x)(1 + \cos x)$.

Example 4

Show that $\dfrac{1+\cos x}{\sin x}$ is equivalent to $\dfrac{\sin x}{1-\cos x}$.

Solution: Notice that the $1 + \cos x$ expression is one of the factors of $(1 - \cos x)(1 + \cos x) = 1 - \cos^2 x = \sin^2 x$.

As a first step, you should reduce the original fraction by multiplying both the numerator and denominator by $1 - \cos x$.

$$\frac{1 + \cos x}{\sin x}$$

$$\frac{(1 + \cos x)(1 - \cos x)}{\sin x(1 - \cos x)}$$

$$\frac{1 - \cos^2 x}{\sin x(1 - \cos x)}$$

$$\frac{\sin^2 x}{\sin x(1 - \cos x)}$$

$$\frac{\cancel{\sin x} \cdot \sin x}{\cancel{\sin x}(1 - \cos x)}$$

$$\frac{\sin x}{1 - \cos x}$$

Check Your Understanding of Section 5.7

A. Multiple-Choice

1. $\csc \theta$ is equal to which of the following?

   (1) $\dfrac{1}{\cos \theta}$ (3) $\dfrac{1}{\sin \theta}$

   (2) $\dfrac{1}{\tan \theta}$ (4) $\dfrac{1}{\sec \theta}$

2. Which is *not* a true trig identity?
   (1) $\sin^2 \theta + \cos^2 \theta = 1$ (3) $1 - \sin^2 \theta = \cos^2 \theta$
   (2) $\sin^2 \theta - \cos^2 \theta = 1$ (4) $1 - \cos^2 \theta = \sin^2 \theta$

Trigonometric Expressions and Equations

3. $\cos \theta \cdot \sec \theta$ is equal to which of the following?
(1) $\tan \theta$ (3) $\sec \theta$
(2) $\cot \theta$ (4) 1

4. $\tan \theta (1 - \sin^2 \theta)$ is equal to which of the following?
(1) $\sin \theta \cos \theta$ (3) $\csc \theta \tan \theta$
(2) $\sin \theta \sec \theta$ (4) $\sec \theta \tan \theta$

5. $\dfrac{1}{\cos \theta} - \cos \theta$ is equal to which of the following?
(1) $\tan \theta \sin \theta$ (3) $\cos \theta \cot \theta$
(2) $\cot \theta \sin \theta$ (4) $\sec \theta \sin \theta$

6. $\cos^2 \theta - \sin^2 \theta$ is equal to which of the following?
(1) $2\cos^2 \theta - 1$ (3) -1
(2) $2\cos^2 \theta + 1$ (4) 1

7. $\dfrac{\cot \theta}{1 + \sin \theta}$ is equal to which of the following?

(1) $\dfrac{1 + \sin \theta}{\cos \theta \sin \theta}$ (3) $\dfrac{1 + \cos \theta}{\cos \theta \sin \theta}$

(2) $\dfrac{1 - \cos \theta}{\cos \theta \sin \theta}$ (4) $\dfrac{1 - \sin \theta}{\cos \theta \sin \theta}$

8. $\cos \theta$ is equal to which of the following?
(1) $\sin(180° - \theta)$ (3) $\sin(\theta - 90°)$
(2) $\sin(\theta + 90°)$ (4) $\sin(90° - \theta)$

9. $-\tan \theta \sin \theta$ is equal to which of the following?
(1) $\sec \theta - \cos \theta$ (3) $\sin \theta - \csc \theta$
(2) $\cos \theta - \sec \theta$ (4) $\csc \theta - \sin \theta$

10. If $0° \le \theta < 90°$ and $\csc \theta = \dfrac{5}{3}$, what is the value of $\cot \theta$?

(1) $\dfrac{5}{4}$ (2) $\dfrac{4}{3}$ (3) $\dfrac{4}{5}$ (4) $\dfrac{3}{4}$

226

B. *Show how you arrived at your answers.*

1. Prove that $\sin\theta, \tan\theta, + \cos\theta = \sec\theta$.

2. Prove that $\dfrac{\cos\theta}{1 + \sin\theta} = \dfrac{1 - \sin\theta}{\cos\theta}$.

3. Prove that $(\cos\theta - \sin\theta)(\cos\theta + \sin\theta) + 2\sin^2\theta = 1$.

4. Prove that $(\cot\theta + \csc\theta)(\cot\theta - \csc\theta) = -1$.

5. What identity do you get if you divide both sides of the identity $\sin^2\theta + \cos^2\theta = 1$ by $\sin^2\theta$?

Chapter Six

SYSTEMS OF EQUATIONS

6.1 TWO EQUATIONS WITH TWO UNKNOWNS

 KEY IDEAS

A *system of two equations with two unknowns* is two equations each containing two variables. The solution set to a system of two equations with two unknowns is usually an *ordered pair* that makes both equations true.

The ordered pair (3, 7) is a solution to the equation $x + y = 10$. If the first number is substituted for the x and if the second number is substituted for the y, the equation becomes $3 + 7 = 10$, which is true. Other ordered pairs that are solutions to this equation are (8, 2), (5, 5), (10, 0), and (12, −2).

An example of a system of two equations with two unknowns is

$$x + y = 10$$
$$x - y = 4$$

The goal is to find, if one exists, an ordered pair that satisfies both equations at the same time. By examining the different ordered pairs that make the first equation true and by testing to see if they also make the second equation true, it is possible through the process of guess-and-check to find that the ordered pair (7, 3) is the solution because $7 + 3 = 10$ and $7 - 3 = 4$. For more complicated examples, though, a method that uses algebra is needed.

The Substitution Method

If one of the variables is already isolated in one of the equations, it is possible to simplify the two equations with two variables into just one equation with one variable.

$$y = 3x - 2$$
$$x + 2y = 24$$

Since the y is isolated in the first equation, the y in the second equation can be replaced with $(3x - 2)$.

$$x + 2(3x - 2) = 24$$
$$x + 6x - 4 = 24$$
$$7x - 4 = 24$$
$$\underline{+4 = +4}$$
$$7x = 28$$
$$\frac{7x}{7} = \frac{28}{7}$$
$$x = 4$$

To get the other number in the ordered pair, substitute the $x = 4$ into the first equation.

$$y = 3 \cdot 4 - 2 = 12 - 2 = 10$$

The ordered pair that satisfies both equations is $(4, 10)$.

Example 1

Find the solution set of the system of equations.

$$y = 2x - 5$$
$$5x - 3y = 9$$

Solution: Substitute $(2x - 5)$ for y in the second equation.

$$5x - 3(2x - 5) = 9$$
$$5x - 6x + 15 = 9$$
$$-x + 15 = 9$$
$$\underline{-15 = -15}$$
$$-x = -6$$
$$\frac{-x}{-1} = \frac{-6}{-1}$$
$$x = 6$$

Substitute $x = 6$ into one of the original equations.

$$y = 2 \cdot 6 - 5 = 12 - 5 = 7$$

The solution is $(6, 7)$.

Systems of Equations

The Elimination Method

If the two equations can be combined in a way that *eliminates* one of the variables, the remaining equation can be solved for one part of the ordered pair. Sometimes this method requires that you multiply both sides of one or of both equations by some constant.

Case 1:

A coefficient of one variable is the opposite of the coefficient of the same variable in the other equation.

Examine the coefficients of the system of equations.

$$2x + 3y = 11$$
$$x - 3y = 1$$

The coefficient of the y in the first equation is 3. The coefficient of the y in the second equation is -3. Remember that $3 + -3 = 0$. If these equations are added together, the result will be an equation with no y-term.

$$2x + 3y = 11$$
$$+ \quad x - 3y = 1$$
$$\overline{3x = 12}$$
$$\frac{3x}{3} = \frac{12}{3}$$
$$x = 4$$

Substitute $x = 4$ into either original equation and solve for y.

$$4 - 3y = 1$$
$$\underline{-4 = -4}$$
$$-3y = -3$$
$$\frac{-3y}{-3} = \frac{-3}{-3}$$
$$y = 1$$

The solution to the system of equations is the ordered pair (4, 1).

Case 2:

A coefficient of one of the variables in one of the equations is equal to the coefficient of the same variable in the other equation.

230

Examine the coefficients in the system of equations.

$$4x + 3y = 29$$
$$4x - 2y = 14$$

The coefficients of the x-terms in both equations are equal to $+4$.

It is permitted to multiply both sides of an equation by the same constant. For this situation, multiply both sides of either equation by -1. Below is what happens if both sides of the second equation are multiplied by -1.

$$4x + 3y = 29$$
$$-1(4x - 2y) = -1(14)$$
$$4x + 3y = 29$$
$$-4x + 2y = -14$$

Now this has become like Case 1.

$$
\begin{array}{r}
4x + 3y = 29 \\
+ \quad -4x + 2y = -14 \\
\hline
5y = 15
\end{array}
$$
$$\frac{5y}{5} = \frac{15}{5}$$
$$y = 3$$

Substitute 3 for y in either of the original equations.

$$4x + 3 \cdot 3 = 29$$
$$4x + 9 = 29$$
$$4x = 20$$
$$x = 5$$

The solution set is (5, 3).

Case 3:

A coefficient of one of the variables in one of the equations is a multiple of the same variable in the other equation.

Examine the coefficients of the system of equations.

$$2x + 6y = 4$$
$$5x + 3y = 34$$

Systems of Equations

The coefficient of the y in the first equation, 6, is a multiple of the coefficient of the y in the second equation, 3. For the equation that has the smaller coefficient, multiply both sides of the equation by the number that would make that coefficient the opposite of the one from the other equation. For this example, multiply both sides of the second equation by -2.

$$2x + 6y = 4$$
$$-2(5x + 3y) = -2(34)$$

$$\begin{array}{r} 2x + 6y = 4 \\ + \quad -10x - 6y = -68 \\ \hline -8x = -64 \end{array}$$

$$\frac{-8x}{-8} = \frac{-64}{-8}$$

$$x = 8$$

Now substitute $x = 8$ into one of the original equations.

$$2 \cdot 8 + 6y = 4$$
$$16 + 6y = 4$$
$$\frac{-16 = -16}{6y = -12}$$

$$\frac{6y}{6} = \frac{-12}{6}$$

$$y = -2$$

The solution is $(8, -2)$.

Case 4:

No coefficient for either variable in either equation is a multiple of the coefficient of the same variable in the other equation.

Examine the coefficients of the system of equations.

$$3x + 4y = 42$$
$$5x - 6y = 32$$

For the x-terms, the 5 is not a multiple of 3. For the y-terms, the -6 is not a multiple of the 4. When this happens, both equations must be changed.

You can choose whether you want to change the equations so that either the x-variables will be eliminated or the y-variables will be eliminated.

To eliminate the y-variables, find the least common multiple of the two coefficients. In this case, the least common multiple of 4 and 6 is 12. The goal is to convert one of the y-coefficients into a +12 and the other into a −12. This can be done by multiplying both sides of the first equation by +3 and both sides of the second equation by +2.

$$3(3x + 4y) = 3(42)$$

$$2(5x - 6y) = 2(32)$$

$$9x + 12y = 126$$
$$+ \quad 10x - 12y = 64$$
$$\overline{ 19x = 190}$$

$$\frac{19x}{19} = \frac{190}{19}$$

$$x = 10$$

Substitute 10 for x into either of the original equations.

$$3 \cdot 10 + 4y = 42$$

$$30 + 4y = 42$$
$$\underline{-30 = -30}$$
$$4y = 12$$

$$\frac{4y}{4} = \frac{12}{4}$$

$$y = 3$$

The solution is (10, 3).

Example 2

Solve the system of equations by eliminating the x-terms.

$$3x + 7y = 4$$

$$4x - 5y = 34$$

Systems of Equations

Solution: The least common multiple of the x-coefficients is 12. Multiply both sides of the first equation by $+4$ and both sides of the second equation by -3.

$$4(3x + 7y) = 4(4)$$
$$-3(4x - 5y) = -3(34)$$
$$12x + 28y = 16$$
$$+\ \underline{-12x + 15y = -102}$$
$$43y = -86$$
$$\frac{43y}{43} = \frac{-86}{43}$$
$$y = -2$$

Substitue $y = -2$ into one of the original equations.

$$3x + 7(-2) = 4$$
$$3x - 14 = 4$$
$$\underline{+14 = +14}$$
$$3x = 18$$
$$\frac{3x}{3} = \frac{18}{3}$$
$$x = 6$$

The solution is $(6, -2)$.

=== **MATH FACTS** ===

Sometimes when solving a system of equations with the elimination, both variables get eliminated and the resulting equation has just numbers. If the two numbers of the resulting equation are not equal, there is no solution to the original system. If the two numbers of the resulting equation are equal, there are an infinite number of solutions to the equation.

Example 3

How many solutions does this system of equations have?

$$2x + 3y = -10$$
$$-2x - 3y = 12$$

(1) No solutions
(2) One solution
(3) Two solutions
(4) More than two solutions

Solution: When the two equations are added together, the result is $0 = 2$, which is never true. There are no solutions to the original system of equations. Choice (1) is correct.

Linear-Quadratic Systems of Equations

If one of the equations has an exponent of 2, the system can be solved with the substitution method. The resulting equation will be a quadratic equation with up to two solutions.

Example 4

What ordered pair(s) satisfy the system?

$$x = y - 1$$
$$y = x^2 - 1$$

Solution:

$$y = (y - 1)^2 - 1$$
$$y = y^2 - 2y + 1 - 1$$
$$y = y^2 - 2y$$
$$0 = y^2 - 3y$$
$$0 = y(y - 3)$$
$$y = 0 \text{ or } y = 3$$

For $y = 0$, $x = -1$. For $y = 3$, $x = 2$. So the two solutions are $(-1, 0)$ and $(2, 3)$.

Systems of Equations

Check Your Understanding of Section 6.1

A. Multiple-Choice

1. Solve the system of equations.

$$y = 2x + 1$$
$$3x + 4y = 15$$

   (1) (3, 1)   (2) (4, 2)   (3) (2, 4)   (4) (1, 3)

2. Solve the system of equations.

$$y = 3x - 2$$
$$7x - 2y = 7$$

   (1) (3, 7)   (2) (7, 3)   (3) (4, 6)   (4) (6, 4)

3. Solve the system of equations.

$$3x + 5y = 37$$
$$2x - 5y = 8$$

   (1) (9, 2)   (2) (2, 9)   (3) (8, 3)   (4) (3, 8)

4. Solve the system of equations.

$$4x + 3y = -11$$
$$-4x + 5y = 35$$

   (1) (5, -3)   (2) (-3, 5)   (3) (-5, 3)   (4) (3, -5)

5. Solve the system of equations.

$$2x + 3y = 11$$
$$3x - 6y = 6$$

   (1) (4, 1)   (2) (1, 4)   (3) (2, 3)   (4) (3, 2)

6. Solve the system of equations.

$$3x - 4y = 7$$
$$5x + 3y = 31$$

   (1) (2, 5)   (2) (5, 2)   (3) (3, 4)   (4) (4, 3)

236

7. Solve the system of equations.
$$5x - 3y = 27$$
$$6x - 7y = 46$$
   (1) (3, –4)   (2) (–4, 3)   (3) (–3, 4)   (4) (4, –3)

8. Solve the system of equations.
$$2x - 5y = 7$$
$$4x - 10y = 12$$
   (1) (6, 1)   (2) (11, 3)   (3) (1, 6)   (4) No solution

9. How many solutions does the system of equations have?
$$5x + 3y = 10$$
$$10x + 6y = 20$$

   (1) No solution   (3) Two solutions
   (2) One solution   (4) Infinite solutions

10. If a system of two equations with two unknowns has no solution, what do the graphs of the two lines representing the two equations look like?
    (1) They intersect at one point.
    (2) They are parallel.
    (3) They are the same line.
    (4) They intersect at two points.

B. *Show how you arrived at your answers.*

1. When will it be easier to use the substitution method rather than the elimination method when solving a system of two equations with two unknowns?

2. Reginald is solving the system of equations.
$$x + y = 12$$
$$x - y = 8$$

Reginald adds the two equations to get $2x = 20$. Madelyn subtracts the two equations to get $2y = 4$. Who is correct?

237

Systems of Equations

3. Solve the system of equations.
$$3x + 5y = -2$$
$$4x + 3y = -10$$

4. Movie tickets cost $7 for children and $12 for adults. If 10 people purchase tickets and the total price is $85, how many adult tickets and how many children tickets were purchased?

5. The following system of equations has the solution (4, 7).
$$2x + y = m$$
$$3x - y = n$$

What are the values of m and n?

6.2 THE ELIMINATION METHOD FOR THREE EQUATIONS WITH THREE UNKNOWNS

KEY IDEAS

With three equations and three unknowns, the solution can be an *ordered triple* (x, y, z). The process of solving a system of equations with three unknowns requires three different elimination steps.

A system of three equations with three unknowns looks like this.
$$-3x - 7y - 2z = -31$$
$$5x + 4y + 2z = 35$$
$$-x + y - z = -4$$

Step 1:

Decide which variable to eliminate first.

The first step in solving a system of three equations with three unknowns is to examine the coefficients of each of the variables to see if the variables for any coefficient are matching numbers or have numbers that are multiples of the other coefficients. Any variable can be eliminated. However, it is easier to eliminate a variable when the coefficients contain multiples of other coefficients.

In this example, the z-variables have the coefficients -2, $+2$, and -1. The z will be the variable we try to eliminate.

Step 2:

Combine the first and second equation to eliminate a variable.

Since -2 is the opposite of $+2$, this is like Case 1 from the previous section. The z can be eliminated by adding the two equations together. This step could require multiplying both sides of one or both equations by a constant before combining them.

$$\begin{array}{r} -3x - 7y - 2z = -31 \\ + \quad 5x + 4y + 2z = 35 \\ \hline 2x - 3y = 4 \end{array}$$

This equation $2x - 3y = 4$ will be used in Step 4.

Step 3:

Combine the second and third equation (or the first and third if it is easier) to eliminate a variable.

Since $+2z + -z$ will not eliminate z, both sides of the second equation must first be multiplied by $+2$ so the z-coefficient will become -2, the opposite of the coefficient of the other z, $+2$.

$$5x + 4y + 2z = 35$$

$$2(-x + y - z) = 2(-4)$$

Now the equations can be combined to eliminate z.

$$\begin{array}{r} 5x + 4y + 2z = 35 \\ + \quad -2x + 2y - 2z = -8 \\ \hline 3x + 6y = 27 \end{array}$$

This equation, together with the one from Step 2, is used in Step 4.

Systems of Equations

Step 4:

Solve the resulting system of two equations with two unknowns

$$2x - 3y = 4$$
$$3x + 6y = 27$$

Since the +6 is a multiple of the −3, multiply both sides of the first equation by +2.

$$2(2x - 3y) = 2(4)$$
$$3x + 6y = 27$$

$$4x - 6y = 8$$
$$+ \quad 3x + 6y = 27$$
$$\overline{\qquad\qquad\qquad}$$
$$7x = 35$$
$$\frac{7x}{7} = \frac{35}{7}$$
$$x = 5$$

Substitute 5 for x into either of the two-variable equations.

$$2(5) - 3y = 4$$
$$10 - 3y = 4$$
$$\frac{-10 = -10}{-3y = -6}$$
$$\frac{-3y}{-3} = \frac{-6}{-3}$$
$$y = 2$$

Step 5:

Substitute the solution of the two-variable system of equations into one of the original equations to find the solution for the third variable.

Any of the original three equations can be used. Since the third equation has the smallest coefficients, it will be simplest to use.

$$-(5) + 2 - z = -4$$
$$-5 + 2 - z = -4$$
$$-3 - z = -4$$
$$\underline{+3 = +3}$$
$$-z = -1$$
$$\frac{-z}{-1} = \frac{-1}{-1}$$
$$z = 1$$

The solution is the ordered triple (5, 2, 1).

This can be checked by substituting $x = 5$, $y = 2$, $z = 1$ into each of the three original equations.

$$-3 \cdot 5 - 7 \cdot 2 - 2 \cdot 1 \overset{?}{=} -31$$
$$-15 - 14 - 2 \overset{?}{=} -31$$
$$-31 \overset{\checkmark}{=} -31$$
$$5 \cdot 5 + 4 \cdot 2 + 2 \cdot 1 \overset{?}{=} 35$$
$$25 + 8 + 2 \overset{?}{=} 35$$
$$35 \overset{\checkmark}{=} 35$$
$$-5 + 2 - 1 \overset{?}{=} -4$$
$$-4 \overset{\checkmark}{=} -4$$

Example

Find the solution to the system of equations.

$$-x + 5y + z = -18$$
$$4x - 5y + 3z = 34$$
$$x + 5y - z = -12$$

Systems of Equations

Solution: Since the y-variables have coefficients of +5, −5, and +5, the y should be eliminated. Combining equations 1 and 2 gets $3x + 4z = 16$. Combining equations 2 and 3 gets $5x + 2z = 22$.

Now solve the system.

$$3x + 4z = 16$$

$$5x + 2z = 22$$

Multiply both sides of the second equation by −2 so the z-coefficient becomes −4.

$$3x + 4z = 16$$
$$+ \quad -10x - 4z = -44$$
$$\overline{\qquad\qquad -7x = -28}$$
$$x = 4$$

Substitute $x = 4$ into one of the two-variable equations.

$$3 \cdot 4 + 4z = 16$$

$$12 + 4z = 16$$

$$4z = 4$$

$$z = 1$$

Now substitute $x = 4$ and $z = 1$ into one of the original equations.

$$-4 + 5y + 1 = -18$$

$$-3 + 5y = -18$$

$$+3 = +3$$
$$\overline{\qquad 5y = -15}$$

$$y = -3$$

The solution is $(4, -3, 1)$.

Check Your Understanding of Section 6.2

B. *Show how you arrived at your answers.*

1. If $x = 2$ and $y = 1$ and if $3x + 2y + 4z = 24$, what is the value of z?

2. Solve the system of equations.

$$3x + 2y + z = 15$$
$$2y + z = 6$$
$$z = 4$$

3. Solve the system of equations.

$$5x + 3y + z = 23$$
$$3x + 4y - z = 21$$
$$4x + 5y - 2z = 26$$

4. Solve the system of equations.

$$4x - 2y + 3z = 27$$
$$2x + 3y - 2z = -1$$
$$5x - 7y + 4z = 39$$

5. Solve the system of equations.

$$3x - 4y + 5z = -38$$
$$4x - 3y + 6z = -41$$
$$5x - 3y - 7z = 9$$

Systems of Equations

6.3 USING A GRAPHING CALCULATOR TO SOLVE SYSTEMS OF EQUATIONS

== Key Ideas ==

The *matrix* function of the graphing calculator can solve systems of equations instantly. If the system of equations is a multiple-choice question, using the calculator method will earn full credit. If the question requires an algebraic solution, using the calculator is still a useful way to check your answer.

Solving Two Equations with Two Unknowns Using Matrices

Matrices is an advanced math topic usually taught in precalculus. Even without understanding what matrices are or how they work, you can use them to solve systems of equations with a graphing calculator.

In Section 6.1, algebra was used to solve the following system of equations.

$$3x + 7y = 4$$
$$4x - 5y = 34$$

To solve the same system with the graphing calculator, do the following.

For the TI-84:

Press [2ND] [x^{-1}] to get to the matrix menu. Move to "EDIT," and press [1] to edit matrix A. Change the numbers before and after the "x" to 2 and 3. Put the numbers 3, 7, and 4 into row 1 and the numbers 4, −5, and 34 into row 2. Press [2ND] [MODE] to quit the matrix menu.

244

Press [2ND] [x⁻¹] and move to "MATH." Scroll to choice "B," called "rref" (Don't ask!) and select it. Then press [2ND] [x⁻¹] [1] [ )] and press [ENTER].

The numbers in the third column are the values for x and y.
The solution is (6, –2).

For the TI-Nspire:

Open the "Calculate" Scratchpad by pressing [A] from the home screen. Press [menu] [7] [5] for the "Reduced Row-Echelon Form" function. Press [template]. Select the matrix icon in the second row of symbols, sixth from the right. Set the number of rows to 2 and the number of columns to 3. Enter 3, 7, and 4 into the first row and 4, –5, and 34 into the second row. Press [enter].

Solving Three Equations with Three Unknowns Using Matrices

Using a graphing calculator to solve a system with three equations and three unknowns is nearly identical to the process for solving two equa-

Systems of Equations

tions with two unknowns. The only difference is that instead of using a matrix with two rows and three columns, use a matrix with three rows and four columns.

This was an example from Section 6.2 that was solved with algebra.

$$-3x - 7y - 2z = -31$$
$$5x + 4y + 2z = 35$$
$$-x + y - z = -4$$

For the TI-84:

Make the matrix with 3 rows and 4 columns.

For the TI-Nspire:

Make the matrix with 3 rows and 4 columns.

Example

Use the graphing calculator to solve the system of equations.

$$2x + 4y + z = 9$$
$$3x + 5y + 6z = -4$$
$$5x + y + 8z = -30$$

Solution: The solution is (−2, 4, −3).

Chapter Seven
FUNCTIONS

7.1 COMPOSITE FUNCTIONS

KEY IDEAS

A *function* is a mathematical rule that generally takes a number as an input and then outputs a number. If the function is called *f*, the notation $f(4) = 11$ means that when the number 4 is put into the function, the number 11 is output from the function. Functions can be used to describe mathematical relationships, including real-world scenarios.

Functions are often defined by a formula. For example, the formula $f(x) = 2x + 3$ means that when a number is put into the function, a number that is three more than twice that number will come out of the function. So $f(4) = 2 \cdot 4 + 3 = 8 + 3 = 11$.

Not only can numbers be put into a function but so can variables or even other functions. Using the function *f* defined above, $f(a) = 2a + 3$ and $f(x^2 + 1) = 2(x^2 + 1) + 3 = 2x^2 + 2 + 3 = 2x^2 + 5$.

If another function *g* is defined as $g(x) = 3x - 2$, it is possible to create a new function $f(g(x))$. By putting the $g(x)$ into the *f* function, the new function becomes $f(g(x)) = 2g(x) + 3 = 2(3x - 2) + 3$, which can be further simplified to $6x - 4 + 3 = 6x - 1$.

When a function is put into another function, the result is called a *composite function*. When working with composite functions, start with the inner function first and then move to the outer function.

Example 1

If $f(x) = 5x + 2$ and $g(x) = 3x - 1$, what is the value of $f(g(4))$?

Solution:

Since $g(4) = 3 \cdot 4 - 1 = 11$, $f(g(4)) = f(11) = 5 \cdot 11 + 2 = 57$.

Example 2

If $f(x) = 2x + 3$ and $g(x) = 3x - 2$, what are $g(f(x))$ and $f(g(x))$?

 Solution:

$$g(f(x)) = g(2x + 3) = 3(2x + 3) - 2 = 6x + 9 - 2 = 6x + 7$$

$$f(g(x)) = f(3x - 2) = 2(3x - 2) + 3 = 6x - 4 + 3 = 6x - 1$$

Notice that in this example, $g(f(x))$ is not equivalent to $f(g(x))$.

Example 3

If $f(x) = \dfrac{x-1}{3}$ and $g(x) = 3x + 1$, what is $f(g(x))$?

 Solution:

$$f(g(x)) = f(3x + 1) = \frac{3x + 1 - 1}{3} = \frac{3x}{3} = x$$

More difficult than finding the composition of two given functions is trying to find two functions whose composition would become some given function. For example
 $h(x) = 5x - 3$ could be *decomposed* into $h(x) = f(g(x))$ where $g(x) = 5x$ and $f(x) = x - 3$. There are other ways to decompose this function into two functions also, but this way is the most useful.

Example 4

Which of the following could be $f(x)$ and $g(x)$ if $f(g(x)) = (x - 2)^2$?

(1) $f(x) = x - 2$, $g(x) = x^2$
(2) $f(x) = x^2$, $g(x) = x - 2$
(3) $f(x) = x + 2$, $g(x) = x^2$
(4) $f(x) = x^2$, $g(x) = x + 2$

 Solution: Choice (2) is correct since $f(x - 2) = (x - 2)^2$. If you test choice (1), it would become $f(x^2) = x^2 - 2$.

Composite Functions on the Graphing Calculator

Graphing calculators can also evaluate functions and composite functions. Using either the TI-84 or the TI-Nspire, this is how to evaluate the function $g(x) = 3x - 1$ at $x = 4$.

Functions

For the TI-84:

Press [Y=] and then enter $3x - 1$ after "Y1=." Press [2ND] [MODE] to exit that menu. Press [VARS] [RIGHT] [1], and select [1] for the Y1 function. Press [(] [4] [)] and [ENTER].

For the TI-Nspire:

From the home screen, press [A] for the calculator scratchpad. Press [menu], [1], and [1] for Define. Type $g(x) = 3x - 1$ and press [enter]. To evaluate the function at the value 4, type $g(3)$ and press [enter].

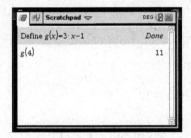

To evaluate a composite function like $f(g(4))$ when $f(x) = 5x + 2$ and $g(x) = 3x - 1$, both functions need to be entered into the calculator. For the TI-84, enter them into Y1 and Y2. For the TI-Nspire, enter them into $f(x)$ and $g(x)$.

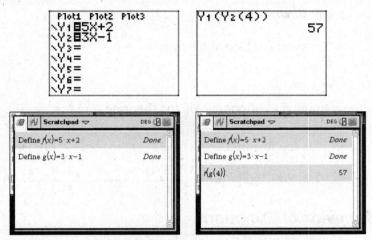

Odd and Even Functions

A function is called an *even* function if its graph has y-axis symmetry. A function is called an *odd* function if its graph has origin symmetry. The graphs of most functions have neither of these symmetries and so are neither odd nor even.

Functions

Average Rate of Change

A function has something called *the average rate of change on an interval*. This is very similar to the concept of slope discussed in Chapter 1.

The formula for the average rate of change of the function $f(x)$ on the interval a to b is shown:

$$\text{Average Rate of Change} = \frac{f(b) - f(a)}{b - a}$$

Example 5

What is the average rate of change of the function $f(x) = x^2$ on the interval $x = 1$ to $x = 3$?

Solution: Use the formula:

$$\text{Average Rate of Change} = \frac{f(3) - f(1)}{3 - 1} = \frac{3^2 - 1^2}{3 - 1} = \frac{9 - 1}{3 - 1} = \frac{8}{2} = 4$$

End Behavior of Functions

The end behavior of a function is a description of what output values happen when very large positive or very large negative values are put into the function. When a large number is put into the function $f(x) = x^2$, for example $f(1,000) = 1,000^2 = 1,000,000$, the function outputs a very large positive number.

Using symbols, we write: As $x \to \infty$, $f(x) \to \infty$, where the ∞ means infinity.

When a large negative number is put into the same function, like $f(-1,000) = (-1,000)^2 = +1,000,000$, the function outputs a very large positive number.

Using symbols, we write: As $x \to -\infty$, $f(x) \to \infty$.

The end behavior can be seen on the graph of the function.

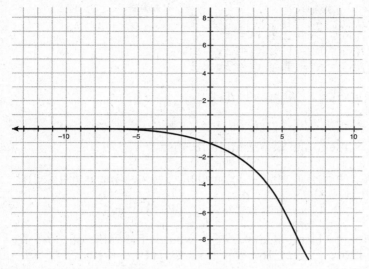

For large positive input values, this function outputs a large negative number. For large negative input values, this function outputs a number close to zero.

The end behavior of this function can be described:

$$\text{As } x \to \infty, f(x) \to -\infty$$
$$\text{As } x \to -\infty, f(x) \to 0$$

Check Your Understanding of Section 7.1

A. *Multiple-Choice*

1. If $f(x) = 2x + 3$ and $g(x) = 3x - 4$, what is $f(g(5))$?
   (1) 35   (2) 30   (3) 25   (4) 20

2. If $f(x) = 2x + 5$ and $g(x) = \dfrac{x-5}{2}$, what is $g(f(7))$?
   (1) 6   (2) 7   (3) 8   (4) 9

Functions

3. Which of the following could be $f(x)$ and $g(x)$ if $f(g(x)) = (3x - 2)^2 + 1$?
   (1) $f(x) = x^2 + 1$, $g(x) = 3x - 2$
   (2) $f(x) = 3x - 2$, $g(x) = x^2 + 1$
   (3) $f(x) = x^2 - 1$, $g(x) = 3x - 2$
   (4) $f(x) = x^2 + 1$, $g(x) = 3x + 2$

4. If $f(g(x)) = 5$, what is $g(f(x))$?

   (1) 5   (2) −5   (3) $\dfrac{1}{5}$

   (4) Not enough information to answer

5. If $f(x) = x + 4$, $g(x) = 3x$, and $h(x) = x^2$, what is $f(h(g(x)))$?
   (1) $(3x + 4)^2$ (3) $3(x^2 + 4)$
   (2) $3x^2 + 4$ (4) $(3x)^2 + 4$

6. If $f(5) = 11$ and $g(x) = x^2 + 5$, what is the value of $g(f(5))$?
   (1) 126   (2) 16   (3) 11   (4) 5

7. If $f(x) = 3x - 7$, what is $f(4x)$?
   (1) $12x - 7$ (2) $12x^2 - 7$ (3) $12x - 28$ (4) $12x^2 - 28$

8. If $g(x) = 3x^2 - 2$, what is $g(f(x))$?
   (1) $2[f(x)]^2 + 3$ (3) $3[f(x)]^2 - 2$
   (2) $2[f(x)]^2 - 3$ (4) $3[f(x)]^2 + 2$

9. If $f(g(x)) = 2(x^2 + 1) - 5$ and $f(x) = 2x - 5$, what is $g(x)$?
   (1) $x + 1$ (2) $2x + 5$ (3) $5x - 2$ (4) $x^2 + 1$

10. If $f(x) = x^2 + 1$, what is the value of $f(f(f(f(0)))))$?
    (1) 2   (2) 5   (3) 26   (4) 677

B. *Show how you arrived at your answers.*

1. $f(x) = 3x - 2$ and $g(x) = 2x + 3$. Griffin says $f(g(3)) = 25$. Ruben says $f(g(3)) = 17$. Who is correct?

2. If $f(x) = x - 4$, $g(x) = x^2$, and $h(x) = 3x$, what is $h(g(f(x)))$ in simplified form?

3. Given the two graphs of $y = f(x)$ and $y = g(x)$, what is the value of $f(g(2))$?

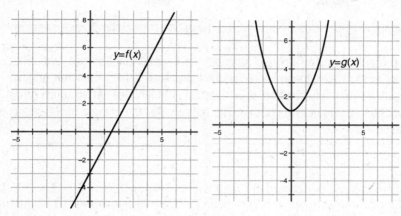

4. If $f(x) = x + 2$ and the graph of $y = g(x)$ is shown below, what does the graph of $f(g(x))$ look like?

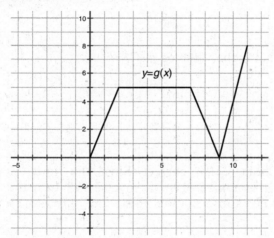

5. If $f(x) = x - 2$ and $g(x) = x^2$, for what value(s) of x does $f(g(x)) = g(f(x))$?

Functions

7.2 INVERSE FUNCTIONS

KEY IDEAS

If a function has an *inverse function*, the inverse function can take what was output from the original function and turn it back into the original input. An inverse function is a function that undoes what some other function has done to a number. The notation for the inverse of a function called f is f^{-1}.

If the function f is defined by $f(x) = x + 3$, then $f(7) = 7 + 3 = 10$. The inverse of function f is $f^{-1}(x) = x - 3$. When the number 10 is put into the inverse function, the output is 7. This value was put into the original function, f, to get the number 10: $f^{-1}(10) = 10 - 3 = 7$. Whatever the function f does to a number, the function f^{-1} "undoes" to that number.

Example 1

If $g(6) = 19$, what is the value of $g^{-1}(19)$?

(1) 6
(2) $\frac{1}{6}$
(3) 19
(4) −19

Solution: Choice (1) is correct. Since the function g turned the number 6 into the number 19, the inverse will turn the 19 back into the 6. With the given information, this is the only value for g^{-1} that can be determined.

MATH FACTS

An exponent of −1 has a different meaning when it is used in functions than when it is used with numbers. Although $3^{-1} = \frac{1}{3}$, $f^{-1}(x)$ is not $\frac{1}{f(x)}$. Instead, $f^{-1}(x)$ represents the inverse of function f.

Some Basic Inverse Functions

If a function has one operation happening to the input value, the inverse will have the opposite operation happening to its input variable. Some examples of basic inverse functions are shown in the table.

$f(x)$	$f^{-1}(x)$
$x + 2$	$x - 2$
$x - 2$	$x + 2$
$2x$	$\dfrac{x}{2}$
$\dfrac{x}{2}$	$2x$
x^2 (for $x \geq 0$)	$\sqrt{x}$
$\sqrt{x}$	x^2 (for $x \geq 0$)
x	x
$-x$	$-x$
$\dfrac{1}{x}$	$\dfrac{1}{x}$

For any of these function/inverse pairs, you can verify that the composition

$$f(f^{-1}(9)) = 9$$

Determining the Inverse Function of a Linear Function

A linear function of the form $f(x) = ax + b$, such as $f(x) = 2x + 3$, has an inverse function that is also a linear function. The process for finding the inverse function is to first rewrite the function but with the $f(x)$ replaced with an x and the x replaced with an $f^{-1}(x)$.

$$x = 2f^{-1}(x) + 3$$

Then solve for $f^{-1}(x)$. In this case, subtract 3 from both sides of the equation and divide both sides of the equation by 2.

Functions

$$x = 2f^{-1}(x) + 3$$

$$\frac{-3 = -3}{x - 3 = 2f^{-1}(x)}$$

$$\frac{x-3}{2} = \frac{2f^{-1}(x)}{2}$$

$$\frac{x-3}{2} = f^{-1}(x)$$

If you pick a number for x like 5, you can see that $f(5) = 2 \cdot 5 + 3 = 10 + 13$ and that $f^{-1}(13) = \frac{13-3}{2} = \frac{10}{2} = 5$. So this inverse function undoes what the function did to the number 5.

Example 2

What is the inverse of the function $f(x) = 3x - 5$?

(1) $f^{-1}(x) = \dfrac{x}{3} + 5$

(2) $f^{-1}(x) = \dfrac{x}{3} - 5$

(3) $f^{-1}(x) = \dfrac{x+5}{3}$

(4) $f^{-1}(x) = \dfrac{x-5}{3}$

Solution: Choice (3) is correct. Replace the $f(x)$ with an x and replace the x with an $f^{-1}(x)$. Solve the equation $x = 3f^{-1}(x) - 5$ for $f^{-1}(x)$ by adding 5 to both sides of the equation and then dividing both sides of the equation by 3.

$$x = 3f^{-1}(x) - 5$$

$$\frac{+ 5 = +5}{x + 5 = 3f^{-1}(x)}$$

$$\frac{x+5}{3} = \frac{3f^{-1}(x)}{3}$$

$$\frac{x+5}{3} = f^{-1}(x)$$

Answering Questions About the Inverse Without Determining the Inverse

A trick question that appears on many Regents exams is one like the following.

If $f(x) = 5x - 2$, what is $f^{-1}(28)$?

The long way to do this question is to use the inverse process to find that the inverse function is $f^{-1}(x) = \dfrac{x+2}{5}$ and then calculate $f^{-1}(28) = \dfrac{28+2}{5} = \dfrac{30}{5} = 6$.

Since $f^{-1}(28)$ is asking what input value for x will output the number 28, a shorter way to do this is to simply solve the equation $28 = 5x - 2$.

$$28 = 5x - 2$$
$$\underline{+2 = +2}$$
$$30 = 5x$$
$$\frac{30}{5} = \frac{5x}{5}$$
$$6 = x$$

Example 3

If $f(x) = x^3 + 4$, what is the value of $f^{-1}(12)$?

Solution: Solve the equation $12 = x^3 + 4$.

$$12 = x^3 + 4$$
$$\underline{-4 = -4}$$
$$8 = x^3$$
$$\sqrt[3]{8} = \sqrt[3]{x^3}$$
$$2 = x$$

Graphs of Inverse Functions

If, for some function f, $f(3) = 9$, then the point $(3, 9)$ will be on the graph for that function. If $f(3) = 9$, then $f^{-1}(9) = 3$. So the point $(9, 3)$ will be on the graph for the inverse function. In general, if (x, y) is a point on the graph of the original function, then (y, x) will be a point on the graph of the inverse function. In terms of transformations, the point $(9, 3)$ is

Functions

the reflection over the line $y = x$ of the point (3, 9). When this transformation is done to every point in the graph of the function, the graph of the inverse will be a reflection of the entire graph of the original function over the line $y = x$.

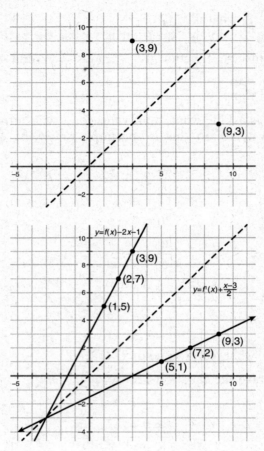

Some more graphs of functions and their inverses are shown in the following table.

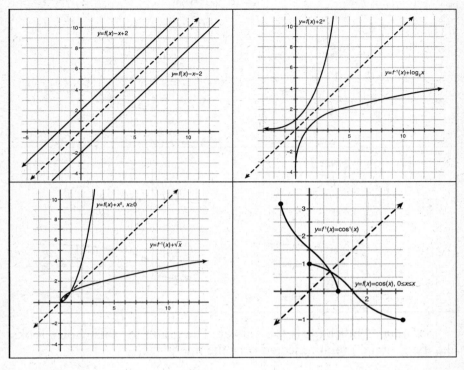

Example 4

Which choice is the graph of the inverse of the function whose graph is below?

Functions

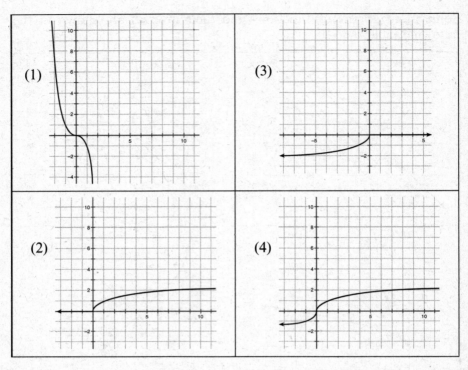

Solution: Choice (4) is the answer. The graph for this choice looks like the original graph reflected over the line $y = x$.

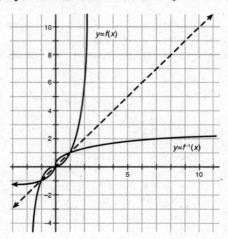

Inverse Functions on the Graphing Calculator

The TI-84 and TI-Nspire can easily graph inverse functions. To graph $y = f(x)$ and $y = f^{-1}(x)$ for the function $f(x) = 2x + 3$, follow these steps.

For the TI-84:

Press [Y1] and enter $2x + 3$ after Y1=. Graph this function by pressing [ZOOM] [6]. To graph the inverse, press [2ND] [PRGM] [8] for DrawInv. Then press [VARS] [RIGHT] [1] [1] and [ENTER].

For the TI-Nspire:

From the homescreen, press [B] for the Graph Scratchpad. Press [tab] and enter $2x + 3$ after $f1(x) =$. Then press enter to see the graph of $f(x) = 2x + 3$. Press and hold [ctrl]. While keeping the [ctrl] key down, press [menu]. Then press [4] for Text. Type "$x = f1(y)$" and then press [enter]. Move the cursor onto this text. Press and hold [ctrl] and then [click]. Drag the text onto the x-axis, and it will graph the inverse.

Functions

More Complicated Inverses

Although they were not called inverse functions when introduced, two inverse functions were already presented in this book.

An exponential function like $f(x) = 2^x$ has an inverse function. Since $f(3) = 2^3 = 8$, the inverse function must make $f^{-1}(8) = 3$. The inverse of $f f^{-1}(x) = \log_2(x)$, and $\log_2 8$ does equal 3.

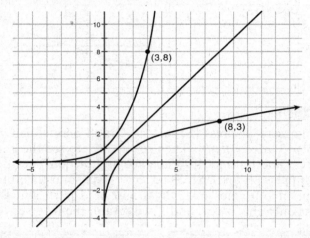

The trig function $f(x) = \cos(x)$, $0° \leq x < 180°$, has the inverse function $f^{-1}(x) = \cos^{-1}(x)$. So $\cos(0°) = 1$ and $\cos^{-1}(1) = 0°$.

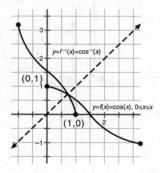

Check Your Understanding of Section 7.2

A. Multiple-Choice

1. If $g(x)$ is the inverse of $f(x)$, which of the following is *not* true?
(1) $f(g(x)) = g(f(x))$
(3) $f(g(x)) = x$
(2) $g(x) = \dfrac{1}{f(x)}$
(4) $g(f(x)) = x$

2. If $g(x)$ is the inverse of $f(x)$, what is the value of $f(g(9))$?
(1) 3
(3) -9
(2) $\dfrac{1}{9}$
(4) 9

3. What is the inverse of $f(x) = x - 3$?
(1) $f^{-1}(x) = x + 3$
(3) $f^{-1}(x) = x - 3$
(2) $f^{-1}(x) = \dfrac{1}{x+3}$
(4) $f^{-1}(x) = \dfrac{1}{x-3}$

4. What is the inverse of $f(x) = 4x$?
(1) $f^{-1}(x) = \dfrac{x}{4}$
(3) $f^{-1}(x) = \dfrac{1}{4x}$
(2) $f^{-1}(x) = 4x$
(4) $f^{-1}(x) = -4x$

5. What is the inverse of $f(x) = 2x - 7$?
(1) $f^{-1}(x) = 2x + 7$
(3) $f^{-1}(x) = \dfrac{x+7}{2}$
(2) $f^{-1}(x) = \dfrac{1}{2x-7}$
(4) $f^{-1}(x) = \dfrac{x}{2} + 7$

6. If $f(3) = 11$, what must also be true?
(1) $f^{-1}(11) = 3$
(3) $f^{-1}(3) = \dfrac{1}{11}$
(2) $f^{-1}(3) = 11$
(4) $f^{-1}(3) = -11$

7. If $f^{-1}(x) = 3x + 2$, what is $f(x)$?
(1) $f(x) = \dfrac{1}{3x+2}$
(3) $f(x) = \dfrac{x}{3} - 2$
(2) $f(x) = \dfrac{x-2}{3}$
(4) $f(x) = -3x - 2$

265

Functions

8. If the graph of $y = f(x)$ is shown below, what is the graph of $y = f^{-1}(x)$?

9. What is the inverse of $f(x) = 3^x$?
   (1) $f^{-1}(x) = x^3$
   (2) $f^{-1}(x) = \dfrac{1}{3^x}$
   (3) $f^{-1}(x) = 3^{-x}$
   (4) $f^{-1}(x) = \log_3 x$

10. What is the inverse of $f(x) = \sin(x)$, $-90° \leq x \leq 90°$?
    (1) $f^{-1}(x) = \dfrac{1}{\sin(x)}$
    (2) $f^{-1}(x) = \csc(x)$
    (3) $f^{-1}(x) = \sin^{-1}(x)$
    (4) $f^{-1}(x) = \cos(x)$

Functions

B. *Show how you arrived at your answers.*

1. If $f(x) = 5x + 4$, what is $f^{-1}(34)$?

2. If the graph of $y = f(x)$ is shown below, sketch on the same axes the graph of $y = f^{-1}(x)$.

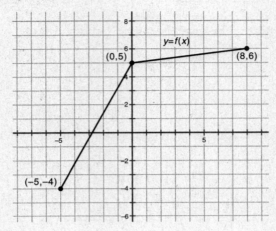

3. If $f(x) = 7x + 3$, what is $f^{-1}(x)$?

4. If $f(x) = \sqrt[3]{5x+19}$, what is $f^{-1}(f(x))$?

5. If $f(x) = 2x$ and $g(x) = x + 7$, show how you could solve $2x + 7 = 17$ by using the inverse functions $f^{-1}(x)$ and $g^{-1}(x)$.

Chapter Eight

SEQUENCES

8.1 SEQUENCES AND SERIES

KEY IDEAS

A *sequence* is a list of numbers that follows some pattern. An example of a sequence is $a = 1, 4, 9, 16, \ldots$. The individual terms of the sequence are called a_1, a_2, a_3, etc., where the small subscript is the term's position on the list. The terms of the sequence can be defined with two types of formulas. An *explicit formula* describes how to calculate the term by knowing the term's position on the list. A *recursive formula* describes how to calculate the term by knowing the previous terms on the list. The explicit formula is generally preferred but often more difficult to find.

Explicit Formula for a Sequence

An explicit formula is a type of function into which only positive integers can be input. Sometimes the explicit formula notation resembles function notation like $a(n) = 2n + 3$. An alternative notation is with a subscript, like $a_n = 2n + 3$. By substituting the values 1, 2, 3, and 4 for n, the first four terms of the sequence can be determined.

$$a_1 = 2 \cdot 1 + 3 = 5$$
$$a_2 = 2 \cdot 2 + 3 = 7$$
$$a_3 = 2 \cdot 3 + 3 = 9$$
$$a_4 = 2 \cdot 4 + 3 = 11$$

The sequence is $a = 5, 7, 9, 11, \ldots$.

MATH FACTS

An *infinite sequence* is one that has no end. The ellipses (...) at the end indicates that the sequence is infinite. A *finite sequence* does have a final term. The sequence 5, 7, 9, 11 is a finite sequence, while the sequence 5, 7, 9, 11, ... is an infinite sequence.

Sequences

Example 1

What are the first three terms of the sequence defined by equation $a_n = 5 + 3(n - 1)$?

Solution: Substitute 1, 2, and 3 for n in the formula to get $a_1 = 5 + 3(1 - 1) = 5 + 3(0) = 5$, $a_2 = 5 + 3(2 - 1) = 5 + 3(1) = 8$, $a_3 = 5 + 3(3 - 1) = 5 + 3(2) = 11$.

Recursive Formula for a Sequence

A recursive formula has two parts. First there is the *base case*, where the first value (or the first few values) of the sequence are given. Second there is the *recursive definition*, where a rule for calculating more terms is described in terms of the previous terms.

For example:

$$a_1 = 5$$

$$a_n = 2 + a_{n-1} \text{ for } n \geq 2$$

To get a_1, there is no calculation needed since a_1 is given as 5 in the base case. For a_2, substitute 2 for n into the recursive part of the definition.

$$a_2 = 2 + a_{2-1} = 2 + a_1 = 2 + 5 = 7$$

Notice that the value of a_1 needs to be known to do this step.

For $n = 3$, $a_3 = 2 + a_{3-1} = 2 + a_2 = 2 + 7 = 9$.

The first three terms are 5, 7, and 9.

Notice that this recursive formula generates the same sequence as does the explicit formula $a_n = 2n + 3$.

Example 2

What is the value of a_3 for the sequence defined by

$$a_1 = 3$$

$$a_n = n \cdot a_{n-1}, \text{ for } n \geq 2?$$

Solution: $a_1 = 3$ is given.
Substitute 2 for n into the recursive part of the definition.

$$a_2 = 2 \cdot a_{2-1} = 2 \cdot a_1 = 2 \cdot 3 = 6$$

Substitute 3 for n into the recursive part of the definition.

$$a_3 = 3 \cdot a_{3-1} = 3 \cdot a_2 = 3 \cdot 6 = 18$$

Example 3

What is the a_7 term of the sequence defined recursively as the following?

$$a_1 = x + 2$$
$$a_n = x \cdot a_{n-1}$$

Solution:

$$a_2 = x \cdot a_1 = x(x + 2) = x^2 + 2x$$
$$a_3 = x \cdot a_2 = x(x^2 + 2x) = x^3 + 2x^2$$
$$a_4 = x \cdot a_3 = x(x^3 + 2x^2) = x^4 + 2x^3$$

Each term becomes $x^n + 2x^{n-1}$. So a_7 will be $x^7 + 2x^6$.

Arithmetic and Geometric Sequences

A sequence like 5, 7, 9, 11, ... is known as an *arithmetic sequence* since each term is equal to 2 more than the previous term. The number 2, in this case, is known as the *common difference*, usually denoted by the variable *d*.

A sequence like 5, 10, 20, 40, ... is known as a *geometric sequence* since each term is equal to 2 times the previous term. The number 2, in this case, is known as the *common ratio*, usually denoted by the variable *r*.

The two formulas for the *n*th term of an arithmetic or of a geometric sequence are provided on the reference sheet given to you in the Regents booklet.

Arithmetic Sequence	Geometric Sequence
$a_n = a_1 + (n - 1)d$	$a_n = a_1 r^{n-1}$

Example 4

Find the explicit formula for the a_n term of sequence 7, 11, 15, 19, ...

Solution: Since each term is equal to the previous term plus 4, this is an arithmetic sequence: $a_1 = 7$ and $d = 4$. The formula is $a_n = 7 + (n - 1)4$.

Example 5

Find the explicit formula for the a_n term of sequence 7, 28, 112, 448, ...

271

Sequences

Solution: Since each term is equal to the previous term times 4, this is a geometric sequence: $a_1 = 7$ and $r = 4$. The formula is $a_n = 7 \cdot 4^{n-1}$.

Example 6

What is the 50th term of the sequence 7, 11, 15, 19, ...?

Solution: The explicit formula is $a_n = 7 + 4(n - 1)$. Substitute 50 for n to get the answer.

$$a_{50} = 7 + 4(50 - 1) = 7 + 4(49) = 203$$

Example 7

The sequence $a = 3, 11, 19, 27, ...$ will eventually reach the number 251. If $a_n = 251$, what is the value of n?

Solution: The formula is $a_n = 3 + 8(n - 1)$. If $a_n = 251$, the equation becomes

$$251 = 3 + 8(n - 1)$$

One way to solve this is first to subtract 3 from both sides of the equation.

$$251 = 3 + 8(n - 1)$$
$$\underline{-3 = -3}$$
$$248 = 8(n - 1)$$

Now divide both sides of the equation by 8.

$$\frac{248}{8} = \frac{8(n-1)}{8}$$

$$31 = n - 1$$

Then add 1 to both sides of the equation.

$$31 = n - 1$$
$$\underline{+1 = +1}$$
$$32 = n$$
$$a_{32} = 251$$

272

Finite Arithmetic and Geometric Series

A *finite series* is like a finite sequence except all the terms are added together. An example of a finite arithmetic series is $5 + 7 + 9 + 11 + 13 + 15$. An example of a finite geometric series is $5 + 10 + 20 + 40 + 80 + 160$.

The formula for calculating the sum of a finite geometric series is given in the reference sheet in the Regents booklet. The formula is shown here.

$$S_n = \frac{a_1 - a_1 r^n}{1-r}$$

To use this formula to calculate $5 + 10 + 20 + 40 + 80 + 160$, substitute 6 for n (since there are 6 terms), 5 for a_1, and 2 for r.

$$S_6 = \frac{5 - 5 \cdot 2^6}{1-2} = \frac{5 - 5 \cdot 64}{-1} = \frac{-315}{-1} = 315$$

The formula for calculating the sum of a finite arithmetic series is not given in the reference sheet in the Regents booklet. That formula is shown here.

$$S_n = \frac{n(a_1 + a_n)}{2}$$

Example 8

What is the sum of the first 100 positive integers in the following series?
$1 + 2 + 3 + \dots + 98 + 99 + 100$

Solution: Substitute 100 for n, 1 for a_1, and 100 for a_n.

$$S_n = \frac{100(1+100)}{2} = \frac{100 \cdot 101}{2} = \frac{10,100}{2} = 5,050$$

Sequences

Check Your Understanding of Section 8.1

A. Multiple-Choice

1. What are the first four terms of the sequence defined by $a_1 = 5$, $a_n = 3 + a_{n-1}$ for $n > 1$?
   (1) 5, 8, 11, 14   (3) 5, 2, −1, −4
   (2) 5, 15, 45, 135   (4) 5, 3, 1, −1

2. The sequence 3, 12, 48, 192, ... can be defined by which of the following?
   (1) $a_1 = 3$, $a_n = 8 + a_{n-1}$ for $n > 1$
   (2) $a_1 = 3$, $a_n = 4a_{n-1}$ for $n > 1$
   (3) $a_1 = 3$, $a_n = 12a_{n-1}$ for $n > 1$
   (4) $a_1 = 3$, $a_n = 12 + a_{n-1}$ for $n > 1$

3. What is the 20th term of the sequence 4, 11, 18, 25, ...?
   (1) 130   (2) 137   (3) 144   (4) 151

4. What position is the number 247 in the sequence 4, 13, 22, 31, ...?
   (1) 26   (2) 27   (3) 28   (4) 29

5. What is the 10th term of the sequence 4, 20, 100, 500, ...?
   (1) 7,812,500   (3) 195,312,500
   (2) 39,062,500   (4) 976,562,500

6. What is the sum of the series $5 + 13 + 21 + 29 + \ldots + 317$?
   (1) 6,380   (2) 6,400   (3) 6,420   (4) 6,440

7. What is the sum of the series $2 + 2 \cdot 3 + 2 \cdot 3^2 + 2 \cdot 3^3 + \ldots + 2 \cdot 3^{15}$?
   (1) $\dfrac{2(1-3^{15})}{1-3}$ (3) $\dfrac{2(1+3^{16})}{1+3}$

   (2) $\dfrac{2(1+3^{15})}{1+3}$ (4) $\dfrac{2(1-3^{16})}{1-3}$

8. A series is defined as $d_1 = 0$, $d_2 = 1$, $d_n = (n-1)(d_{n-2} + d_{n-1})$ for $n > 2$. What is the value of d_5?
   (1) 2   (2) 9   (3) 44   (4) 265

274

9. What is the sum of the first 20 terms of $2 - 6 + 18 - 54 + 162 - 486 + ...$?
   (1) $-1,743,392,000$ (3) $-1,743,392,200$
   (2) $-1,743,392,100$ (4) $-1,743,392,300$

10. $\dfrac{1}{2} + \dfrac{1}{4} + \dfrac{1}{8} + \dfrac{1}{16} + ... + \dfrac{1}{2^{20}}$ is closest to which of the following?
    (1) 0.99995   (2) 0.999995   (3) 0.9999995   (4) 1

B. *Show how you arrived at your answers.*

1. For 20 years, Journey puts \$1,000 into the bank every January 1. The bank pays 3% interest every December 31. If she starts on January 1, 2000, the amount of money she will have on January 2, 2020 can be shown by the sequence:

   $1,000 + 1,000(1.05) + 1,000(1.05^2) + ... + 1,000(1.05)^{20}$. How much money is this?

2. What is greater: the 20th term of the sequence 10,000; 20,000; 30,000; 40,000; ... or the 20th term of the sequence 1, 2, 4, 8, ...?

3. The Fibonacci sequence is defined by $F_1 = 1$, $F_2 = 1$, $F_n = F_{n-1} + F_{n-2}$ for $n > 2$. What is F_{10}?

4. If $a_1 = 3$ and $a_n = n^2 \cdot a_{n-1} + n$, what is a_4?

5. 2,657,205 is what term number in the sequence 5, 15, 45, 135, ...?

Chapter Nine PROBABILITY

9.1 SAMPLE SPACES

Key Ideas

A *sample space* is a list of all the possible *outcomes* of some experiment. When it is possible to create an accurate sample space, calculating the *probability* of different types of outcomes can be found by counting. The probability of something happening can range from as low as 0% = 0.00 (if it is impossible) to as high as 100% = 1.00 (if it must happen). Most probabilities are somewhere between 0 and 1.

When a fair coin is flipped, there are two possible *outcomes*: heads and tails. Since each outcome is equally likely, the two outcomes can be written as a set called the *sample space*.

{Heads, Tails}

Anything you can try to calculate the probability of is known as an *event*. The sample space can be used to calculate the *probability* of different events that can happen with the coin.

To calculate the probability of the coin landing on heads, first count the number of possible outcomes in the sample space. Make that the denominator of your answer. Then count the number of *favorable* outcomes, meaning outcomes in the sample space that are heads. Make that the numerator of your answer. Since there are 2 possible outcomes and 1 outcome with heads, the probability of the coin landing on heads is $\frac{1}{2}$.

Math Facts

If there is a sample space, the probability of some event happening can be calculated with the fraction $\frac{\text{Favorable outcomes in the sample space}}{\text{Total outcomes in the sample space}}$.

Below is a picture of a circle divided into 6 equal slices, numbered 1 though 6. A spinner is put into the middle and spun.

The sample space for one spin is {1, 2, 3, 4, 5, 6}.
With this sample space, various questions can be answered.

Basic Probability Questions

The most basic question is one like: What is the probability that the spinner will land on the number 4? This question is abbreviated as P(4).

Since there is one 4 in the sample space and there are 6 total numbers in the sample space, $P(4) = \frac{1}{6}$.

Probability Questions Involving the Word "Not"

If the question is to find the probability that the spinner will not land on the number 4, or P(not 4), count the number of outcomes in the sample space that are not 4. Since 5 of the numbers in the sample space are not the number 4, $P(\text{not } 4) = \frac{5}{6}$.

Probability Questions Involving the Word "And"

Questions involving the word "and" require you to analyze each element in the sample space to see if it meets two different conditions. Here's a typical question involving the spinner with 6 equal slices, numbered 1 through 6.

"On one spin, what is the probability that the spinner will land on a number that is greater than 3 and even, or P(greater than 3 and even)?"

The denominator of the solution is still 6. For the numerator, examine all 6 numbers in the sample space to see how many of them are both greater than 3 and also even.

Probability

- 1 is neither even nor greater than 3.
- 2 is even but is not greater than 3.
- 3 is neither even nor greater than 3.
- 4 is both even and greater than 3.
- 5 is greater than 3 but not even.
- 6 is both even and greater than 3.

Exactly two of the numbers, 4 and 6, satisfy both conditions. So the numerator of the probability fraction is 2, and the solution is $P(\text{even and greater than 3}) = \dfrac{2}{6} = \dfrac{1}{3}$.

Probability Questions Involving the Word "Or"

Imagine there is a game where you spin the spinner and you win a prize if the spinner lands on an even number or on a number that is greater than 3. To find the probability of winning this game, go through the 6 outcomes to find how many of them are either even, greater than 3, or both.

Of the 6 numbers, the numbers 2, 4, 5, and 6 all are either even, greater than 3, or both. The probability of getting a number that is even or greater than 3 is $P(\text{even or greater than 3}) = \dfrac{4}{6} = \dfrac{2}{3}$.

Probability Questions Involving the Word "Given"

Imagine the spinner is spun. Somebody looks at it before you get a chance to and tells you a hint about what number it landed on, such as "It landed on an even number." Then a question is asked like, "What is the probability that it landed on a number greater than 3?" Had you not known about the hint, the solution would be $\dfrac{3}{6}$ since 3 of the 6 numbers (4, 5, and 6) are greater than 3. With the hint, though, you can get a more accurate answer.

This question could be described as "What is the probability of getting a number greater than 3 given that it landed on an even number?" or P(greater than 3 given even). You can instead use the "given" symbol, |, to write P(greater than 3|even).

To calculate the probability using the sample space, first use the given information to reduce the sample space, crossing out anything that cannot have happened. For this example, the sample space is now $\{\cancel{1}, 2, \cancel{3}, 4, \cancel{5}, 6\} = \{2, 4, 6\}$. Now the question can be answered using this new, more accurate sample space. There are 3 outcomes in this

278

sample space, and 2 of them (4 and 6) are greater than 3. So

$P(\text{greater than 3}|\text{even}) = \dfrac{2}{3}$.

Example 1

A fair coin is flipped, and a fair 6-sided die is rolled. The sample space has 12 outcomes {H1, H2, H3, H4, H5, H6, T1, T2, T3, T4, T5, T6}. Using this sample space, calculate the probability of getting:

A) An outcome with tails on the coin
B) An outcome with tails on the coin and 2 on the die
C) An outcome with tails on the coin or 2 on the die
D) An outcome with tails on the coin given that there is a 2 on the die
E) An outcome with a 2 on the die given that there is a tails on the coin

Solution:

A) Since 6 out of the 12 outcomes have a T in them, the probability is $\dfrac{6}{12} = \dfrac{1}{2}$.

B) Only one of the 12 outcomes has a T and a 2, so the probability is $\dfrac{1}{12}$.

C) The outcomes H2, T1, T2, T3, T4, T5, T6 each have a 2 or a T (T2 has both). This is 7 out of the 12 outcomes in the sample space, so the probability is $\dfrac{7}{12}$.

D) Since it is given that 2 is on the die, the sample space gets reduced to {H2, T2}. Only one of the two outcomes has a T, so the probability is $\dfrac{1}{2}$.

E) Since it is given that tails is on the coin, the sample space set gets reduced to {T1, T2, T3, T4, T5, T6}. Of the 6 possible outcomes in this reduced sample space, only one has a 2 on the die. So the probability is $\dfrac{1}{6}$.

Probability

Sample Spaces Represented as Tables

For certain surveys, the results can be represented in a table. This table can serve as a sample space for probability questions.

Imagine that 100 people are surveyed and asked two questions "Are you more than 14 years old?" and "Are you more than 5 feet tall?" The results of the survey can be shown with a table like this.

	Under 14 Years	Over 14 Years	Total
Under 5 feet	21	12	33
Over 5 feet	9	58	67
Total	30	70	100

A probability question that can be answered with this table is "If one person is taken from the 100 at random, what is the probability the person will be both under 14 years old and over 5 feet tall?"

The denominator of the fraction will be 100 since there are 100 people in total. For the numerator, the number of people who are both under 14 years old and over 5 feet tall, based on information in the table, is 21.

So the solution is $\frac{21}{100} = .21$.

Example 2

Using the above table, what is the probability that a randomly chosen person will be under 14 years old?

Solution: The denominator of the solution is 100. The numerator is 21 + 9 = 30 since there are 21 people under 14 years old who are under 5 feet tall and 30 people under 14 years old who are over 5 feet tall. The solution is $\frac{30}{100} = .30$.

Example 3

Using the above table, find the probability that a randomly chosen person is over 5 feet tall or over 14 years old.

Solution: The total number of people is 100, which will be the denominator of the solution. For the numerator, look at the four possibilities and see which count people who are either over 5 feet tall, over 14 years old, or both. Three of the four possibilities have at least one of

280

those characteristics. The only one that doesn't is the people who are both under 14 years old and under 5 feet tall. Add 9 + 12 + 58 = 79, and make that the numerator. The solution is $\frac{79}{100} = .79$.

Example 4
Using the above table, find the probability of a randomly chosen person being over 5 feet tall given that the person is over 14 years old.

Solution: Because of the word "given," the denominator of the fraction is no longer going to be 100. Since it is known already that the person is over 14 years old, the 30 people under 14 years old are no longer relevant. The number of people over 14 years old is 70, and this will be the denominator of the solution. Of those 70 people, 58 of them are over 5 feet tall. So the solution is $\frac{58}{70} \approx .83$.

Sample Spaces Represented as Venn Diagrams

Another way to represent the data from the age/height survey is with a Venn diagram.

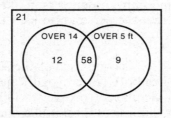

The two circles represent the people who are over 14 years old and the people who are over 5 feet tall. Just as there are four different categories of people in the table (both under 14 and under 5 feet, under 14 but over 5 feet, over 14 but under 5 feet, and both over 14 and over 5 feet), this Venn diagram has 4 regions (in neither circle, in the circle on the left but not in the circle on the right, in the circle on the right but not in the circle on the left, and in both circles).

Use the Venn diagram to answer a question like "If a person is chosen randomly from the 100 surveyed, what is the probability that the person is under 14 years old but over 5 feet tall?" The circle on the right represents the people who are over 5 feet tall. This is composed of the 58, which are the people who are both over 5 feet tall and over 14 years old, and of the 9, which are the people who are over 5 feet tall but not over

Probability

14 years old (since they are not inside the over 14 years old circle on the right). The solution is $\frac{9}{100} = .09$.

Example 5

The Venn diagram below is based on two survey questions "Do you like frozen yogurt?" and "Do you like mustard?"

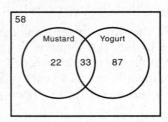

Use this Venn diagram to answer the following questions:

A) What is the probability that a person chosen randomly among those surveyed likes frozen yogurt but does not like mustard?
B) What is the probability that a person chosen randomly among those surveyed likes mustard?
C) Given that a randomly chosen person likes mustard, what is the probability that the person also likes frozen yogurt?

Solution:

A) The circle on the left is composed of the 33, which is the number of people who like both frozen yogurt and mustard, and of the 87, which is the number of people who like frozen yogurt (they are inside the Y circle) but do not like mustard (they are outside the M circle). The solution is $\frac{87}{200} = .435$.

B) The M circle has 33 + 22 = 55. So the solution is $\frac{55}{200} = .275$.

C) Since it is given that the person likes mustard, the denominator of the fraction will not be the total surveyed but just those surveyed who like mustard. From part (b), this is 55. Since 33 of those 55 like frozen yogurt, the solution is $\frac{33}{55} = .60$.

Check Your Understanding of Section 9.1

A. Multiple-Choice

1. If a fair penny and a fair dime are flipped, the sample space of possible outcomes is {HH, HT, TH, TT}. What is the probability of getting tails on both coins?

   (1) $\dfrac{1}{2}$　　　　(2) $\dfrac{1}{4}$　　　　(3) $\dfrac{3}{4}$　　　　(4) 1

2. A fair 6-sided die is rolled. The sample space is {1, 2, 3, 4, 5, 6}. What is the probability that the number that comes up is both even and greater than 4?

   (1) $\dfrac{1}{6}$　　　　(2) $\dfrac{2}{6}$　　　　(3) $\dfrac{3}{6}$　　　　(4) $\dfrac{4}{6}$

3. A spinner with equal-size sections and with the numbers 1 to 8 on it is spun. The sample space is {1, 2, 3, 4, 5, 6, 7, 8}. What is the probability that the number the spinner lands on is either even or greater than 5?

   (1) $\dfrac{3}{8}$　　　　(2) $\dfrac{4}{8}$　　　　(3) $\dfrac{7}{8}$　　　　(4) $\dfrac{5}{8}$

4. A fair coin is tossed, and a fair 6-sided die is rolled. The sample space of possible outcomes is {H1, H2, H3, H4, H5, H6, T1, T2, T3, T4, T5, T6}. If it is known that the die landed on a number greater than 4, what is the probability that the coin landed on heads?

   (1) $\dfrac{1}{2}$　　　　(2) $\dfrac{1}{3}$　　　　(3) $\dfrac{1}{4}$　　　　(4) $\dfrac{1}{6}$

Questions 5, 6, and 7 use the following information:

40 people are surveyed about whether they like iOS or Android. The results are collected on this table.

	Like Android	Don't Like Android	Total
Like iOS	7	20	27
Don't like iOS	12	1	13
Total	19	21	40

283

Probability

5. If one of the 40 people is selected at random, what is the probability the person likes both iOS and Android?
   (1) $\dfrac{39}{40}$ (2) $\dfrac{7}{40}$ (3) $\dfrac{46}{40}$ (4) $\dfrac{1}{40}$

6. What is the probability that a randomly selected person likes Android if it is known that the person likes iOS?
   (1) $\dfrac{7}{27}$ (2) $\dfrac{19}{40}$ (3) $\dfrac{7}{40}$ (4) $\dfrac{19}{27}$

7. What is the probability that a randomly selected person likes either iOS or Android?
   (1) $\dfrac{39}{40}$ (2) $\dfrac{37}{40}$ (3) $\dfrac{35}{40}$ (4) $\dfrac{32}{40}$

Questions 8, 9, and 10 use the following information:

50 students are surveyed about how they like the red and blue Barron's books. The results are collected on this Venn diagram.

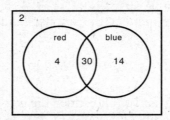

8. What is the probability that a randomly selected person likes both the red and the blue book?
   (1) .30   (2) .40   (3) .50   (4) .60

9. What is the probability that a randomly chosen person will not like the blue book?
   (1) .06   (2) .12   (3) .18   (4) .24

10. From the group of people who like the red book, a person is randomly chosen. What is the probability that the person also likes the blue book?
    (1) .72   (2) .80   (3) .88   (4) .96

B. *Show how you arrived at your answers.*

1. A fair 6-sided die with the faces numbered 1 to 6 is rolled. A spinner with the letters A, B, and C, with equal chances of occurring, is spun. What is the sample space of possible outcomes?

2. A family with three children can be one of eight possibilities:

    {BBB, BBG, BGB, BGG, GBB, GBG, GGB, GGG}.

    What is the probability of the family having 2 boys and a girl in any order?

3. 80 students are asked two questions: (1) Do you like the Rangers? and (2) Do you like the Islanders? 40 people answered yes to question 1, and 55 people answered yes to question 2. 10 people answered no to both. How many people answered yes to both? Use this Venn diagram to assist you.

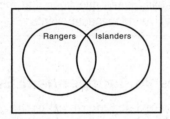

4. 200 students get to choose an ice cream cone. They can pick one flavor from the three choices: chocolate, vanilla, and strawberry. They can pick one topping, either sprinkles or peanuts. The results are collected on this table.

	Sprinkles	**Peanuts**	**Total**
Chocolate	40	50	90
Vanilla	25	5	30
Strawberry	65	15	80
Total	130	70	200

   a) What is the probability that a randomly selected person chose chocolate ice cream?
   b) What is the probability that a person who chose chocolate ice cream also chose peanuts?
   c) What is the probability that a person who chose sprinkles also chose strawberry ice cream?

5. A family has 4 children.
   a) Make a sample space of all the boy/girl combinations; there are 16 of them.
   b) What is more likely: having a 2/2 split (2 boys and 2 girls) or a 3/1 split (either 3 boys and 1 girl or 3 girls and 1 boy)?

9.2 CALCULATING PROBABILITIES INVOLVING INDEPENDENT EVENTS

Two events are *independent* if one happening (or not happening) has nothing to do with whether or not the other happens (or doesn't happen). When events are not independent, they are *dependent*. It is much simpler to calculate the probability when questions that include the words "or," "and," and "given" involve independent events.

Dependent Events vs. Independent Events

Many things in the real word depend on other things. If you ask someone, "Are you going to the beach next Saturday?" that person could say, "It depends." If you follow up with "Depends on what?" the person could respond, "On what the weather is like," "On whether or not my friend with a car is working," or all kinds of other possibilities.

If you ask the same person, "Is your birthday next Saturday?" he or she will not likely say, "It depends," since there isn't anything else that will make it more likely or less likely that the person's birthday is next Saturday. Next Saturday will be their birthday or it will not, regardless, for example, of what the weather is like.

The events "person will go to the beach on Saturday" and "the weather is nice on Saturday" are dependent events. The events "person's birthday is Saturday" and "the weather is nice on Saturday" are independent events.

The table shows some events and whether or not they qualify as dependent or as independent when combined.

Event A	Event B	Dependent or Independent
A coin is flipped	A die is rolled	Independent
What happens in the Yankees baseball game	What happens in the Rangers hockey game	Independent
Winning the lottery	Buying an airplane	Dependent
Passing the Algebra II Regents	Studying this book	Dependent
Rooting for the Mets	The Mets winning	Probably Independent, but maybe if you're at the game and cheering really hard, it could help a little!

MATH FACTS

It is sometimes not very clear whether two events are dependent or independent. Questions involving coin tosses, spinners, and dice are generally about independent events. For events involving human behavior, an argument can sometimes be made for either dependent or independent.

Example 1

Do you think these events are dependent or independent? Explain your reasoning.

Event A: The groundhog sees his shadow on Groundhog's Day.

Event B: There are six more weeks of cold weather.

Solution: As this is an opinion question, either independent or dependent is correct as long as your reasoning is clear. For independent, you could say that the weather does not in any way know whether the groundhog saw his shadow or not. So winter will come whenever it does, regardless. For dependent, you could say that maybe the groundhog seeing his shadow means that it is sunny on Groundhog's Day and that the sunny day is an indication that there will be more sunny days in the future so the warm weather will come sooner.

Probability

Probability Questions Involving the Word "And"

When a coin is flipped and a 6-sided die is rolled, the outcome of the coin flip is independent of the outcome of the die roll. The coin does not know (or care, for that matter) what happened with the die.

In Section 9.1, sample spaces were used to answer probability questions involving the word "and." With independent events, there is a shortcut for this. The shortcut is partly justified by the fact that when one fraction is multiplied by another fraction, the result will be a fraction that is smaller than either of the fractions. Likewise, the probability of two things happening is smaller than either one of them occurring individually.

When two events are independent, the probability of the first event and of the second event both happening is equal to the product of the probabilities of each happening separately.

===== **MATH FACTS** =====

If event A and event B are independent events, the probability of

$$P(A \text{ and } B) = P(A) \cdot P(B).$$

If A is the event "the coin lands on tails," then $P(A) = \dfrac{1}{2}$. If B is the event "the die shows a 2," then $P(B) = \dfrac{1}{6}$. (Small sample spaces {H, T} and {1, 2, 3, 4, 5, 6} could be used to determine these probabilities.) Since these are independent events, the probability that the coin lands on heads and that the die shows a 2 is $P(A \text{ and } B) = P(A) \cdot P(B) = \dfrac{1}{2} \cdot \dfrac{1}{6} = \dfrac{1}{12}$. This is the same answer as was obtained in Section 9.1 with the large sample space.

Example 2

If the probability of event A, that it will rain in Boston on Friday, is .6 and the probability of event B, that the Knicks will win the game they play in Los Angeles on Friday, is .3, what is the probability that it both rains in Boston on Friday and that the Knicks win the game they play in Los Angeles on Friday?

Solution: These are independent events. Whether or not it rains in Boston on Friday has no impact on whether or not the Knicks win in Los Angeles on Friday. (Some people think that all things affect each other in tiny cosmic ways, even events like these. However, for the Regents, these are implied to be independent events.)

$$P(A \text{ and } B) = P(A) \cdot P(B) = .6 \cdot .3 = .18$$

Probability Questions Involving the Word "Or"

For independent events, the probability of either A or B (or both) happening can be calculated with the formula $P(A \text{ or } B) = P(A) + P(B) - P(A \text{ and } B)$. Since the events are independent, $P(A \text{ and } B) = P(A) \cdot P(B)$. So the formula becomes

$$P(A \text{ or } B) = P(A) + P(B) - P(A) \cdot P(B).$$

====================== **MATH FACTS** ======================

If events A and B are independent, then $P(A \text{ or } B) = P(A) + P(B) - P(A) \cdot P(B)$

Example 3

If the probability of event A, that it will rain in Boston on Friday, is .6 and the probability of event B, that the Knicks will win the game they play in Los Angeles on Friday, is .3, what is the probability that it rains in Boston on Friday or that the Knicks win the game they play in Los Angeles on Friday (or both)?

Solution: Since they are independent events, $P(A \text{ or } B)$ can be calculated with the formula $P(A) + P(B) - P(A) \cdot P(B) = .6 + .3 - .6 \cdot .3 = .72$.

Probability Questions Involving the Word "Given"

For independent events A and B, the probability of A happening is not affected by whether or not B happened. So when A and B are independent, $P(A|B) = P(A)$.

Example 4

If the probability of event A, that it will rain in Boston on Friday, is .6 and the probability of event B, that the Knicks will win the game they play in Los Angeles on Friday, is .3, what is the probability that the Knicks win on Friday given that it will rain in Boston on Friday?

Solution: Since the events are independent, $P(A|B) = P(A) = .3$.

Probability

Check Your Understanding of Section 9.2

A. Multiple-Choice

1. Which two events are independent of one another?
   (1) A = It rains in Nevada on January 1, 1975.
       B = A New York resident goes to the movies on May 17, 2016.
   (2) A = It is Valentine's Day.
       B = Flower sales have increased.
   (3) A = A student eats a healthy breakfast.
       B = A student has energy at 10:00 A.M.
   (4) A = A teacher wins the lottery.
       B = The same teacher quits his job.

2. Which two events are *not* independent?
   (1) A = A child has brown hair.
       B = The child's biological parents both have brown hair.
   (2) A = A coin lands heads side up.
       B = A 6-sided die lands with the 6 face up.
   (3) A = A student's birthday is on May 1.
       B = Another student seated next to that student is born on May 1.
   (4) A = A person wears a blue shirt.
       B = The same person likes the TV show "Seinfeld."

3. A and B are independent events. If $P(A)$ = .4 and $P(B)$ = .5 what is $P(A$ and $B)$?
   (1) .4   (2) .5   (3) .2   (4) .9

4. A and B are independent events. If $P(A)$ = .4 and $P(B)$ = .5 what is $P(A$ or $B)$?
   (1) .7   (2) .9   (3) .5   (4) .45

   Questions 5, 6, and 7 use the following information:

   A fair coin is flipped, and a spinner with the letters A, B, C, D, and E, all equally likely, is spun.

5. What is the probability the coins will land heads and the spinner will land on D?
   (1) $\dfrac{1}{2}$ (2) $\dfrac{1}{5}$ (3) $\dfrac{1}{10}$ (4) $\dfrac{7}{10}$

290

6. What is the probability that the coin will land on heads or the spinner will land on D?

(1) $\dfrac{4}{10}$ (2) $\dfrac{5}{10}$ (3) $\dfrac{6}{10}$ (4) $\dfrac{7}{10}$

7. What is the probability that the coin will land on heads given that the spinner landed on D?

(1) $\dfrac{1}{10}$ (2) $\dfrac{3}{10}$ (3) $\dfrac{7}{10}$ (4) $\dfrac{1}{2}$

8. If the probability that the Rangers win the Stanley Cup is .05 and the probability that the Yankees win the World Series is .15, what is the probability that both will happen?

(1) .20   (2) .0075   (3) .05   (4) .15

9. Anne will go out to dinner with a group of friends if the restaurant is nearby or if it is a sushi restaurant. If P(nearby) = .7 and P(sushi) = .2 and these are independent events, what is the probability of Anne going out with the group?

(1) .76   (2) .83   (3) .90   (4) .14

10. *A* and *B* are independent events. If P(*A*) = .6 and P(*B*) = .9, what is P(*A*|*B*)?

(1) .3   (2) .6   (3) .9   (4) .54

B. *Show how you arrived at your answers.*

1. Write an event that would be dependent with the event *A* = "It rains this Saturday."

2. If *A* is the event "I eat pizza on Tuesday" and *B* is the event "I eat roast beef on Saturday" and if P(*A*) = .4 and P(*B*) = .3, what is the probability that I eat pizza on Tuesday and I do *not* eat roast beef on Saturday?

3. The probability of it raining on Saturday is 70%. The probability of Nayeli going to the movies on Saturday is 30%. Nayeli reasons that it will definitely either rain or she will go to the movies since .70 + .30 = 1. Is this reasoning correct?

4. *A* and *B* are independent events. P(*A*) = .6 and P(*B*) = .2. What is P(*B*|*A*)?

Probability

5. The Mets are playing the Royals in baseball. The probability that the Mets will win is .45. The probability that the Royals will win is .55. What is the probability that the Mets will win *and* that the Royals will win?

9.3 CALCULATING PROBABILITIES INVOLVING DEPENDENT EVENTS

When events are dependent, rather than independent, calculating probabilities is more complicated. If in a problem different probabilities are already given, it is possible to use them to determine if certain events are independent or dependent.

Dependent events can be complicated. If the probability of Camden going to a party is .8 and the probability of Greyson going to a party is .7, what is the probability of them both going to the party?

Well, it depends. If these were independent events, the probability of both going to the party would be the product of $.8 \cdot .7 = .56$. However, what if Camden and Greyson are rivals? What if when Camden got there and saw Greyson, he would turn around and leave? Then the probability of both going is not .56 but 0.

A simpler example of dependent events is choosing a doughnut from a box of doughnuts for event A, eating it, and then choosing a second doughnut for event B. Suppose the box of 12 doughnuts has 3 glazed, 5 jelly filled, 3 powdered sugar, and 1 custard filled.

Probability Questions Involving the Word "Given"

In the doughnut example, it is possible to calculate the probability that the second doughnut was glazed given that the first doughnut is glazed. If the first doughnut was glazed, after it is eaten there are only 11 doughnuts left in the box. Of those 11, only 2 are glazed. So the probability of the first doughnut being glazed is $P(A) = \frac{3}{12} = \frac{1}{4}$. However, the probability of the second doughnut being glazed given that the first doughnut was glazed is $P(B|A) = \frac{2}{11}$.

Example 1

If event *A* is "the first doughnut is custard filled" and event *B* is "the second doughnut is custard filled," what is the probability P(*B*|*A*)?

Solution: The first doughnut is custard filled. After it is eaten, there are no custard-filled doughnuts left. So it is impossible to get a custard filled for the second doughnut, P(*B*|*A*) = 0.

One Way of Determining if Events Are Independent or Dependent

If the probability of an event is known and the probability of that event given that another event happened is also known, it is possible to determine whether or not the two events are dependent events.

MATH FACTS

For dependent events, *A* and *B*, P(*A*|*B*) ≠ P(*A*). In other words, if *B* has happened, the probability of *A* happening is different than it would have been if *B* had not happened.

Example 2

If the probability of Joshua being late is .3 and the probability of Joshua being late given that he missed the bus is .9, what can be said about the events "Joshua is late to school" and "Joshua missed the bus"?

(1) They are independent events.

(2) They are dependent events.

(3) There is not enough information to determine if they are independent or dependent events.

(4) Joshua needs to wake up earlier.

Solution: Choice (2) is correct. Since P(Joshua is late) ≠ P(Joshua is late given that Joshua missed the bus), these are dependent events. If these probabilities were equal, the events would be independent.

Probability Questions Involving the Word "And"

With dependent events *A* and *B*, the probability of *A* and *B* both happening is not just P(*A*) · P(*B*) as it was with independent events.

With the box of doughnuts from before, the probability of getting a glazed doughnut as the first doughnut and getting a glazed doughnut as the second doughnut is $\frac{3}{12}$ (the probability of getting a glazed

Probability

doughnut first) times $\dfrac{2}{11}$ (the probability of getting a glazed doughnut second given that the first one was glazed).

P(glazed doughnut first and glazed doughnut second)

$$= \dfrac{3}{12} \cdot \dfrac{2}{11} = \dfrac{6}{132} = .05$$

=== MATH FACTS ===

For dependent events A and B, $P(A \text{ and } B) = P(A) \cdot P(B|A)$. When the events are independent, this formula also works since for independent events $P(B|A) = P(B)$. So $P(A) \cdot P(B|A) = P(A) \cdot P(B)$, which is the formula for $P(A \text{ and } B)$ when A and B are independent.

Another Way of Determining if Events Are Independent or Dependent

If A and B are independent events, $P(A \text{ and } B) = P(A) \cdot P(B)$. If they are not independent events, then $P(A \text{ and } B) \neq P(A) \cdot P(B)$. If the values of $P(A)$, $P(B)$, and $P(A \text{ and } B)$ are known, comparing $P(A \text{ and } B)$ to $P(A) \cdot P(B)$ will quickly reveal if the two events are dependent or independent.

Example 3

The probability of Genevieve having pizza for lunch on Wednesday is .4. The probability of her having pizza for dinner on Wednesday is .3. If the probability of Genevieve having pizza for both lunch and for dinner on Wednesday is .05, what can be concluded about event A, Genevieve will have pizza for lunch on Wednesday, and about event B, Genevieve will have pizza for dinner on Wednesday?

(1) The events are independent since $P(A \text{ and } B) = P(A) \cdot P(B)$.

(2) The events are dependent since $P(A|B) \neq P(A)$.

(3) The events are dependent since $P(A \text{ and } B) \neq P(A) \cdot P(B)$.

(4) The events are independent since $P(A|B) = P(A)$.

Solution: Since $.3 \cdot .4 \neq .05$, $P(A \text{ and } B) \neq P(A) \cdot P(B)$. This means A and B are dependent events. Choice (3) is correct. Although choice (2) also concludes they are dependent events, the value of $P(A|B)$ is not known, so it can't be compared to $P(A)$.

Dependent or Independent Events with Tables or Venn Diagrams

In Section 9.1 the results of a survey were presented in this table.

	Under 14 Years	Over 14 Years	Total
Under 5 feet	21	12	33
Over 5 feet	9	58	67
Total	30	70	100

Using the topics from this section, it can be determined whether or not the event A, "The person is over 14 years old" and event B, "The person is over 5 feet tall" are dependent or independent.

One way to check is to see if $P(A) = P(A|B)$.

$P(A) = \frac{70}{100} = .7$. However, $P(A|B) = \frac{58}{67} = .87$. Since these are not equal, the events are not independent.

A second way to check is to see if $P(A \text{ and } B) = P(A) \cdot P(B)$.

$P(A \text{ and } B) = \frac{58}{100} = .58$, $P(A) = \frac{70}{100} = .7$, and $P(B) = \frac{67}{100} = .67$. Since $P(A) \cdot P(B) = .469$, which does not equal $P(A \text{ and } B) = .58$, the events are not independent.

These two tests can also be applied to a Venn diagram.

Example 4

The Venn diagram below is based on two survey questions, "Do you like frozen yogurt?" and "Do you like mustard?"

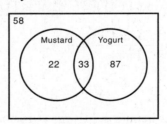

Based on the information in this Venn diagram, determine if event Y, "The person likes frozen yogurt" and event M, "The person likes mustard" are independent or dependent.

Solution: One way to check is to see if $P(Y \text{ and } M) = P(Y) \cdot P(M)$. For this example

Probability

$$P(Y \text{ and } M) = \frac{33}{200} = .165 . \quad P(Y) = \frac{120}{200} = .60 . \quad P(M) = \frac{55}{200} = .275 .$$

Since $P(Y) \cdot P(M) = .60 \cdot .275 = .165$, these are independent events.
The other way this could also be checked is by comparing
$P(Y) = \frac{120}{200} = .6$ to $P(Y|M) = \frac{33}{55} = .6$. Since $P(Y) = P(Y|M)$, they are
independent events.

Calculating the Probability of A Given B

The formula $P(A \text{ and } B) = P(A) \cdot P(B|A)$ can be rearranged into a formula that enables you to calculate $P(B|A)$ if the right information is known.

MATH FACTS

For any events A and B, $P(B|A) = \dfrac{P(A \text{ and } B)}{P(A)}$. For independent events,

this becomes $P(B|A) = \dfrac{P(A \text{ and } B)}{P(A)} = \dfrac{P(A) \cdot P(B)}{P(A)} = P(B) .$

To use this formula, very specific information must be provided in the question.

Example 5

If the $P(A \text{ and } B)$ is .4 and $P(A)$ is .7, what is $P(B|A)$?

Solution: $P(B|A) = \dfrac{P(A \text{ and } B)}{P(A)} = \dfrac{.4}{.7} \approx .57$

Example 6

Event B is "patient has hepatitis." Event A is "patient tests positive for hepatitis." If $P(A) = .01098$ and $P(A \text{ and } B) = .00099$, what is the probability that the patient has hepatitis given that the person tests positive for hepatitis?

Solution: $P(B|A) = \dfrac{P(A \text{ and } B)}{P(A)} = \dfrac{.00099}{.01098} \approx .09 = 9\%$

Check Your Understanding of Section 9.3

Chapter Nine   PROBABILITY

A. Multiple-Choice

1. If $P(A) = .4$ and $P(A|B) = .5$, what conclusion can be made?
   (1) $P(B) = .6$
   (2) A and B are independent events.
   (3) A and B are not independent events.
   (4) No conclusion can be made.

2. If $P(A) = .7$ and if A and B are independent events, what must be true?
   (1) $P(B) \neq .7$ (3) $P(B|A) = .7$
   (2) $P(A|B) = .7$ (4) No conclusion can be made.

3. If $P(A) = .4$, $P(B) = .7$, and $P(B|A) = .9$, what is $P(A$ and $B)$?
   (1) 1.3   (2) .28   (3) .36   (4) .63

4. If $P(A) = .3$ and $P(A$ and $B) = .12$, what is $P(B|A)$?
   (1) .2   (3) .4
   (2) .3   (4) Not enough information.

5. A dish has 20 blue candies and 30 yellow candies. One candy is chosen at random and eaten. Then another one is chosen at random and eaten. What is the probability that both candies were blue?
   (1) $\dfrac{20}{50} + \dfrac{19}{49}$ (2) $\dfrac{20}{50} \cdot \dfrac{20}{50}$ (3) $\dfrac{20}{50} \cdot \dfrac{19}{50}$ (4) $\dfrac{20}{50} \cdot \dfrac{19}{49}$

6. If $P(A) = .4$, $P(B) = .6$, and $P(A$ and $B) = .3$, what conclusion can you make about A and B?
   (1) $P(B|A) = .6$
   (2) A and B are independent events.
   (3) A and B are not independent events.
   (4) No conclusion is possible.

7. If $P(A) = .7$, $P(B) = .6$, and $P(A$ and $B) = .42$, what conclusion can you make?
   (1) $P(A|B) = .6$
   (2) A and B are independent events.
   (3) A and B are not independent events.
   (4) No conclusion can be made.

Probability

8. A group of 40 adults are asked if they like classical music, rock music, both, or neither. The results are collected in this Venn diagram.

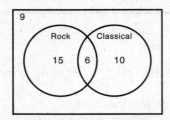

Based on this data, what conclusion can be made about liking classical music A and liking rock music B?
(1) $P(B) \neq P(B|A)$
(2) They are not independent of each other.
(3) They are independent of each other.
(4) No conclusion can be made.

9. 60 students are surveyed and asked if they use Facebook A, Instagram B, both, or neither. The results of the survey are collected in this Venn diagram below.

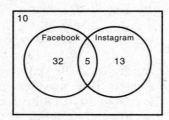

Based on this data, what conclusion can be made about using Facebook and using Instagram?
(1) $P(A|B) = P(A)$
(2) They are not independent of each other.
(3) They are independent of each other.
(4) No conclusion can be made.

10. 200 people are surveyed and asked two questions. (1) Which drink do you prefer: water, juice, or milk? (2) Which snack do you prefer: fruit or chips? The data are shown on the table below.

	Water	**Juice**	**Milk**	**Total**
Fruit	50	70	10	130
Chips	20	10	40	70
Total	70	80	50	200

Based on the data in the table, which of the following is *not* true?

(1) P(prefers juice or prefers chips) $= \dfrac{150}{200}$

(2) P(prefers fruit and prefers water) $= \dfrac{50}{200}$

(3) P(prefers fruit given that they prefer water) $= \dfrac{50}{70}$

(4) P(prefers milk given that they prefer chips) $= \dfrac{40}{70}$

B. *Show how you arrived at your answers.*

1. *A* is the event "Audrina smokes a pack of cigarettes a day." What is an event that is likely *not* to be independent of *A*?

2. P(*A*) = .2, P(*B*) = .7. How can the value of P(*A* and *B*) help you determine if *A* and *B* are independent events?

3. The probability that Charlotte goes to the party is P(Charlotte goes to party) = .7. The probability that Layla goes to the party is P(Layla goes to party) = .6. Explain how it could be that P(Layla goes to the party given that Charlotte goes to the party) = .9?

4. 104 students were polled about whether they like soccer, basketball, both, or neither. Based on this data, determine if liking basketball and liking soccer are independent events.

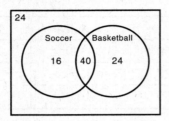

5. 150 people are surveyed and asked these two questions. (1) Which subject do you prefer: math, English, or science? (2) Is your hair curly or straight? The data are shown on the chart below.

	Math	English	Science	Total
Curly hair	25	30	35	90
Straight hair	45	5	10	60
Total	70	35	45	150

Based on the data in the chart, is having straight hair independent of preferring science? Explain your reasoning.

Chapter Ten

NORMAL DISTRIBUTION

10.1 STANDARD DEVIATION

The *standard deviation* of a set of numbers is a measure of how "spread out" the numbers in the set are. If all the numbers are equal, the standard deviation is zero. The more spread out the numbers are, the larger the standard deviation.

Comparing Standard Deviations for Two Data Sets with the Same Mean

Two rooms each have 100 people in them. The mean (average) height of the people in each room is approximately 66 inches. Even though the rooms have the same number of people and the same mean, there is something very different about the collection of people in each room.

Below are the dot plots for room 1 and room 2.

Normal Distribution

The people in room 1 are mostly very close to the 66-inch average. There are people in room 2, though, who are much shorter than the average and people who are much taller than the average.

In statistics, we say that the set of people in the second room have a higher standard deviation. The standard deviation for the heights of the people in room 1 was approximately 2, while the standard deviation for the heights of the people in room 2 was 6.

Example 1

Each of the dot plots below has a mean of approximately 40. Which of the dot plots below corresponds to the set with the lowest standard deviation?

Normal Distribution

Solution: Choice (4) is correct. The dots are bunched together much more closely in choice (4) than in the other choices. Choice (1) has a standard deviation of about 7. Choice (2) has a standard deviation of about 5. Choice (3) has a standard deviation of about 3. Choice (4) has a standard deviation of about 1.

Calculating Standard Deviation with a Calculator

On 10 Algebra II quizzes, Willian gets scores of 72, 76, 81, 81, 79, 71, 85, 79, 85, and 89. Calculating the standard deviation for the quiz scores can be done very quickly with a graphing calculator.

For the TI-84:

Press [STAT] [1] to get to the list editor. Go to the L1 cell, and press [CLEAR] [ENTER]. Enter the 10 scores into L1. Press [STAT], move to "CALC," and select [1]. The first value, 81.1 is the mean. The fifth value, next to the σx=, is the standard deviation. The standard deviation for this data set is 5.4.

For the TI-Nspire:

From the home screen, select the Add Lists & Spreadsheet icon. Name column A x, and fill in cells A1 though A10 with the 10 scores. Press [menu] [4] [1] [1] for One-Variable Statistics. Select the [ok] button. Set the X1 List to a[ ]. Set Frequency List to 1. In cell E2 will be the mean, 81.1. In cell C6 will be the standard deviation next to the σx=. The standard deviation for this data set is 5.4.

Example 2

Find the standard deviation of the data set 4, 6, 8, 8, 9, 12, 15, rounded to the nearest tenth.

Solution: Using the graphing calculator, the standard deviation of this set is 3.4.

Standard Deviation for Sets with Repeated Numbers

Entering all the numbers for a large data set into the calculator can be very time consuming. If, however, there are many repeated numbers, you can use a shortcut. Below the data for 55 numbers are summarized in a *frequency table.* The 7 in the frequency column after the number 42 means that there are seven 42s in the data set.

Number	Frequency
42	7
53	9
61	12
74	8
86	4
127	10
150	5

305

Normal Distribution

Though this could be done by entering 42, 42, 42, 42, 42, 42, 42, 53, 53, 53, etc., into the calculator, there is a shortcut for finding the standard deviation of the complete 55-number set.

For the TI-84:

Press [STAT] [1] and enter the numbers 42, 53, 61, 74, 86, 127, and 150 into L1, and enter the numbers 7, 9, 12, 8, 4, 10, 5 into L2. Press [STAT], move to "CALC" and press [1]. In the next menu, put L1 into the List field by pressing [2ND][1] and L2 into the FreqList field by pressing [2ND][2]. Then select Calculate. The standard deviation is 35.2.

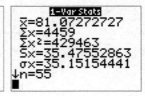

For the TI-Nspire:

From the home screen, select the Add Lists & Spreadsheet icon. Name column A x, and fill in cells A1 though A7 with the 7 scores. Name column B y, and fill in cells B1 through B7 with the 7 frequencies. Press [menu] [4] [1] [1] for One-Variable Statistics. Set the number of lists to 1. Select the [ok] button. Set the X1 List to a[ ] and the Frequency List to b[ ]. In cell E2 will be the mean, 81.1. In cell E6 will be the standard deviation next to the $\sigma x=$. The standard deviation for this data set is 35.2.

306

z-scores

In a data set with a mean of 60 and a standard deviation of 4, the number 64 is one standard deviation above the mean since it is equal to the mean plus the standard deviation, 60 + 4 = 64. The number 68 is two standard deviations above the mean since it is equal to the mean plus two times the standard deviation 60 + 2 · 4 = 68.

In statistics, we say that a number that is one standard deviation above the mean has a *z-score* of 1, while a number that is two standard deviations above the mean has a *z*-score of 2.

MATH FACTS

The *z-score* of a number is how many standard deviations above or below the mean that number is. If the number is greater than the mean, the *z*-score is positive. If the number is less than the mean, the *z*-score is negative.

A formula for calculating *z*-score is $z = \dfrac{\text{number} - \text{mean}}{\text{standard deviation}}$.

Example 3

If the mean is 60 and the standard deviation is 4, what is the *z*-score of the number 52?

Solution: Since 52 is less than 60, the *z*-score will be negative. For this example, it is possible to subtract the standard deviation, 4, from 60 to get to 56, which has a *z*-score of −1. Then subtract it again to get to 52, which has a *z*-score of −2.

This can also be calculated with the formula $z = \dfrac{\text{number} - \text{mean}}{\text{standard deviation}} = \dfrac{52 - 60}{4} = \dfrac{-8}{4} = -2$.

Example 4

If the mean is 40 and the standard deviation is 7, what is the *z*-score of the number 50?

Solution: The *z*-score must be between 1 and 2 since 40 + 7 = 47 would have a *z*-score of 1 while 40 + 2 · 7 = 54 would have a *z*-score of 2.

Use the formula: $z = \dfrac{50 - 40}{7} = \dfrac{10}{7} \approx 1.4$

Normal Distribution

Example 5

If the mean is 73 and the standard deviation is 8, what number has a z-score of −1.5?

Solution: Multiply the z-score by the standard deviation to get $-1.5 \cdot 8 = -12$. Add this to the mean to get $73 + (-12) = 61$, which is the number that has a z-score of −1.5.

This can also be done with the formula with z as a known variable and the number as the unknown.

$$-1.5 = \frac{x - 73}{8}$$
$$8 \cdot (-1.5) = x - 73$$
$$-12 = x - 73$$
$$\underline{+73 = +73}$$
$$61 = x$$

Check Your Understanding of Section 10.1

A. Multiple-Choice

1. Each of these data sets has a mean of 5. Which has the smallest standard deviation?
   (1) 1, 2, 5, 7, 10 (3) 4, 5, 5, 5, 6
   (2) 2, 3, 5, 5, 10 (4) 3, 4, 5, 5, 8

2. What is the standard deviation of the data set 16, 19, 14, 12, 10, 15, 15, 16?
   (1) 2.5   (2) 2.6   (3) 2.7   (4) 2.8

3. What is the standard deviation for this data set?

x_i	f_i
18	4
19	7
20	2
21	3
22	6

   (1) 1.41   (2) 1.51   (3) 1.54   (4) 1.61

4. If the mean of a data set is 60 and the standard deviation is 8, what is the z-score for the number 72?
   (1) .67   (2) 1.00   (3) 1.25   (4) 1.50

5. If the mean of a data set is 40 and the standard deviation is 6, what number has a z-score of -2.5?
   (1) 25   (2) 35   (3) 45   (4) 55

6. Which number has a z-score closest to -1 in the data set 45, 38, 47, 41, 42, 34, 46, 37?
   (1) 42   (2) 41   (3) 38   (4) 37

7. If the mean of a data set is 41 and the standard deviation is 6, which of the following has a z-score between 1.5 and 2.0?
   (1) 45   (2) 47   (3) 49   (4) 51

8. How many of these 8 numbers are more than two standard deviations from the mean?

                   18, 40, 44, 45, 48, 50, 50, 68

   (1) 1   (2) 2   (3) 3   (4) 0

9. How many of these 8 numbers are more than two standard deviations from the mean?

                   11, 49, 52, 52, 53, 53, 55, 90

   (1) 1   (2) 2   (3) 3   (4) 0

10. What percent of these 14 numbers have a z-score between -1 and $+1$?

           59, 62, 54, 58, 62, 62, 55, 56, 65, 64, 58, 53, 59, 61

    (1) 68%   (2) 71%   (3) 78%   (4) 93%

10.2 NORMAL DISTRIBUTION

KEY IDEAS

The dot plot of real-world data often has the shape of a *bell curve*, which is also known as a *normal distribution curve*. When this happens, it is possible to answer certain questions about the data that you would not be able to had the shape of the dot plot not been a bell curve.

Normal Distribution

The Normal Curve

Below is a dot plot with 400 data points representing the weights of 400 children. The mean value is approximately 60, and the standard deviation is approximately 4.

Height

Since the mean is 60 and the standard deviation is 4, a score of 64 is said to be *one standard deviation above the mean*. A score of 68 is said to be *two standard deviations above the mean*. A score of 56 is said to be *one standard deviation below the mean*. A score of 52 is said to be *two standard deviations below the mean*.

Percentages in the Normal Distribution Curve

In a normal distribution with a lot of data points, approximately 68% of the values will be between one standard deviation below the mean and one standard deviation above the mean. 95% of the values will be between two standard deviations below the mean and two standard deviations above the mean.

The dot plot about the weights of the 100 children was approximately *normal* because 70% of the 100 points are between 56 and 64 while 93% of the 100 points are between 52 and 68.

Using a Calculator to Find the Percentage of Numbers Between Two Values

If the mean and the standard deviation of a data set are known, the graphing calculator can quickly determine what percent of the numbers in the data set will be between any two values.

The heights of 400 people are measured. The mean is approximately 60, and the standard deviation is approximately 4.

Since 68% of the numbers are expected to fall within one standard deviation of the mean, then approximately $.68 \cdot 400 = 272$ should be between $60 - 4 = 56$ inches and $60 + 4 = 64$ inches. Since 95% of the numbers are expected to fall within two standard deviations of the mean, then approximately $.95 \cdot 400 = 380$ should be between $60 - 8 = 52$ inches and $60 + 8 = 68$ inches.

Knowing the 68% and 95% rule is useful if a question asks how many members of a set are within one or two standard deviations of the mean. Most questions, though, are about finding how many members of a set are between two numbers that are not exactly one or two standard deviations from the mean.

In the scenario above, a graphing calculator can be used to determine how many out of the 400 people can be expected to be between 57.5 and 62 inches tall. Between 56 and 64 inches was 68% since those were the numbers exactly one standard deviation above and below the mean. The percent of people between 57.5 and 62 inches will be lower than 68% and can be found with the graphing calculator.

Normal Distribution

For the TI-84:
Press [2ND] [VARS] [2] for normacdf. Fill in the fields with 57.5 for lower, 62 for upper, 60 for µ, and 4 for σ Then select "PASTE."

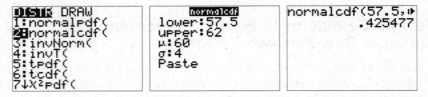

For the TI-Nspire:
From the home screen, press [A]. Press [menu] [6] [5] [2] for Normal Cdf. Set the Lower Bound to 57.5, the Upper Bound to 62, the mean (μ) to 60, and the standard deviation (σ) to 4.

The percent is approximately 43%, which is approximately $.43 \cdot 400 = 72$ people between 57.5 inches and 62 inches tall.

Example 1

If the mean of a set of 500 Regents scores is 80, the standard deviation is 15, and the data are approximately normally distributed, approximately how many scores can be expected to be between 74 and 89?

Solution: Using the Normal cdf function of the calculator, the percent of scores between 74 and 89 will be approximately 38%. Out of 500 scores, this will be approximately $.38 \cdot 500 = 190$ scores.

Example 2

If the mean temperature in New York throughout the year is 55 degrees and the standard deviation is 15 degrees, what percent of the days can be expected to have temperatures greater than 80 degrees?

Solution: In this example, there is no upper bound. When this happens, make the upper bound a very large number, at least four standard deviations above the mean. (The number 9,999 is generally safe to use unless there is a very large mean and/or standard deviation.) When using 9,999 as the upper bound, the calculator determines the percent to be approximately 5%.

For examples where there is no lower bound, make the lower bound a very small negative number like –9,999.

Check Your Understanding of Section 10.2

A. *Multiple-Choice*

1. If the mean of a set of normally distributed 200 numbers is 70 and the standard deviation is 6, approximately what percent of the numbers will be between 64 and 76?
   (1) 68%   (2) 70%   (3) 76%   (4) 80%

2. The weights of 400 children are collected, and the data are normally distributed. What percent of the children were between 42 and 52 pounds if the mean was 50 pounds and the standard deviation was 3?
   (1) 57%   (2) 60%   (3) 63%   (4) 74%

3. On the Algebra II Regents, the mean score was 75 and the standard deviation was 9. If the data are normally distributed, what percent of the scores were over 90?
   (1) 20%   (2) 15%   (3) 10%   (4) 5%

4. The amount of time that 300 adults exercise per week was collected. If the data were normally distributed, what percent of the people exercised under 20 minutes a week if the mean was 40 minutes and the standard deviation was 15?
   (1) 9%   (2) 11%   (3) 13%   (4) 15%

Normal Distribution

5. In a normally distributed data set, what percent of the samples will have a z-score between 1 and 2?
   (1) 14%　　　(2) 16%　　　　　(3) 18%　　　　　(4) 20%

6. In a set of 500 samples, the mean is 90 and the standard deviation is 17. If the data are normally distributed, how many of the 500 are expected to have a value between 93 and 101?
   (1) 82　　　　(2) 84　　　　　　(3) 86　　　　　(4) 88

7. In a high school, 1,000 students run a mile. The mean time is 9 minutes with a standard deviation of 2 minutes. What running time would put a student in the top 5 percentile rank?
   (1) 4.3 minutes　　　　　　　(3) 5.3 minutes
   (2) 4.8 minutes　　　　　　　(4) 5.8 minutes

8. A group of baseball players had a mean of 15.3 home runs with a standard deviation of 9.9. If the data are normally distributed, what percent of players can be expected to hit over 30 home runs?
   (1) 3%　　　　(2) 5%　　　　　　(3) 7%　　　　　(4) 9%

9. The average snowfall in Regents Town is 19 inches per year with a standard deviation of 5 inches. If the data are normally distributed, what is the probability that it will snow between 18.5 and 19.5 inches in a year?
   (1) 6%　　　　(2) 8%　　　　　　(3) 20%　　　　　(4) 27%

10. In a normally distributed data set with a mean of 60, what is more likely: a number greater than 80 or a number less than 30?
    (1) A number greater than 80
    (2) A number less than 30
    (3) They are equally likely
    (4) It depends on what the standard deviation is

B. *Show how you arrived at your answers.*

1. A data set of ten numbers is shown below:

   $$28, 83, 73, 88, 81, 54, 75, 66, 66, 91$$

   a) Calculate the mean and the standard deviation for the data set.
   b) How many of the numbers fall within one standard deviation of the mean?
   c) Based on your answer to part (b), do you think this data set is approximately normal or not? Explain your reasoning.

2. Which, if any, of these numbers is not within 2 standard deviations of the mean?

    46, 47, 38, 38, 36, 41, 44, 40, 33, 38, 40, 39?

3. The heights of 200 adults are measured. The average height is 66 inches, and the standard deviation is 8 inches. If 14 of the people are over 82 inches, what does this suggest about whether or not the data are normally distributed? Explain your reasoning.

4. The ages of 20 people are shown:

    66, 72, 70, 74, 78

    80, 73, 68, 84, 76

    76, 78, 70, 62, 75

    76, 79, 73, 73, 70

    Which, if any, of these ages are not within 2 standard deviations of the mean?

5. A data set has 68.3% of the population of samples within one standard deviation of the mean. Is this a normal distribution? Explain your reasoning.

Chapter Eleven

STATISTICS

11.1 TYPES OF STATISTICAL STUDIES

Three ways to collect data for a statistical study are *survey*, *observational study*, and *experiment*. *Bias* in a statistical study is when something about how the data are collected may have caused inaccurate results. Causes of this bias should be identified and, if possible, eliminated before collecting the data.

Statistical Survey

A survey is a question or series of questions that participants, also called *subjects*, in the study are asked to answer. The survey can have a simple yes/no question like "Do you like pistachio ice cream?" or a question with a numerical answer like "How much time did you spend on homework last night?" A survey is the simplest way of collecting data, but there are many ways that it can lead to bias.

The wording of the survey question can cause bias. For example, if the question says "Do you not hate pistachio ice cream?" rather than "Do you like pistachio ice cream?" the negative phrasing of the question could change the results even if the questions are supposed to mean the same thing.

The participants of the survey should be randomly selected, otherwise the results can also be distorted. The results of a survey about pop music, for example, will not be accurate if most of the people asked to complete the survey are under 20 years old. If the participants in the survey do not accurately represent the total population, the survey has *selection bias*.

If the survey is voluntary, there is a chance that people who respond to the survey are more likely to answer the questions a certain way. A survey that is conducted by text messaging, for example, might be answered more by younger people, causing bias in the results.

316

Observational Study

An *observational study* is like a survey. Instead of subjects being asked to answer questions, the person conducting the study observes the behavior of the participants and records the results. To learn about whether or not people like pistachio ice cream, the observer could go to an ice cream shop and watch what different people order. An important aspect of an observational study is that the person conducting the study cannot do anything that could interfere with or control what the subjects do.

An observational study could have bias if the subjects are not randomly selected. For example, if the ice cream shop observations happen during school hours, people under 18 will not be adequately represented.

Experimental Study

In an *experimental study*, the person conducting the study randomly chooses some of the subjects and exposes them to some kind of *treatment*. An example is a study to see if taking vitamin C prevents colds. Fifty people are randomly selected and then, from those fifty, twenty-five are randomly selected to take vitamin C pills daily while the other twenty-five are not given the pills.

It is important in an experimental study to choose the group of people getting the treatment randomly. If there is some kind of bias in who receives the treatment, the experiment might lead to an inaccurate conclusion.

Check Your Understanding of Section 11.1

B. Show how you arrived at your answers.

1. You want to find out what type of movie is the favorite among the residents of New York City. If you want to do this by conducting a survey, what are some ways that you can reduce bias?

2. You want to study how the amount of sleep a student gets the night before the Regents relates to his or her Regents score. How can this be studied as an observational study? How can this be studied as an experiment?

3. There are 500 10th graders at Regentsville High School. You want to do an experiment where 20 students study for the Regents with just

the red Barron's book and 20 other students study for the Regents with both the blue and the red Barron's books. How can the groups be chosen to reduce the chance of skewed data through bias?

4. A survey is conducted to learn what subjects students enjoy most at school. One question reads, "What is your most boring subject?" Could the wording of this question skew the results of the survey? Explain.

5. Is this describing a survey, an experiment, or an observational study? Explain.

A teacher teaches two Algebra II math classes, one during period 5 and another during period 6. In the period 5 class, the temperature in the room is set to 70 degrees for a two-week period. In the period 6 class, the temperature in the room is set to 85 degrees for the same two-week period. At the end of the two weeks, the students are given a test to see how well they learned the math. The results are collected and analyzed.

11.2 INFERENTIAL STATISTICS

KEY IDEAS

If much of the data in a data set is unknown, it is still possible to estimate different information about it by using a smaller collection of the total set known as a *sample*. Statistics about the sample can provide valuable information about the complete data set.

A complete data set can contain thousands, if not millions, of data points. The complete set of data is known as the *population*. A much smaller subset of the population is known as a *sample*.

Two types of measures about the population that can be estimated with samples are the *population proportion* and the *population mean*. The population proportion is the percent of the total number of things in the population that has a certain characteristic. The population mean is the average value of some characteristic of all the things in the population. The percent of some sample of things that has a certain characteristic is called the *sample proportion*. The *sample mean* is the average value of some characteristic of all the things in that sample.

To estimate the population proportion or the population mean, you generally find a sample proportion or sample mean and then add and subtract the *standard error*.

For example, if the sample mean is 20 and the standard error is 7, the population mean will be between $20 - 7 = 13$ and $20 + 7 = 27$. If the sample proportion is 0.40 and the standard error is 0.15, the population mean estimate will be between $0.40 - 0.15 = 0.25$ and $0.40 + 0.15 = 0.55$.

Calculating the standard error depends on the type of question. Four different types are presented below.

Using One Sample Set to Approximate the Population Proportion

In an auditorium with 800 students, a certain percent of them will be boys and a certain percent will be girls. If 450 of those students are boys, then the percent of total students who are boys is $\frac{450}{800} = 0.5625 = 56.25\%$.

This percent is the population proportion, represented by the symbol p, of boys for the entire population of students.

If, for whatever reason, we are only able to survey a subset of 75 students, we can calculate the sample proportion for that sample. If 45 out of those 75 students were boys, the sample proportion for the percent of boys, represented by the symbol $\hat{p}$, would be $\hat{p} = \frac{45}{75} = 0.60$.

For this type of problem, the standard error, SE, can be calculated with the following formula.

$$\text{SE} = 2 \cdot \sqrt{\frac{\hat{p}(1-\hat{p})}{n}} = 2 \cdot \sqrt{\frac{0.60(1-0.60)}{75}} \approx 0.11$$

The sample proportion is different from the population proportion in this example. However, there is a formula that allows you to find a range of likely values for the population proportion based on this one sample of 75 students.

$$\hat{p} - \text{SE} \le p \le \hat{p} + \text{SE}$$

The population proportion is, *almost surely*, between $0.60 + 0.11 = 0.71 = 71\%$ and $0.60 - 0.11 = 0.49 = 49\%$.

This is a pretty large range of values. However, the actual value of the population proportion, 56.25%, is between 49% and 71% as the formula predicted.

Using Multiple Sample Sets to Approximate the Population Proportion

For a more accurate estimate of the population proportion, more than one sample set is needed. In the auditorium example, one sample set of 75 students with a sample mean of 0.60 was used. To get a better approximation, repeat this process 50 times. The first sample of 75 had 45 boys, which was 60%. The second sample of 75 had 35 boys, which was approximately 47%. A computer is used to do this process 50 times. When the percent of boys for all 50 sets of 75 students are plotted on a dot plot, it looks like this.

This plot is called the *sampling distribution of the sample proportion*.

Two statistics about this plot are also provided. The mean value for the numbers in this plot is approximately 0.54, and the standard deviation for the numbers in this plot is approximately 0.05.

These two numbers can now be used to approximate the population proportion of the complete data set.

For this type of problem, the standard error, SE, can be calculated with the following formula.

$$SE = 2 \cdot \sigma = 2 \cdot 0.05 = 0.10$$

The estimated population proportion is then mean of plot $-$ SE $\leq p \leq$ mean of plot $+$ SE.

For this example the actual value of the population proportion is, *almost surely*, between $0.54 - 0.10 = 0.44$ and $0.54 + 0.10 = 0.64$. The actual proportion is 56.25%, which is, indeed, between 44% and 64%.

On the Regents, the information will be provided so you can do the analysis. Creating the dot plot and finding the mean and the standard deviation of the numbers in the dot plot is something that requires a computer.

Making this plot required 50 · 75 = 3,750 numbers. Even though this is a lot of numbers for calculating the percent of boys out of just 800 people, this method of using a computer to do this analysis is very powerful either when the population is either extremely large or when, for various reasons, it is not possible to get access to all the data.

Using One Sample Set to Approximate the Population Mean

At a high school graduation party are 600 people, including parents and students. There are more parents than students at the party, and a dot plot of the ages of all the party goers looks like this.

The approximate average age of the 600 people at the party can be approximated by first taking a random sample of 30 people at the party. Here is one random sampling of 30 people from the total population of 600 people.

51, 45, 42, 13, 40, 20, 45, 49, 41, 15, 18, 45, 41, 14, 35, 41, 17, 40, 42, 45, 51, 18, 31, 11, 19, 47, 20, 14, 17, 47

To use these 30 numbers to approximate the mean age of the 600 people, first find the mean and the standard deviation of these 30 values. For this sample set, the mean value, represented by $\bar{x}$, is approximately 32.4. The standard deviation of the sample, represented by s, is approximately 13.8.

For this type of problem the standard error, SE, can be calculated with the following formula.

$$\text{SE} = 2 \cdot \frac{s}{\sqrt{n}} = 2 \cdot \frac{13.8}{\sqrt{30}} \approx 5.04$$

Statistics

The formula for approximating the population mean, represented by μ, is $\bar{x} - SE \leq \mu \leq \bar{x} + SE$.

So according to this formula, the population mean is *almost surely* between 32.4 − 5.04 = 27.4 years old and 32.4 + 5.05 = 37.4 years old. The actual population mean is 34.4 years old, so the predicted range was correct.

Using Multiple Samples Set to Approximate the Population Mean

By using just one sample set of 30 ages, it was possible to find an approximate value of the population mean. To get an even more accurate approximation, more sample sets are needed.

By using a computer, 100 sample sets of 30 ages each were taken. The mean for the first set, used in the other approximation, was 32.4. The mean for the second set was 36.5. The mean for each of these 100 sets was then plotted on this dot plot or histogram.

Measures from Sample of Graduation Party

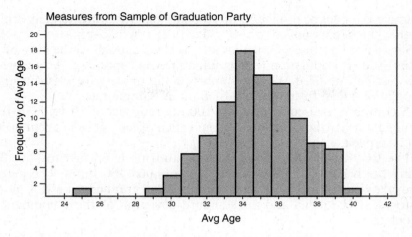

This is called the *sampling distribution of the sample mean*.

Two more statistics will be needed and likely given on the Regents. The mean value for the numbers in the plot is approximately 34.5. The standard deviation is approximately 2.5.

For this scenario, the standard error, SE, is 2 times the standard deviation of all the different means.

$$SE = 2 \cdot 2.5 = 5$$

With these numbers, the population mean can be calculated with the formula:

$$\text{mean of plot} - SE \leq \mu \leq \text{mean of plot} + SE.$$

For this example, $34.5 - 5 \leq \mu \leq 34.5 + 5$. So the population mean is between 29.5 and 39.5. This is more accurate than the approximation that used just one set of 30. The mean of the samples, 34.5, is extremely close to the actual population mean of 34.4.

If the standard deviation of the sample means is not given, it can be approximated with the formula $\sigma = \dfrac{s}{\sqrt{n}}$, where s is the standard deviation of any one of the sample sets and n is the number of samples in each of the sample sets. When using the first sample set, which had a standard deviation of 13.8, $\sigma = \dfrac{13.8}{\sqrt{30}} \approx 2.5$. Since the standard error is two times the standard deviation, it is also true that $SE = 2 \cdot \dfrac{s}{\sqrt{n}}$.

Using a Simulation for a Randomization Test

A *randomization test* is a way to check if the results of a small random sample are near what you would expect (or hope) them to be.

In this example, a focus group of 30 students are asked to give their opinion on a new style of backpack. The manufacturer is willing to mass produce the backpack if 40% of students like it. Unfortunately, only 9 of the 30 students polled, or 30%, liked the new backpack.

There is a way to check if that 30% is close enough to the 40% of the entire student population the manufacturer had specified. A *randomization test* can be used. This is similar to the process for estimating the population mean from a bunch of random sample sets.

A computer is used to generate 100 random sets of 30 students from an imaginary group of students from which exactly 40% of them do like the backpack.

The first set of these imaginary 30 students happened to have 37% liking the backpack. This process is continued 100 times. The percent who like the backpack for each sample set is graphed on a dot plot or histogram. The mean and the standard deviation of those numbers are calculated.

In this case, the mean value is 0.404 and the standard deviation is 0.089.

Even without being told the mean and standard deviation of this plot of the sampling distribution for the imaginary set, it is possible to approximate them. The mean will be 0.40, since that is the proportion that was used to create the plot. The standard deviation can be calculated with the formula $\sigma = \dfrac{\sqrt{p(1-p)}}{n} = \dfrac{\sqrt{0.4 \cdot 0.6}}{30} \approx 0.089$, which was exactly what it was for the data on the dot plot. On the dot plot, it can be seen that 30% or less happened 16 times. So it does not seem so unusual for this to happen.

To determine if 30% is something that is *likely* to happen even though the real percent was 40% for this imaginary population, check to see if the 30% is within two standard deviations of the mean.

For this example, $0.404 - 2 \cdot 0.089 = 0.226$ and $0.404 + 2 \cdot 0.089 = 0.582$. Since 30% is between 22.6% and 58.2%, the company can justify producing the backpacks even though the sample of 30 people had fewer than 40% liking it.

Check Your Understanding of Section 11.2

A. *Multiple-Choice*

1. An ecologist wants to know what percent of the 10,000 fish in a lake are cod. She conducts a study in which 50 fish are randomly caught. Of those 50 fish, 15 are cod. Based on this data, using a 95% confidence interval, the researcher can determine that the percent of fish that are cod in the lake is which of the following?
   (1) Between 17% and 43%   (3) Between 24% and 36%
   (2) Exactly 30%   (4) Between 20% and 40%

2. Out of 30,000 households in a town, 100 are selected at random and asked if they have a landline or not. 70 out of the 100 households chosen replied that they do have a landline. Based on this data, using a 95% confidence interval, you can determine that the percent of households in the town that have a landline is which of the following?
   (1) Exactly 70%   (3) Between 64% and 76%
   (2) Between 68% and 72%   (4) Between 61% and 79%

3. A sports physician conducts an observational study to learn the average amount of time that 3,000 swimmers in the town can hold their breath underwater. 60 swimmers are chosen at random and tested. The average amount of time those 60 swimmers can hold their breath is 70 seconds with a standard deviation of 6. Based on this data, using a 95% confidence interval, the researcher can determine that for the entire 3,000 swimmers, the average amount of time they can hold their breath underwater is which of the following?
   (1) Between 71.5 seconds and 68.5 seconds
   (2) Exactly 70 seconds
   (3) Between 69.5 seconds and 70.5 seconds
   (4) Between 69 seconds and 71 seconds

Statistics

4. A concert promoter wants to estimate the average age of the 20,000 people attending a Taylor Swift concert. At the concert, 80 people are selected at random. The average age of the 80 people is 24.3 with a standard deviation of 11.9. Based on this data, using a 95% confidence interval, the researcher can determine that the average age for all 20,000 people at the concert is which of the following?
   (1) Exactly 24.3
   (2) Between 21.64 and 26.96
   (3) Between 23.3 and 25.3
   (4) Between 23.8 and 25.1

5. Use the same population as from question 1 except take 500 samples of size 50. The average for all these samplings is 0.24 with a standard deviation of 0.06.

    This is a histogram of the sampling distribution of the sample proportion.

Using this data, with a 95% confidence interval, we can determine that the percent of fish in the lake that are cod is which of the following?
   (1) Between 0.12 and 0.36
   (2) Between 0.14 and 0.34
   (3) Between 0.20 and 0.28
   (4) Exactly 0.24

6. Using the same population as in question 2 but take 300 samplings of size 100. The average of all the means is 0.73, and the standard deviation is 0.05.

   This is a histogram of the sampling distribution of the sample proportion.

   Using this data, with a 95% confidence interval, we can determine that the percent of homes in the town that have a land line is which of the following?
   (1) Between 0.63 and 0.83
   (2) Between 0.68 and 0.78
   (3) Between 0.70 and 0.76
   (4) Exactly 0.73

7. The sports physician uses the same population as in question 3 but uses 150 samplings of 60 people. The average of the means of all the samplings is 72.7, and the standard deviation is 0.92.

   This is a histogram of the sampling distribution of the sample mean.

Statistics

Based on this data, with a 95% confidence interval, the researchers can determine that the actual average amount of time the entire population can hold its breath under water is which of the following?
(1) Exactly 72.7
(2) Between 72 and 73.4
(3) Between 71.28 and 73.12
(4) Between 70.86 and 74.54

8. The concert promoter uses the same population as in question 4 but takes 150 samplings of 80 people. The average of the means of all the samplings is 25.5, and the standard deviation is 1.5.

This is a histogram of the sampling distribution of the sample mean.

Based on this data, with a 95% confidence interval, the researchers can determine that the average age of the entire 20,000 person population is which of the following?
(1) Exactly 25.5
(2) Between 22.5 and 28.5
(3) Between 23.5 and 27.5
(4) Between 24.5 and 26.5

9. Amy Hogan rolls a six-sided die 12 times and lands on 6 on four of the rolls. She wants to examine if this might suggest that the die is defective since, on average, the number of expected 6s in 12 rolls is 2. She uses a computer to simulate 500 times of rolling a fair die 12 times. A histogram of the results is shown below.

Based on this histogram, Amy concludes which of the following?
(1) Yes, the die is defective because two 6s is most common.
(2) Yes, the die is defective because four 6s is not common.
(3) No, the die is not necessarily defective because two 6s is not very common.
(4) No, the die is not necessarily defective because four 6s is fairly common.

10. A new plant food is developed that the inventors claim helps plants grow faster. They cite an experiment they did where out of 40 randomly chosen plants, 20 of them were randomly chosen to receive the new plant food while the other 20 plants received a different kind of plant food.

The 20 plants that received the new plant food grew an average of 2 inches more than the plants that received a different plant food. To test if this difference is statistically significant, the data for the 40 plants are entered into a computer that randomly separates them into two groups of 20. This is done 50 times. The results are shown in the histogram below, where the number is the difference between the first group and the second group of the randomly chosen groupings.

Based on this data, can you conclude that the new plant food causes plants to grow faster?
(1) No, the difference is not statistically significant. +2 is uncommon.
(2) No, the difference is not statistically significant. +2 is fairly common.
(3) Yes, the difference is statistically significant. +2 is uncommon.
(4) Yes, the difference is statistically significant. +2 is fairly common.

| Chapter Twelve | # TEST-TAKING STRATEGIES |

Knowing the material is only part of the battle in acing the Algebra Regents exam. Things like not managing your time, making careless errors, and struggling with the calculator can cost valuable points. In this chapter some test-taking strategies are explained to help you perform your best on test day.

12.1 TIME MANAGEMENT

Don't Rush

The Algebra Regents exam is three hours long. Even though you are permitted to leave after one and a half hours, to get the best grade possible, stay until the end of the exam. Just as it wouldn't be wise to come to the test an hour late, it is almost the same thing if you leave a test an hour early.

Do the Test Twice

The best way to protect against careless errors is to do the entire test twice and compare the answers you got the first time to the answers you got the second time. For any answers that don't agree, do a "tie breaker" third time. Redoing the test and comparing answers is much more effective than simply "looking over" your work. People tend to skim by careless errors when looking over their work. Redoing the questions, you are less likely to make the same careless error.

Bring a Watch

In some classrooms the clock is broken. Without knowing how much time is left, you might rush and make careless errors. Though the proctor should write the time on the board and update it every so often, it is best if you have your own watch.

Test-Taking Strategies

The TI-84 graphing calculator has a built in clock. Press the [MODE] to see it. If the time is not right, go to SET CLOCK and set it correctly. The TI-Nspire does not have a built-in clock.

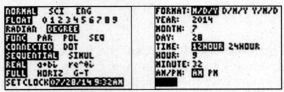

12.2 KNOW HOW TO GET PARTIAL CREDIT

Know the Structure of the Exam

The Algebra Regents exam has 37 questions. The first 24 of those questions are multiple-choice worth two points each. There is no partial credit if you make a mistake on one of those questions. Even the smallest careless error, like missing a negative sign, will result in no credit for that question.

Parts Two, Three, and Four are free-response questions with no multiple-choice. Besides giving a numerical answer, you may be asked to explain your reasoning.

Part Two has eight free-response questions worth two points each. The smallest careless error will cause you to lose one point, which is half the value of the question.

Part Three has four free-response questions worth four points each. These questions generally have multiple parts.

Part Four has one free-response question worth six points. This question will have multiple parts.

Explaining Your Reasoning

When a free-response question asks to "Justify your answer," "Explain your answer," or "Explain how you determined your answer," the grader is expecting a few clearly written sentences. For these, you don't want to write too little since the grader needs to see that you understand why you did the different steps you did to solve the equation. You also don't want to write too much because if anything you write is not accurate, points can be deducted.

Here is an example followed by two solutions. The first would not get full credit, but the second would.

Example 1

Use algebra to solve for x in the equation $\frac{2}{3}x + 1 = 11$. Justify your steps.

Solution 1 (part credit):

$\frac{2}{3}x + 1 = 11$ $\underline{\quad -1 = -1 \quad}$ $\frac{2}{3}x = 10$ $x = 15$	I used algebra to get the x by itself. The answer was $x = 15$.

Solution 2 (full credit):

$\frac{2}{3}x + 1 = 11$ $\underline{\quad -1 = -1 \quad}$ $\frac{2}{3}x = 10$ $\frac{3}{2} \cdot \frac{2}{3}x = \frac{3}{2} \cdot 10$ $1x = 15$ $x = 15$	I used the subtraction property of equality to eliminate the +1 from the left-hand side. Then to make it so the x had a 1 in front of it, I used the multiplication property of equality and multiplied both sides of the equation by the reciprocal of $\frac{2}{3}$, which is $\frac{3}{2}$. Then since $1 \cdot x = x$, the left-hand side of the equation just became x and the right-hand side became 15.

Computational Errors vs. Conceptual Errors

In the Part Three and Part Four questions, the graders are instructed to take off one point for a "computational error" but half credit for a "conceptual error." This is the difference between these two types of errors.

If a four point question was $x - 1 = 2$ and a student did it like this,

$$x - 1 = 2$$
$$\underline{\quad +1 = +1 \quad}$$
$$x = 4$$

the student would lose one point out of 4 because there was one computational error since $2 + 1 = 3$ and not 4.

333

Had the student done it like this,

$$x - 1 = 2$$
$$\underline{-1 = -1}$$
$$x = 1$$

the student would lose half credit, or 2 points, since this error was conceptual. The student thought that to eliminate the -1, he should subtract 1 from both sides of the equation.

Either error might just be careless, but the conceptual error is the one that gets the harsher deduction.

12.3 KNOW YOUR CALCULATOR

Which Calculator Should You Use?

The two calculators used for this book are the TI-84 and the TI-Nspire. Both are very powerful. The TI-84 is somewhat easier to use for the functions needed for this test. The TI-Nspire has more features for courses in the future. The choice is up to you. This author prefers the TI-84 for the Algebra II Regents.

Graphing calculators come with manuals that are as thick as the book you are holding. There are also plenty of video tutorials online for learning how to use advanced features of the calculator. To become an expert user, watch the online tutorials or read the manual.

Clearing the Memory

You may be asked at the beginning of the test to clear the memory of your calculator. When practicing for the test, you should clear the memory too so you are practicing under test-taking conditions.

This is how you clear the memory.

For the TI-84:

Press [2ND] and then [+] to get to the MEMORY menu. Then press [7] for Reset.

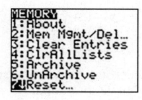

Use the arrows to go to [ALL] for All Memory. Then press [1].

Press [2] for Reset.

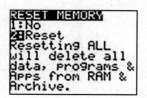

The calculator will be reset as if brand new condition. The one setting that you may need to change is to turn the diagnostics on if you need to calculate the correlation coefficient.

For the TI-Nspire:

The TI-Nspire must be set to Press-To-Test mode when taking the Algebra Regents. Turn the calculator off by pressing [ctrl] and [home]. Press and hold [esc] and then press [home].

While in Press-to-Test mode, certain features will be deactivated. A small green light will blink on the calculator so a proctor can verify the calculator is in Press-to-Test mode.

To exit Press-to-Test mode, use a USB cable to connect the calculator to another TI-Nspire. Then from the home screen on the calculator in Press-to-Test mode, press [doc], [9] and select Exit Press-to-Test.

Use Parentheses

The calculator always uses the order of operations where multiplication and division happen before addition and subtraction. Sometimes, though, you may want the calculator to do the operations in a different order.

Test-Taking Strategies

Suppose at the end of a quadratic equation, you have to round $x = \frac{-1+\sqrt{5}}{2}$ to the nearest hundredth. If you enter (–) (1) (+) (2ND) (x^2) (5) (/) (2), it displays

 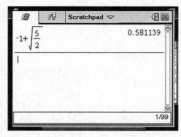

which is not the correct answer.

One reason is that for the TI-84 there needs to be a closing parentheses (or on the TI-Nspire, press [right arrow] to move out from under the radical sign) after the 5 in the square root symbol. Without it, it calculated $-1 + \sqrt{\frac{5}{2}}$.

More needs to be done, though, since

 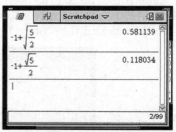

still is not correct. This is the solution to $-1 + \frac{\sqrt{5}}{2}$.

To get this correct, there also needs to be parentheses around the entire numerator, $-1 + \sqrt{5}$.

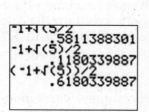

This is the correct answer.

336

On the TI-Nspire, fractions like this can also be done with [templates].

 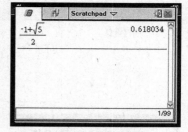

Using the ANS Feature

The last number calculated with the calculator is stored in something called the ANS variable. This ANS variable will appear if you start an expression with a +, −, ×, or ÷. When an answer has a lot of digits in it, this saves time and is also more accurate.

If for some step in a problem you need to calculate the decimal equivalent of $\frac{1}{7}$, it will look like this on the TI-84:

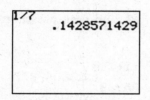

For the TI-Nspire, if you try the same thing, it leaves the answer as $\frac{1}{7}$. To get the decimal approximation, press [ctrl] and [enter] instead of just [enter].

 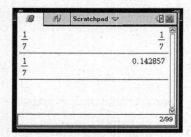

Test-Taking Strategies

Now if you have to multiply this by 3, just press [×], and the calculator will display "Ans*"; press [3] and [enter].

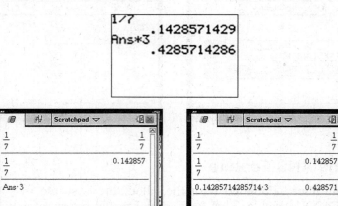

The ANS variable can also help you do calculations in stages. To calculate $x = \frac{-1+\sqrt{5}}{2}$ without using so many parentheses as before, it can be done by first calculating $-1 + \sqrt{5}$ and then pressing [÷] and [2] and Ans will appear automatically.

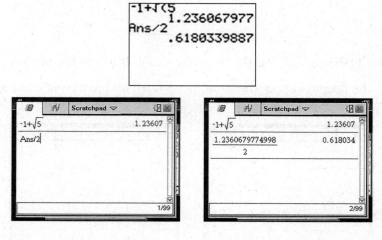

The ANS variable can also be accessed by pressing [2ND] and [−] at the bottom right of the calculator. If after calculating the decimal equivalent of $\frac{1}{7}$ you wanted to subtract $\frac{1}{7}$ from 5, for the TI-84 press [5], [−], [2ND], [ANS], and [ENTER].

For the TI-Nspire press [5], [–], [ctrl], [ans], and [enter].

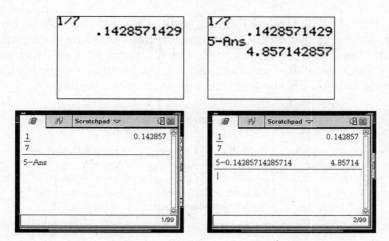

Test-Taking Strategies

12.4 USE THE REFERENCE SHEET

In the back of the Algebra Regents booklet is a reference sheet with 17 conversion facts, like inches to centimeters and quarts to pints, and also 17 formulas. Many of these conversion facts and formulas will not be needed for an individual test, but the quadratic formula and the arithmetic sequence formula are the two that will come in the handiest.

High School Math Reference Sheet

1 inch = 2.54 centimeters	1 kilometer = 0.62 mile	1 cup = 8 fluid ounces
1 meter = 39.37 inches	1 pound = 16 ounces	1 pint = 2 cups
1 mile = 5280 feet	1 pound = 0.454 kilogram	1 quart = 2 pints
1 mile = 1760 yards	1 kilogram = 2.2 pounds	1 gallon = 4 quarts
1 mile = 1.609 kilometers	1 ton = 2000 pounds	1 gallon = 3.785 liters
		1 liter = 0.264 gallon
		1 liter = 1000 cubic centimeters

Triangle	$A = \frac{1}{2}bh$	Pythagorean Theorem	$a^2 + b^2 = c^2$
Parallelogram	$A = bh$	Quadratic Formula	$x = \dfrac{-b \pm \sqrt{b^2 - 4ac}}{2a}$
Circle	$A = \pi r^2$	Arithmetic Sequence	$a_n = a_1 + (n - 1)d$
Circle	$C = \pi d$ or $C = 2\pi r$	Geometric Sequence	$a_n = a_1 r^{n-1}$
General Prisms	$V = Bh$	Geometric Series	$S_n = \dfrac{a_1 - a_1 r^n}{1 - r}$ where $r \neq 1$
Cylinder	$V = \pi r^2 h$	Radians	$1 \text{ radian} = \dfrac{180}{\pi}$ degrees
Sphere	$V = \frac{4}{3}\pi r^3$	Degrees	$1 \text{ degree} = \dfrac{\pi}{180}$ radians
Cone	$V = \frac{1}{3}\pi r^2 h$	Exponential Growth/Decay	$A = A_0 e^{k(t - t_0)} + B_0$
Pyramid	$V = \frac{1}{3}Bh$		

12.5 HOW MANY POINTS DO YOU NEED TO PASS?

The Algebra II Regents exam is scored out of a possible 86 points. Unlike most tests given in the year by your teacher, the score is not then turned into a percent out of 86. Instead each test has a conversion sheet that varies from year to year. For the June 2016 test, the conversion sheet looked like this.

Raw Score	Scale Score	Raw Score	Scale Score	Raw Score	Scale Score	Raw Score	Scale Score
88	100	65	83	42	61	19	33
87	99	64	82	41	60	18	31
86	99	63	81	40	59	17	30
85	98	62	80	39	58	16	28
84	97	61	80	38	56	15	27
83	97	60	79	37	55	14	25
82	96	59	78	36	54	13	24
81	95	58	77	35	53	12	22
80	95	57	76	34	52	11	21
79	94	56	75	33	51	10	19
78	93	55	74	32	49	9	18
77	92	54	73	31	48	8	16
76	92	53	72	30	47	7	14
75	91	52	71	29	46	6	13
74	90	51	70	28	44	5	11
73	89	50	69	27	43	4	9
72	88	49	68	26	42	3	7
71	88	48	67	25	41	2	5
70	87	47	66	24	39	1	3
69	86	46	65	23	38	0	0
68	86	45	64	22	37		
67	85	44	63	21	35		
66	84	43	62	20	34		

On this test, 30 points became a 65, 57 points became a 75, and 73 points became an 85. This means that for this examination a student who got 30 out of 86, which is just 35% of the possible points, would get a 65 on this exam. 57 out of 86 is 66%, but this scaled to a 75. 73 out of 86, however, is actually 85% and became an 85. So in the past there has been a curve on the exam for lower scores, though the scaling is not released until after the exam.

341

Answers and Solution Hints to Practice Exercises

CHAPTER 1

Section 1.1

A

1. (1)   4. (1)   7. (1)   10. (4)
2. (2)   5. (4)   8. (2)   11. (3)
3. (4)   6. (4)   9. (1)

B

1. $10x^2 - 15x + 20$
2. Zahra did not distribute the $-$ through. The correct answer is $3x^2 + 4x + 6$.
3. $a = 2$
4. $4x^2 + 10x + 25$ is a perfect square trinomial because $\left(\dfrac{10}{2}\right)^2 = 25$. It factors to $\left(x + \dfrac{10}{2}\right)^2 = (x+5)^2$. So $a = 5$.
5. $x^3 + 9x^2 + 27x + 27$

Section 1.2

A

1. (1)   3. (3)   5. (4)   7. (2)   9. (1)
2. (1)   4. (4)   6. (2)   8. (1)   10. (3)

B

1. $(2x + 1)(x + 3)$
2. $(x + 17)(x + 39)$
3. 81
4. $29 \cdot 19$, because $24^2 - 5^2 = (24 + 5)(24 - 5)$
5. $(x^2 - 9)(x^2 - 4) = (x - 3)(x + 3)(x - 2)(x + 2)$

Section 1.3

A

1. (1)   3. (2)   5. (3)   7. (3)   9. (4)
2. (2)   4. (4)   6. (4)   8. (1)   10. (3)

342

Answers and Solution Hints to Practice Exercises

B
1. $4^3 + (5)(4^2) - (7)(4) + 3 = 119$
2. 7   The remainder equals the value of the function at $x = 3$.
3. 100   The second polynomial is 6 greater than the first polynomial. Thus, the remainder of the second polynomial must be $6 + 94$.
4. Jose checked to see if $7^3 - 12 \cdot 7^2 + 43 \cdot 7 - 56 = 0$.
5. $(x + 3)$

Section 1.4

A

1. (2)	3. (4)	5. (1)	7. (3)	9. (3)
2. (1)	4. (4)	6. (1)	8. (4)	10. (2)

B
1. Both are right and can complete the question to get $x = \pm 5$.
2. This factors by grouping to become $(x + 4)(x^2 - 9) = 0$, which then becomes $(x + 4)(x - 3)(x + 3) = 0$. The solutions are -4, 3, and -3.
3. The solutions are $(x - 2)(x - 4)(x - 5) = 0$ or $x^3 - 11x^2 + 38x - 40 = 0$.
4. $x = \dfrac{-b \pm \sqrt{b^2 + 16}}{2}$
5. Delilah noticed that $x^2 + 10x + 16$ factors to $(x + 2)(x + 8)$.

Section 1.5

A

1. (4)	3. (2)	5. (1)	7. (2)	9. (3)
2. (1)	4. (1)	6. (1)	8. (3)	10. (1)

343

Answers and Solution Hints to Practice Exercises

B
1.
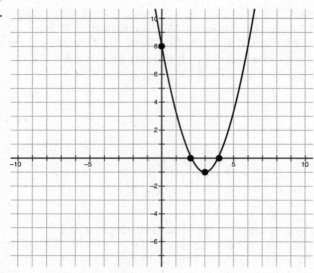

2. 8
3. The x-coordinate of the vertex is 5. The y-coordinate is $5^2 - 10(5) + 23 = -2$. In vertex form, the equation is $y = (x - 5)^2 - 2$.
4. The vertex is halfway between the focus and directrix. So the vertex is at (4, 2). The equation is $y = a(x - 4)^2 + 2$. Since $p = \dfrac{3-1}{2} = 1$, $a = \dfrac{1}{4p} = \dfrac{1}{4 \cdot 1} = \dfrac{1}{4}$. So the complete equation is $y = \dfrac{1}{4}(x-4)^2 + 2$.
5. $\overline{PF} = \overline{PD}$ because any point on the parabola is equidistant from the focus and the directrix so $\overline{PF} = \overline{PD} = 8$.

Section 1.6

A
1. (1)   3. (3)   5. (4)   7. (1)   9. (4)
2. (1)   4. (3)   6. (2)   8. (2)   10. (2)

Answers and Solution Hints to Practice Exercises

B
1. Intercepts at (2, 0), (−2, 0), (−3, 0), (0, −2)

2. $y = \dfrac{1}{3}(x-3)^2(x+1)$
3. $y = a(x-2)^2(x+1)$ where a is any real number
4. The a-value of the first graph is positive. The a-value of the second graph is negative.
5. (−5, 0) and (−1, 0). Divide by $(x-2)$, and factor the quotient $x^2 + 6x + 5$ into $(x+5)(x+1)$.

CHAPTER 2

Section 2.1

A
1. (1)   3. (1)   5. (4)   7. (2)   9. (2)
2. (3)   4. (3)   6. (4)   8. (1)   10. (3)

B
1. Braylon is right. $\dfrac{2x+5}{2} = x + \dfrac{5}{2}$

2. $\dfrac{1}{x} - \dfrac{1}{x+1} = \dfrac{x+1}{x(x+1)} - \dfrac{x}{x(x+1)} = \dfrac{1}{x(x+1)}$

3. $\left(\dfrac{x}{x(x+h)} - \dfrac{x+h}{x(x+h)}\right) \div h = \dfrac{-h}{x(x+h)} \cdot \dfrac{1}{h} = -\dfrac{1}{x(x+h)}$

4. Talia forgot to distribute the minus sign. It should be $5x - 15 - 2x - 4$ in the numerator to get $\dfrac{3x - 19}{(x+2)(x-3)}$.

345

Answers and Solution Hints to Practice Exercises

5. $\dfrac{2x^2 + 4x + 1}{(x + 1)(x + 2)}$

6. $\dfrac{(x + 4)(x + 5)}{(x - 3)}$. Since the denominator has the factors $(x - 3)$ and $(x + 3)$, the numerator must be divisible by one of these factors. The numerator factors to $(x + 3)(x + 4)(x + 5)$.

Section 2.2

A

1. (3)   3. (3)   5. (3)   7. (1)   9. (4)
2. (4)   4. (3)   6. (2)   8. (1)   10. (1)

B

1. Originally, 6 people pay $4 each. Then 8 people pay $3 each.
2. Paxton is correct. The first and third terms are undefined when $x = 2$.
3. They drove 60 mph. The equation $\dfrac{240}{x} + \dfrac{4}{5} = \dfrac{240}{x - 10}$, where x = rate from New York to Boston.
4. The equation simplifies to $2x^2 + 11x - 30 = 0$. So $x = -7.5$ or $x = 2$.
5. The equation simplifies to $7x + 5 = 26$, so $x = 3$.

Section 2.3

A

1. (2)   2. (4)   3. (3)   4. (1)   5. (2)

CHAPTER 3

Section 3.1

A

1. (2)   3. (3)   5. (3)   7. (2)   9. (3)
2. (1)   4. (4)   6. (1)   8. (3)   10. (2)

B

1. Ashlynn is right because $5^3 \cdot 5^4 = 5^{3+4} = 5^7$.
2. $5.2 \times 10^{-4} = 5.2 \times \dfrac{1}{10^4} = 5.2 \times \dfrac{1}{10,000} = 0.00052$.

346

Answers and Solution Hints to Practice Exercises

3. $2^{11} = 2^{10} \cdot 2^1 = 1{,}024 \cdot 2 = 2{,}048.$

4. $9^{1.5} = 9^{\frac{3}{2}} = \left(\sqrt[2]{9}\right)^3 = 3^3 = 27$

5. $\dfrac{1}{64} \cdot 125 \cdot \dfrac{1}{9} = \dfrac{125}{576}$

Section 3.2

A

1. (1) 3. (4) 5. (3) 7. (4) 9. (4)
2. (2) 4. (1) 6. (2) 8. (4) 10. (3)

B

1. They intersect at (2.7, 20).

2. They are both right. $2^{-x} = (2^{-1})^x = \left(\dfrac{1}{2}\right)^x.$

3. The equation simplifies to $-2x + 6 = -6x + 18$. Thus, $x = 3$.
4. Between 3 and 4 since $4^3 = 64$ and $4^4 = 256$.
5. $3 \cdot 2^x$ is not equivalent to 6^x.

Section 3.3

A

1. (3) 3. (3) 5. (3) 7. (3) 9. (1)
2. (4) 4. (1) 6. (4) 8. (1) 10. (2)

B

1. $\log_3 1{,}000$ is greater. Since 3 is less than 4, it would have to be raised to a larger power to become 1,000.
2. As an exponential equation, it would become $7^x = 7^{123456789}$. So $x = 123456789$.
3. Subtract 7 from 262, and then divide by 3. You get $5^x = 85$. So $x = \log_5 85 \approx 2.76$.

4. They are both correct: $\dfrac{\log 263}{\log 7} = \dfrac{\ln 263}{\ln 7} = \log_7 263.$

347

Answers and Solution Hints to Practice Exercises

5.

Section 3.5

A

1. (3) 3. (1) 5. (2) 7. (2) 9. (2)
2. (1) 4. (4) 6. (3) 8. (4) 10. (1)

B

1. a) 61 million   b) $T = 37$, year 2037
2. a) 57 degrees   b) 4 minutes
3. $A = 200(1.01)^{12T}$
4. $100 \cdot {}^e(0.49^2) = 110.296$ Reagan will have \$110.25, and David will have \$110.30. So David will have more money.
5. Since $240{,}000 = \dfrac{1}{1{,}267{,}200} \cdot 2^N$, $N = 41$ folds.

CHAPTER 4

Section 4.1

A

1. (2) 3. (1) 5. (1) 7. (2) 9. (2)
2. (4) 4. (2) 6. (4) 8. (3) 10. (1)

B

1. Mia is correct. There is no distributive property for radicals.
2. No. $\sqrt{49} + \sqrt{576} = 7 + 24 = 31$
3. $\sqrt{8^2 - 4^2} = \sqrt{64 - 16} = \sqrt{48} = 4\sqrt{3}$
4. $\dfrac{6\sqrt{3}}{3} = 2\sqrt{3}$
5. $25 - 10\sqrt{3} + 10\sqrt{3} - 4 \cdot 3 = 13$

Section 4.2
A

1. (1)   3. (1)   5. (4)   7. (2)   9. (1)
2. (4)   4. (1)   6. (2)   8. (1)   10. (3)

B

1.

The absolute value is $\sqrt{(-6)^2 + (8)^2} = \sqrt{100} = 10$.

2. $(x^2 + 4)(x^2 + 9) = 0$
   $x^2 = -4$ or $x^2 = -9$
   $x = \pm 2i$ or $x = \pm 3i$

3. $4i$

4. $\dfrac{7}{4} + \dfrac{3}{2}i$

5. $\sqrt{1.9^2 + 0.8^2} \approx 2.06 > 2$

Section 4.3
A

1. (2)   3. (2)   5. (2)   7. (3)   9. (4)
2. (3)   4. (3)   6. (1)   8. (1)   10. (2)

B

1. $x = 13$
2. $x = 4$
3. $x^2 = 9$ has the solution set $\{-3, 3\}$.
4. $\left(\sqrt{x-4} + \sqrt{x+4}\right)^2$ does not equal $x - 4 + x + 4$. Instead, it equals $x - 4 + 2\sqrt{x-4}\sqrt{x+4} + x + 4$.
5. $x = 3$ $\sqrt{x+6} = 2 + \sqrt{4-x}$, $x + 6 = 4 + 4\sqrt{4-x} + 4 - x$, $2x - 2 = 4\sqrt{4-x}$, $4x^2 - 8x + 4 = 16(4-x)$, $4x^2 + 8x - 60 = 0$, $4(x^2 + 2x - 15) = 0$, $4(x-3)(x+5) = 0$.

Answers and Solution Hints to Practice Exercises

Section 4.4

A
1. (3)   3. (1)   5. (2)   7. (1)   9. (3)
2. (2)   4. (4)   6. (1)   8. (1)   10. (1)

B
1.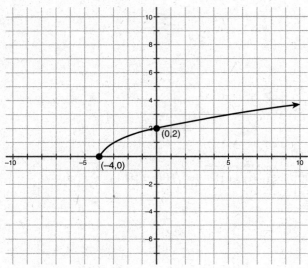

2. $y = \sqrt{x-2} - 4$
3. They are both right. $\sqrt{4x} = 2\sqrt{x}$.

Answers and Solution Hints to Practice Exercises

4. They both go through (0, 0) and (1, 1). $y = \sqrt[3]{x}$ is in quadrants I and III, while $y = \sqrt{x}$ is in only quadrant I.

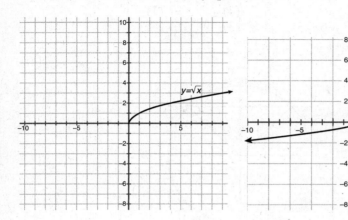

5. $\sqrt{x+2} + 3 = 6$

CHAPTER 5

Section 5.1

A

1. (2)   3. (4)   5. (2)   7. (4)   9. (2)
2. (1)   4. (3)   6. (4)   8. (4)   10. (1)

B
1. a) By the Pythagorean theorem, $AC \approx 0.8$.
   b) By similar triangles, $DF \approx 2.43$.
2. $FG = 8 \cdot \sin 40° \approx 5.14$, $EF = 8 \cdot \cos 40° \approx 6.13$.
3. The $\sin \angle AOB$ is the y-coordinate of point B, which is 0.6, $\sin^{-1}(0.6) = 37°$. Therefore, $\angle AOB = 180° - 37°$.
4. $180° - 25° = 155°$, $180° + 25° = 205°$, $360° - 25° = 335°$.
5. $\sin^{-1}.4848 \approx 29°$. $\sin \theta < 0$ in quadrants III and IV so $\theta = 180° + 29° = 209°$. $\cos 209° \approx -.8746$

351

Section 5.2

A

1. (4)   3. (1)   5. (4)   7. (4)   9. (1)
2. (2)   4. (4)   6. (3)   8. (3)   10. (4)

B

1. 45° and 135°

2. 115° and 245°

3. The horizontal line $y = 2$ does not intersect the circle.
4. $x = 180° + \sin^{-1}(0.8192) \approx 235°$ and $x = 360° - \sin^{-1}(0.8192) \approx 305°$
5. When $45° < x < 225°$, the y-coordinate of the point on the unit circle is greater than the x-coordinate.

Section 5.3

A

1. (3)   3. (4)   5. (4)   7. (4)   9. (2)
2. (3)   4. (3)   6. (1)   8. (1)   10. (4)

Answers and Solution Hints to Practice Exercises

B
1. One radian is smaller than 60°. It is approximately 57.3°. Also triangle AOB cannot be equilateral since $\overline{OA} = \overline{OB} = \widehat{AB} > \overline{AB}$.

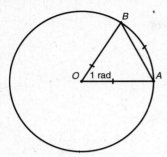

2. $\dfrac{5\pi}{3}$ radians = 300° and sin 300° = $-\dfrac{\sqrt{3}}{2} \approx -0.8660$.
3. 1 degree, 1 radian, 1 right angle
4. $\measuredangle AOB$ is bigger. Since BC is sin $\measuredangle AOB$ and arc $\widehat{BA}$ is the same as $\measuredangle AOB$ in radians, $\measuredangle AOB = \widehat{BA} > \overline{BC} = \sin \measuredangle AOB$.

5. 100 gradians. $\dfrac{\pi}{2}$ radians is 90°, which is $\dfrac{1}{4}$ of a circle. So $\dfrac{1}{4}$ of 400 gradians is 100 gradians.

Section 5.4

A
1. (3)   3. (2)   5. (3)   7. (1)   9. (1)
2. (2)   4. (4)   6. (4)   8. (4)   10. (4)

Answers and Solution Hints to Practice Exercises

B
1.

2.

3. $y = -3\sin(2x) + 4$
4. $y = 4\cos\left(\dfrac{\pi}{30}x\right)$
5.

Section 5.5
A
1. (4) 3. (1) 5. (2) 7. (1) 9. (1)
2. (1) 4. (3) 6. (2) 8. (3) 10. (4)

Section 5.6
B
1. a) 30 seconds
   b) 44 feet
   c)

2. a)

   b) $h = -30\cos\left(\dfrac{\pi}{40}t\right) + 35$

   c) 46.5 feet
   d) 23 seconds and 58 seconds

3. a) $y = -30\cos\left(\dfrac{\pi}{6}x\right) + 55$

   b) 40 degrees
   c) $x = 8$ (September) and $x = 7$ (August)

Section 5.7
A

1. (3)   3. (4)   5. (1)   7. (4)   9. (2)
2. (2)   4. (1)   6. (1)   8. (4)   10. (2)

B

1. $\sin\theta\tan\theta + \cos\theta = \sin\theta\dfrac{\sin\theta}{\cos\theta} + \dfrac{\cos^2\theta}{\cos\theta} = \dfrac{\sin^2\theta + \cos^2\theta}{\cos\theta} = \dfrac{1}{\cos\theta} = \sec\theta$

2. $\dfrac{\cos\theta}{1+\sin\theta} = \dfrac{\cos\theta(1-\sin\theta)}{(1+\sin\theta)(1-\sin\theta)} = \dfrac{\cos\theta(1-\sin\theta)}{\cos^2\theta} = \dfrac{1-\sin\theta}{\cos\theta}$

3. $(\cos\theta - \sin\theta)(\cos\theta + \sin\theta) + 2\sin^2\theta = \cos^2\theta - \sin^2\theta + 2\sin^2\theta$
   $\qquad\qquad\qquad\qquad\qquad\qquad = \cos^2\theta + \sin^2\theta = 1$

Answers and Solution Hints to Practice Exercises

4. $(\cot\theta + \csc\theta)(\cot\theta - \csc\theta) = \left(\dfrac{\cos\theta}{\sin\theta} + \dfrac{1}{\sin\theta}\right)\left(\dfrac{\cos\theta}{\sin\theta} - \dfrac{1}{\sin\theta}\right)$

$$= \dfrac{\cos^2\theta}{\sin^2\theta} - \dfrac{1}{\sin^2\theta} = \dfrac{\cos^2\theta - 1}{\sin^2\theta}$$

$$= \dfrac{-1(1 - \cos^2\theta)}{\sin^2\theta} = \dfrac{-\sin^2\theta}{\sin^2\theta} = -\sin\theta$$

5. $\dfrac{\sin^2\theta + \cos^2\theta}{\sin^2\theta} = \dfrac{1}{\sin^2\theta}$ becomes $1 + \dfrac{\cos^2\theta}{\sin^2\theta} = \csc^2\theta$ becomes

$1 + \cot^2\theta = \csc^2\theta$.

CHAPTER 6

Section 6.1

A

1. (4)   3. (1)   5. (1)   7. (1)   9. (4)
2. (1)   4. (3)   6. (2)   8. (4)   10. (2)

B

1. The substitution method is easier to use if one of the equations has either x or y isolated.
2. Both are correct.
3. (−4, 2)
4. Tickets were bought for 3 adults and 7 children.
5. $m = 15$, $n = 5$

Section 6.2

B

1. 4
2. (3, 1, 4)
3. (2, 4, 1)
4. (4, −1, 3)
5. (−2, 3, −4)

356

CHAPTER 7

Section 7.1

A

1. (3) 3. (1) 5. (4) 7. (1) 9. (4)
2. (2) 4. (4) 6. (1) 8. (3) 10. (3)

B

1. Griffin is correct. $g(f(3)) = 17$
2. $3x^2 - 24x + 48$
3. $g(2) = 4$, $f(g(2)) = f(4) = 5$
4. It is $y = f(x)$ shifted up by 2 units.

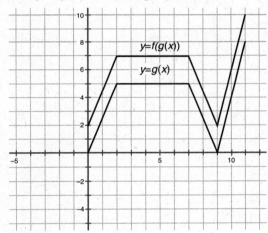

5. $(x-2)^2 = x^2 - 2$, $x = \dfrac{3}{2}$

Section 7.2

A

1. (2) 3. (1) 5. (3) 7. (2) 9. (4)
2. (4) 4. (1) 6. (1) 8. (3) 10. (3)

Answers and Solution Hints to Practice Exercises

B
1. 6
2.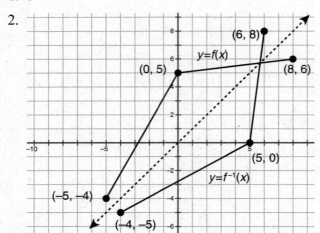

3. $f^{-1}(x) = \dfrac{x-3}{7}$
4. x
5. $f^{-1}(x) = \dfrac{x}{2}, g^{-1}(x) = x - 7$

$$2x + 7 = g(f(x)) = 17$$
$$g^{-1}(g(f(x))) = g^{-1}(17)$$
$$f(x) = 10$$
$$f^{-1}(f(x)) = f^{-1}(10)$$
$$x = 5$$

CHAPTER 8

Section 8.1

A

1. (1) 3. (2) 5. (1) 7. (4) 9. (3)
2. (2) 4. (3) 6. (4) 8. (3) 10. (3)

B

1. $\dfrac{1{,}000(1-1.05^{21})}{1-1.05} = \$35{,}719$

2. The 20th term of the second sequence is $2^{19} = 524{,}288$ while the 20th term of the first sequence is 200,000.

Answers and Solution Hints to Practice Exercises

3. 55 The first ten terms are 1, 1, 2, 3, 5, 8, 13, 21, 34, 55.
4. $a_4 = 2,356$
5. $5 \cdot 3^{n-1} = 2,657,205$
   $3^{n-1} = 531,441$
   $n - 1 = \log_3 531,441$
   $n - 1 = 12$
   $n = 13$

CHAPTER 9

Section 9.1

A

1. (2) 3. (4) 5. (2) 7. (1) 9. (2)
2. (1) 4. (1) 6. (1) 8. (4) 10. (3)

B

1. {1A, 2A, 3A, 4A, 5A, 6A, 1B, 2B, 3B, 4B, 5B, 6B, 1C, 2C, 3C, 4C, 5C, 6C}
2. $\dfrac{3}{8}$
3. 25

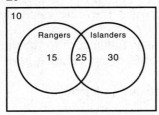

4. a) $\dfrac{90}{200}$ b) $\dfrac{50}{90}$ c) $\dfrac{65}{130}$
5. a) {BBBB, BBBG, BBGB, BBGG, BGBB, BGBG, BGGB, BGGG, GBBB, GBBG, GBGB, GBGG, GGBB, GGBG, GGGB, GGGG}

   b) There is a $\dfrac{6}{16}$ chance of 2 boys 2 girls, $\dfrac{4}{16}$ chance of 3 boys 1 girl, and a $\dfrac{4}{16}$ chance of 3 girls 1 boy. So there is an $\dfrac{8}{16}$ chance of 3 of one gender and 1 of the other, which is greater than $\dfrac{6}{16}$ for 2 of each gender.

Answers and Solution Hints to Practice Exercises

Section 9.2

A

1. (1)   3. (3)   5. (3)   7. (4)   9. (1)
2. (1)   4. (1)   6. (3)   8. (2)   10. (2)

B

1. Various answers. One could be B = "I have an outdoor picnic this Saturday."
2. $.4 \times (1 - .3) = .28$
3. No. P(A or B) = P(A) + P(B) − P(A and B). She did not subtract P(A and B).
4. If events are independent, $P(B|A) = P(B)$ so $P(B|A) = .2$.
5. 0. It is impossible for both to win since they are playing against each other. These are not independent events.

Section 9.3

A

1. (3)   3. (3)   5. (4)   7. (2)   9. (2)
2. (2)   4. (3)   6. (3)   8. (2)   10. (1)

B

1. Various answers are acceptable such as, "She gets lung cancer," "She has yellow teeth," "She has bad breath," etc.
2. For independent events, P(A and B) = P(A) $\cdot$ P(B). If P(A and B) = .14, these are independent events. Otherwise, they are dependent events.
3. These are not independent events, otherwise P(Layla goes to the party given that Charlotte goes to the party) would equal P(Layla goes to the party).
4. There are three ways to verify if two events are independent:
   1) If $P(S) = P(S|B)$,
   2) If $P(B) = P(B|A)$, or
   3) If P(A and B) = P(A) $\cdot$ P(B). $P(S) = P(S|B) = .4$, $P(B) = P(B|S) = .5$, and $P(S$ and $B) = P(S) \cdot P(B) = .20$. By either of the three tests, these are independent events.

360

Answers and Solution Hints to Practice Exercises

5. There are three ways to check for independence:

1) Does P(science) = P(science|straight hair)? No, since $\dfrac{45}{150} \neq \dfrac{10}{60}$.

2) Does P(straight hair) = P(straight hair | science)? No, since $\dfrac{60}{150} \neq \dfrac{10}{45}$.

3) Does P(straight hair and science) = P(straight hair) · P(science)? No, since $\dfrac{10}{150} \neq \dfrac{2700}{22,500}$. None of these equalities work, so these are dependent events.

CHAPTER 10

Section 10.1

A

1. (3)   3. (2)   5. (1)   7. (4)   9. (1)
2. (1)   4. (4)   6. (4)   8. (1)   10. (2)

Section 10.2

A

1. (1)   3. (4)   5. (1)   7. (4)   9. (2)
2. (4)   4. (1)   6. (3)   8. (3)   10. (1)

B

1. a) mean = 70.5, standard deviation = 17.7
   b) 70.5 − 17.7 = 52.8 and 70.5 + 17.7 = 88.2. So of the ten numbers, 8 or 80% are within one standard deviation of the mean. (c) If the set was perfectly normal, 68% would be within one standard deviation of the mean. So this set could be considered approximately normal.
2. The mean is 40, and the standard deviation is 3.9. All of the numbers are between 40 − 2(3.9) = 32.2 and 40 + 2(3.9) = 47.8. So all are within two standard deviations of the mean.
3. 14 people is 7%. If this was a normal distribution, it would be expected for just 2.2% to be 82 inches or more. This suggests that the data are not normally distributed.

Answers and Solution Hints to Practice Exercises

4. The mean is 73.65, and the standard deviation is 5.01. To be within two standard deviations of the mean, numbers would need to be between $73.65 - 2(5.01) = 63.6$ and $73.65 + 2(5.01) = 83.7$. The numbers 62 and 84 are not within two standard deviations of the mean. The rest are.

5. 68.3% within one standard deviation of the mean suggest that it could be a normal distribution. However, more information would need to be known to be sure.

CHAPTER 11

Section 11.1

B

1. One way to reduce bias is to choose the subjects randomly from a list rather than choosing people who are in a certain place or at a certain time. The wording of the questions should also avoid language that could interfere with the way the subjects answer the questions.

2. As an experiment, you would somehow have to force one group of students to have a certain amount of sleep. This could be done by having some kind of late meeting or event that causes the group to get less sleep than the other group. As an observation, you would not do anything to affect the amount of sleep that people get and then record what you observe about people who have different amounts of sleep.

3. One way is to assign each student in the grade a random number. Then have a computer generate random numbers to choose the groups without any chance of bias.

4. When the question is worded with language like "boring," it can cause responders to get a negative feeling about the question. Questions should use more neutral language.

5. This is an experiment since the researcher is controlling who is exposed to different temperatures.

Section 11.2

A

1. (1) 3. (1) 5. (1) 7. (4) 9. (3)
2. (4) 4. (2) 6. (1) 8. (2) 10. (2)

362

Glossary of Algebra II Terms

A

Algebraic identity Two expressions usually involving polynomials that can be shown to be equal to one another.

Amplitude The distance between the middle of a sine or cosine curve to the top of that sine or cosine curve. When the equation is $y = a\sin(bx) + d$ or $y = a\cos(bx) + d$, the amplitude is a.

Arithmetic sequence A sequence of numbers like 3, 7, 11, 15, ... where each number is the same amount more (or less) than the previous number. That difference is called d. In this case, $d = 4$.

Arithmetic series The sum of the terms in an arithmetic sequence. $3 + 7 + 11 + 15$ is an arithmetic series.

Asymptote A line that a curve gets closer and closer to without ever reaching.

B

Base In an exponential expression, the base is the number being raised to a power. In the expression $5 \cdot 2^x$, 2 is the base. In a logarithmic expression, the base is the small subscript number after the word "log." In the expression $\log_3 81 = 4$, the base is 3.

Binomial A polynomial with two terms linked by either a plus (+) or minus (−) sign. The expression $2x + 5$ is a binomial.

C

Coefficient The number that gets multiplied by a variable in a mathematical expression. In the expression $5x^2$, the coefficient of the x^2-term is 5.

Complex number A number of the form $a + bi$ where a and b are real numbers. $5 + 3i$ is a complex number.

Composite function When the expression that defines one function is put into another function. If $f(x) = 3x$ and $g(x) = x^2$, then $g(f(x)) = (3x)^2$ is a composite function.

Compound interest When money in a bank gets interest on an initial deposit and then later gets more interest based on the original deposit and on the interest already earned.

Constant A number. The term in a polynomial that has no variable part. In the polynomial $2x + 3$, the constant is 3.

Cosine The ratio between the adjacent side and the hypotenuse of a right triangle. Also the x-coordinate of a point on the unit circle.

Cubic equation An equation of the form $ax^3 + bx^2 + cx + d = 0$.

Cubic function A function of the form $f(x) = ax^3 + bx^2 + cx + d$.

D

Degree The highest exponent in a polynomial. The degree is $x^3 + 5x^2 + 7x + 2$ is 3.

363

Glossary of Algebra II Terms

Degree A unit of angle measurement that is $\dfrac{1}{360}$ of a circle.

Dependent events When the probability of one event is affected by whether or not some other event happened, the two events are called dependent events. The event "The baseball game will be canceled" and the event "It will rain" are dependent events.

Difference of perfect squares A way of factoring a polynomial like $x^2 - 3^2$ into $(x - 3)(x + 3)$. In general, an expression of the form $a^2 - b^2$ can be factored into $(a - b)(a + b)$.

Directrix Every point on a parabola is equidistant from a point and a line. That line is called the directrix.

E

e A mathematical constant that is approximately 2.72. It is involved in many questions related to modeling growth.

Elimination method Multiplying one or more equations in a system of equations by a constant and then adding the equations in a way that makes one of the variables cancel out. The system $x + 2y = 5$ and $3x - 2y = 7$ becomes $4x = 12$ when the two equations are added together.

Even function A function whose graph is symmetric with respect to the y-axis.

Experimental study When a researcher has one group of subjects do one thing and another group do another thing and then studies how the two groups react.

An example is giving half of the subjects one type of medicine and the rest of them another type of medicine.

Explicit sequence formula A way of describing a sequence based on relating the position of a term to the number in that position. $a_n = 5n + 8$ is an explicit sequence formula.

Exponential equation An equation that involves an exponential expression. The variable is in the exponent. $3 \cdot 2^x = 96$ is an exponential equation.

Exponential expression An expression that has a variable as an exponent. $3 \cdot 2^x$ is an exponential expression.

Exponential function A function that has a variable as an exponent. $f(x) = 3 \cdot 2^x$ is an exponential function.

F

Factor by grouping A way of factoring that involves separating a polynomial into two or more groups, finding a common factor in each group, and, if possible, factoring a common factor out of what remains. The expression $x^3 + 3x^2 + 2x + 6 = x^2(x + 3) + 2(x + 3) = (x + 3)(x^2 + 2)$ is an example of factoring by grouping.

Factoring Finding two factors whose product is a given number or polynomial. $15 = 3 \cdot 5$ is a way of factoring 15. $x^2 - 5x + 6 = (x - 2)(x - 3)$ is a way of factoring $x^2 - 5x + 6$.

Factors When a number or polynomial is divided by a number or polynomial and there is no

364

Glossary of Algebra II Terms

remainder, the divisor is a factor of the dividend. Since $\frac{15}{3} = 5$ with remainder 0, 3 is a factor of 15.

Factor theorem A theorem that states that if the remainder when you divide $P(x)$ by $x - a$ is 0, then $x - a$ is a factor of $P(x)$. For example, if you divide $P(x) = x^2 + 5x + 6$ by $x + 2$, you get $x + 3$ with remainder 0, so $x + 2$ is a factor of $P(x)$.

Finite arithmetic series formula The formula $S_n = \frac{(a_1 + a_n)n}{2}$ for summing the terms of a finite arithmetic series.

Finite geometric series formula The formula $S_n = \frac{a_1(1 - r^n)}{1 - r}$ for summing the terms of a finite geometric series.

Focus Every point on a parabola is equidistant from a point and a line. That point is called the focus of the parabola.

FOIL A way to remember the four multiplications when multiplying a binomial by another binomial. When multiplying $(2x + 3)(4x + 5)$, F (for firsts) represents $2x \cdot 4x$, O (for outers) represents $2x \cdot 5$, I (for inners) represents $3 \cdot 4x$, and L (for lasts) represents $3 \cdot 5$.

Frequency The number of sine or cosine curves that can fit into a 360-degree interval.

Function A rule that takes a number or an expression as an input and converts it into a number or an expression in a predicable way. The function $f(x) = x^2$ takes a number or expression and turns it into the

number or expression squared, for example $f(3) = 3^2 = 9$.

G

Geometric sequence A sequence of numbers like 3, 6, 12, 24, ... where each number is equal to the previous number in the sequence multiplied by the same number. That number, the common ratio, is called r. In this case, $r = 2$.

Geometric series The sum of the terms in a geometric sequence. $3 + 6 + 12 + 24$ is a geometric series.

Given When calculating the probability of B given A, the given means that event A has already happened and may, or may not, affect the probability that event B will also happen.

Greatest common factor Of all the common factors between two numbers or polynomials, the ones that are largest. The numbers 12 and 18 have the common factors 1, 2, 3, and 6, so 6 is the greatest common factor between 12 and 18.

Growth rate In the exponential expression $a(1 + r)^x$, the r is the growth rate. For example, in $100(1 + 0.07)^x$, the growth rate is 0.07.

H

Horizontal shift When all the points of a graph move right or left the same number of units, it is called a horizontal shift.

Horizontal shrink When the x-coordinates of every point in a graph are divided by the same number, it is called a horizontal shrink.

365

Glossary of Algebra II Terms

I

i A shorthand way of writing $\sqrt{-1}$.

Imaginary number A number that is a multiple of the imaginary unit i. $5i$ is an imaginary number.

Independent events When the probability of one event is not affected by whether or not some other event happened, the two events are called independent events. The event "I wake up at 7:00 A.M." and "It will rain today" are independent events.

Inverse function If a function $f(x)$ has an inverse $f^{-1}(x)$, then $f^{-1}(f(x)) = x$. For example, if $f(x) = x + 1$, $f^{-1}(x) = x - 1$. An inverse function undoes whatever the original function did.

L

Like terms When two monomials with the same variable part are added or subtracted, it is called combining like terms. In the expression $2x^2 + 5x + 3x^2 + 7$, the $2x^2$ and the $3x^2$ are like terms and can be combined to form $5x^2$.

Linear function A function whose largest exponent is 1. $f(x) = 3x + 7$ is a linear function.

ln x A shorthand for writing $\log_e x$.

$\log_b x$ The number that b would have to be raised to in order to get a result of x. For example, $\log_5 125 = 3$ since $5^3 = 125$.

Logarithm An exponent that tells what a base needs to be raised to in order to become another number. The logarithm of 25 base 5 or $\log_5 25 = 2$ since $5^2 = 25$.

M

Monomial A polynomial that has just one term. The polynomial $3x^5$ is a monomial.

N

Normal distribution When a set of data points resembles a bell curve with 68% being within one standard deviation, 95% being within two standard deviations, it is known as a normal distribution.

O

Observational study When a statistical researcher collects data by studying the behavior of the subjects without interfering with their environment.

Odd function A function whose graph is symmetric with respect to the origin.

P

Perfect square trinomial A polynomial of the form $x^2 + 2ax + a^2$ or $x^2 - 2ax + a^2$ that can be factored into $(x + a)^2$ or $(x - a)^2$. For example, $x^2 + 6x + 9$ can be factored into $(x + 3)^2$.

Period The number of degrees (or radians) in one cycle of a sine or cosine curve.

Polynomial An expression like $2x^3 + 5x^2 - 3x + 7$ that consists of one or more terms, each term having a coefficient and a variable raised to a power.

Population mean The average of an entire population for some characteristic. For example, if out of 10,000 people studied their average height is 64 inches, 64 is the population mean.

Population proportion The percent of an entire population that has some characteristic. For example, if out of 10,000 people studied 52% are boys, 0.52 is the population proportion.

366

Glossary of Algebra II Terms

Probability The chance of something happening, ranging from 10 (impossible) to 1 (certain).

Q

Quadratic equation An equation of the form $ax^2 + bx + c = 0$.

Quadratic formula The formula $x = \dfrac{-b \pm \sqrt{b^2 - 4ac}}{2a}$ used to find the solutions to the equation $ax^2 + bx + c = 0$.

Quadratic function A function whose largest exponent is 2. $f(x) = x^2 - 3x + 7$ is a quadratic function.

R

Radian A unit of angle measurement that is equal to approximately 57 degrees.

Radical equation An equation involving a radical expression. $\sqrt{x+2} = 5$ is a radical equation.

Radical expression An expression involving a radical symbol. $\sqrt{x+2}$ is a radical expression.

Randomization test When a computer is used to simulate an ideal experiement with the goal of seeing if something that already happened was likely to happen or not.

Rational equation An equation that involves one or more rational expressions. $\dfrac{3}{x-2} + \dfrac{4}{x-2} = 5$ is a rational equation.

Rational expression An expression that has a fraction with a polynomial of degree at least one in the denominator. $\dfrac{3}{x+1}$ is a rational expression.

Rational Function A function that has a fraction with a polynomial of degree at least one in the denominator. $f(x) = \dfrac{3}{x+1}$ is a rational function.

Recursive sequence formula A way of describing a sequence based on how each term relates to the preceding term. $a_1 = 3$, $a_n = 5 + a_{n-1}$ is a recursive sequence formula.

Remainder theorem A theorem that states that when you divide a polynomial $P(x)$ by $x - a$, the remainder will always equal $P(a)$. For example, if you divide the polynomial $P(x) = x^2 + 5x + 6$ by $x - 3$, you will get a remainder of $P(3) = 30$.

S

Sample mean The average of a sample that has some characteristic. For example, if out of 10,000 people 100 people are randomly chosen and of those 100 people their average height is 61 inches, then 61 is the sample mean for that sample.

Sample proportion The percent of a sample that has some characteristic. For example, if out of 10,000 people 100 people are randomly chosen and of those 100 people 38% are boys, then 0.38 is the sample proportion.

Sample space A list of all the possible outcomes of a probability experiment.

367

Sampling distribution of the sample mean When multiple groups of samples are taken from a larger population and the average of some attribute of each sample group is calculated and graphed in a histogram.

Sampling distribution of the sample proportion When multiple groups of samples are taken from a larger population and the percent of some attribute of each sample group is calculated and graphed in a histogram.

Simplified radical form When the number inside a square root (or cube root) sign has no factors that are perfect squares (or perfect cubes). The expression $\sqrt{18}$ is not in simplified radical form since $18 = 2 \cdot 9$. $3\sqrt{2}$ is equal to $\sqrt{18}$ but is in simplified radical form.

Sine The ratio between the opposite side and the hypotenuse of a right triangle. Also the y-coordinate of a point on the unit circle.

Standard deviation A way of measuring how spread out a data set is. The smallest standard deviation is 0. It occurs when all the elements of a set are equal.

Standard form of a quadratic equation An equation is the form $y = ax^2 + bx + c$.

Statistical significance When the data from an experiment are randomly mixed together multiple times with the goal of seeing if a difference between two groups is likely to happen.

Substitution method A system of equations where one of the variables is solved in terms of the other variable and then substituted into the other equation. For example, in the system of equations $y = 2x + 1$, $3x + 2y = 10$, the y in the second equation can be replaced by $2x + 1$ to become $3x + 2(2x + 1) = 10$.

Survey A question or set of questions answered by a subject in a statistical study.

System of equations Two or more equations that have two or more variables. A system of equations generally has one solution that is an ordered pair (or triple) with numbers that satisfy each of the equations.

T

Tangent The ratio between the opposite side and the adjacent side of a right triangle.

Trigonometric function A function that involves a trigonometry ratio, like sine, cosine, or tangent. $f(x) = 3\sin(x) + 1$ is a trigonometric function.

Trigonometric identity Two expressions involving trigonometric ratios that can be shown to be equal to one another.

Trinomial A polynomial with three terms linked by plus (+) and/or minus (–) signs. The expression $x^2 + 5x + 6$ is a trinomial.

Two-way frequency table A chart used for listing the different ways that things can either have or not have certain attributes.

U

Unit circle A circle with a radius of 1 and center at (0, 0).

Glossary of Algebra II Terms

V

Variable A letter that is used to represent a number. Often this letter represents the thing to be solved for in an equation. In the expression $2x + 5$, the variable is x.

Venn diagram A way of organizing data that has two or more intersecting circles surrounded by a rectangle.

Vertex form of a quadratic equation An equation of the form $y = a(x - h)^2 + k$. When graphed, it is a parabola with vertex (h, k).

Vertical shift When all the points of a graph move up or down the same number of units, it is called a vertical shift.

Vertical stretch When the y-coordinates of every point in a graph are multiplied by the same number, it is called a vertical stretch.

Z

z-score A measure of how many standard deviations above (positive z-score) or below (negative z-score) a number is from the mean.

THE ALGEBRA II REGENTS EXAMINATION

The Regents Examination in Algebra II is a three-hour exam that is divided into four parts with a total of 37 questions. All 37 questions must be answered. Part I consists entirely of regular multiple-choice questions. Parts II, III, and IV each contain a set of questions that must be answered directly in the question booklet. You are required to show how you arrived at the answers for the questions in Parts II, III, and IV. The acompanying table shows how the exam breaks down.

Question Type	Number of Questions	Credit Value
Part I: Multiple choice	24	$24 \times 2 = 48$
Part II: 2-credit open ended	8	$8 \times 2 = 16$
Part III: 4-credit open ended	4	$4 \times 4 = 16$
Part IV: 6-credit open ended	1	$1 \times 6 = 6$
	Total = 37 questions	Total = 86 points

The Algebra II Regents Examination

How Is the Exam Scored?

- Each of your answers to the 24 multiple-choice questions in Part I will be scored as either right or wrong.
- Solutions to questions in Part II, III, and IV that are not completely correct may receive partial credit according to a special rating guide that is provided by the New York State Education Department. In order to receive full credit for a correct answer to a question in Parts II, III, or IV, you must show or explain how you arrived at your answer by indicating the key steps taken, including appropriate formula substitutions, diagrams, graphs, and charts. A correct numerical answer with no work shown will receive only 1 credit.
- The raw scores for the four parts of the test are added together. The maximum total raw score for the Algebra II Regents Examination is 86 points. Using a special conversion chart that is provided by the New York State Education Department, your total raw score will be equated to a final test score that falls within the usual 0 to 100 scale.

What Type of Calculator Is Required?

Graphing calculators are *required* for the Algebra II Regents Examination. During the administration of the Regents exam, schools are required to make a graphing calculator available for the exclusive use of each student. You will need to use your calculator to work with trigonometric functions of angles, find roots of numbers, and perform routine calculations.

Knowing how to use a graphing calculator gives you more options when deciding how to solve a problem. Rather than solving a problem algebraically with pen and paper, it may be easier to solve the same problem using a graph or table created by a graphing calculator. A graphical or numerical solution using a calculator can also be used to help confirm an answer obtained by solving the problem algebraically.

Are Any Formulas Provided?

The Algebra II Regents Examination test booklet will include a reference sheet containing the formulas in the accompanying table. Keep in mind that you may be required to know other formulas that are not included in this sheet. Please see page 373 for these formulas.

The Algebra II Regents Examination

What Else Should I Know?

- Do not omit any questions from Part I. Since there is no penalty for guessing, make certain that you record an answer for each of the 24 multiple-choice questions.
- If the method of solution is not stated in the problem, choose an appropriate method (numerical, graphical, or algebraic) with which you are most comfortable.
- If you solve a problem in Parts II, III, or IV using a trial-and-error approach, show the work for at least three guesses with appropriate checks. Should the correct answer be reached on the first trial, you must further illustrate your method by showing that guesses below and above the correct guess do not work.
- Avoid rounding errors when using a calculator. Unless otherwise directed, the (pi) key on a calculator should be used in computations involving the constant π rather than the common rational approximation of 3.14 or $\frac{22}{7}$. When performing a sequence of calculations in which the result of one calculation is used in a second calculation, do not round off. Instead, use the full power/display of the calculator by performing a "chain" calculation, saving intermediate results in the calculator's memory. Unless otherwise specified, rounding, if required, should be done only when the *final* answer is reached.
- Check that each answer is in the requested form. If a specific form is not required, answers may be left in any equivalent form, such as $\sqrt{75}$, $5\sqrt{3}$, or 8.660254038 (the full power/display of the calculator).
- If a problem requires using a formula that is not provided in the question, check the formula reference sheet in the test booklet to see if it is listed. Clearly write any formula you use before making any appropriate substitutions. Then evaluate the formula in step-by-step fashion.
- For any problem solved in Parts II, III, and IV using a graphing calculator, you must indicate how the calculator was used to obtain the answer such as by copying graphs or tables created by your calculator together with the equations used to produce them. When copying graphs, label each graph with its equation, state the dimensions of the viewing window, and indentify the intercepts and any points of intersection with their coordinates. Whenever appropriate, indicate the rationale of your approach.

Examination
June 2016
Algebra II

HIGH SCHOOL MATH REFERENCE SHEET

Conversions

1 inch = 2.54 centimeters

1 meter = 39.37 inches

1 mile = 5280 feet

1 mile = 1760 yards

1 mile = 1.609 kilometers

1 kilometer = 0.62 mile

1 pound = 16 ounces

1 pound = 0.454 kilogram

1 kilogram = 2.2 pounds

1 ton = 2000 pounds

1 cup = 8 fluid ounces

1 pint = 2 cups

1 quart = 2 pints

1 gallon = 4 quarts

1 gallon = 3.785 liters

1 liter = 0.264 gallon

1 liter = 1000 cubic centimeters

Formulas	
Triangle	$A = \dfrac{1}{2}bh$
Parallelogram	$A = bh$
Circle	$A = \pi r^2$
Circle	$C = \pi d$ or $C = 2\pi r$
General Prisms	$V = Bh$

Examination June 2016

Cylinder	$V = \pi r^2 h$
Sphere	$V = \frac{4}{3}\pi r^3$
Cone	$V = \frac{1}{3}\pi r^2 h$
Pyramid	$V = \frac{1}{3}Bh$
Pythagorean Theorem	$a^2 + b^2 = c^2$
Quadratic Formula	$x = \dfrac{-b \pm \sqrt{b^2 - 4ac}}{2a}$
Arithmetic Sequence	$a_n = a_1 + (n-1)d$
Geometric Sequence	$a_n = a_1 r^{n-1}$
Geometric Series	$S_n = \dfrac{a_1 - a_1 r^n}{1 - r}$ where $r \neq 1$
Radians	$1 \text{ radian} = \dfrac{180}{\pi}$ degrees
Degrees	$1 \text{ degree} = \dfrac{\pi}{180}$ radians
Exponential Growth/Decay	$A = A_0 e^{k(t - t_0)} + B_0$

Examination June 2016

PART I

Answer all **24 questions** in this part. Each correct answer will receive 2 credits. No partial credit will be allowed. For each statement or question, write in the space provided the numeral preceding the word or expression that best completes the statement or answers the question. [48 credits]

1 When $b > 0$ and d is a positive integer, the expression $(3b)^{\frac{2}{d}}$ is equivalent to

(1) $\dfrac{1}{\left(\sqrt[d]{3b}\right)^2}$

(3) $\dfrac{1}{\sqrt{3b^d}}$

(2) $\left(\sqrt{3b}\right)^d$

(4) $\left(\sqrt[d]{3b}\right)^2$

1 _____

2 Julie averaged 85 on the first three tests of the semester in her mathematics class. If she scores 93 on each of the remaining tests, her average will be 90. Which equation could be used to determine how many tests, T, are left in the semester?

(1) $\dfrac{255 + 93T}{3T} = 90$

(3) $\dfrac{255 + 93T}{T + 3} = 90$

(2) $\dfrac{255 + 90T}{3T} = 93$

(4) $\dfrac{255 + 90T}{T + 3} = 93$

2 _____

3 Given i is the imaginary unit, $(2 - yi)^2$ in simplest form is

(1) $y^2 - 4yi + 4$

(3) $-y^2 + 4$

(2) $-y^2 - 4yi + 4$

(4) $y^2 + 4$

3 _____

375

4 Which graph has the following characteristics?
- three real zeros
- as $x \to -\infty$, $f(x) \to -\infty$
- as $x \to \infty$, $f(x) \to \infty$

(1)

(3)

(2)

(4)

4 _____

5 The solution set for the equation is $\sqrt{56-x} = x$ is
(1) {−8, 7}   (3) {7}
(2) {−7, 8}   (4) { }

5 _____

6 The zeros for $f(x) = x^4 - 4x^3 - 9x^2 + 36x$ are
(1) {0, ±3, 4}   (3) {0, ±3, −4}
(2) {0, 3, 4}   (4) {0, 3, −4}

6 _____

7 Anne has a coin. She does not know if it is a fair coin. She flipped the coin 100 times and obtained 73 heads and 27 tails. She ran a computer simulation of 200 samples of 100 fair coin flips. The output of the proportion of heads is shown below.

Given the results of her coin flips and of her computer simulation, which statement is most accurate?

(1) 73 of the computer's next 100 coin flips will be heads.
(2) 50 of her next 100 coin flips will be heads.
(3) Her coin is not fair.
(4) Her coin is fair.

8 If $g(c) = 1 - c^2$ and $m(c) = c + 1$, then which statement is *not* true?

(1) $g(c) \bullet m(c) = 1 + c - c^2 - c^3$
(2) $g(c) + m(c) = 2 + c - c^2$
(3) $m(c) - g(c) = c + c^2$
(4) $\dfrac{m(c)}{g(c)} = \dfrac{-1}{1-c}$

Examination June 2016

9 The heights of women in the United States are normally distributed with a mean of 64 inches and a standard deviation of 2.75 inches. The percent of women whose heights are between 64 and 69.5 inches, to the *nearest whole percent*, is

(1) 6   (3) 68

(2) 48   (4) 95   9 _____

10 The formula below can be used to model which scenario?

$$a_1 = 3000$$
$$a_n = 0.80a_{n-1}$$

(1) The first row of a stadium has 3000 seats, and each row thereafter has 80 more seats than the row in front of it.

(2) The last row of a stadium has 3000 seats, and each row before it has 80 fewer seats than the row behind it.

(3) A bank account starts with a deposit of $3000, and each year it grows by 80%.

(4) The initial value of a specialty toy is $3000, and its value each of the following years is 20% less.   10 _____

11 Sean's team has a baseball game tomorrow. He pitches 50% of the games. There is a 40% chance of rain during the game tomorrow. If the probability that it rains given that Sean pitches is 40%, it can be concluded that these two events are

(1) independent   (3) mutually exclusive

(2) dependent   (4) complements   11 _____

378

Examination June 2016

12 A solution of the equation $2x^2 + 3x + 2 = 0$ is

(1) $-\dfrac{3}{4} + \dfrac{1}{4}i\sqrt{7}$

(3) $-\dfrac{3}{4} + \dfrac{1}{4}\sqrt{7}$

(2) $-\dfrac{3}{4} + \dfrac{7}{4}i$

(4) $\dfrac{1}{2}$

12 _____

13 The Ferris wheel at the landmark Navy Pier in Chicago takes 7 minutes to make one full rotation. The height, H, in feet, above the ground of one of the six-person cars can be modeled by $H(t) = 70\sin\left(\dfrac{2\pi}{7}(t-1.75)\right) + 80$, where t is time, in minutes. Using $H(t)$ for one full rotation, this car's minimum height, in feet, is

(1) 150

(3) 10

(2) 70

(4) 0

13 _____

14 The expression $\dfrac{4x^3 + 5x + 10}{2x+3}$ is equivalent to

(1) $2x^2 + 3x - 7 + \dfrac{31}{2x+3}$

(3) $2x^2 + 2.5x + 5 + \dfrac{15}{2x+3}$

(2) $2x^2 - 3x + 7 - \dfrac{11}{2x+3}$

(4) $2x^2 - 2.5x - 5 - \dfrac{20}{2x+3}$

14 _____

15 Which function represents exponential decay?

(1) $y = 2^{0.3t}$

(3) $y = \left(\dfrac{1}{2}\right)^{-t}$

(2) $y = 1.2^{3t}$

(4) $y = 5^{-t}$

15 _____

379

Examination June 2016

16 Given $f^{-1}(x) = -\dfrac{3}{4}x + 2$, which equation represents $f(x)$?

(1) $f(x) = \dfrac{4}{3}x - \dfrac{8}{3}$ (3) $f(x) = \dfrac{3}{4}x - 2$

(2) $f(x) = -\dfrac{4}{3}x + \dfrac{8}{3}$ (4) $f(x) = -\dfrac{3}{4}x + 2$ 16 _____

17 A circle centered at the origin has a radius of 10 units. The terminal side of an angle, θ, intercepts the circle in quadrant II at point C. The y-coordinate of point C is 8. What is the value of $\cos \theta$?

(1) $-\dfrac{3}{5}$ (3) $\dfrac{3}{5}$

(2) $-\dfrac{3}{4}$ (4) $\dfrac{4}{5}$ 17 _____

18 Which statement about the graph of $c(x) = \log_6 x$ is *false*?

(1) The asymptote has equation $y = 0$.
(2) The graph has no y-intercept.
(3) The domain is the set of positive reals.
(2) The range is the set of all real numbers.   18 _____

19 The equation $4x^2 - 24x + 4y^2 + 72y = 76$ is equivalent to

(1) $4(x - 3)^2 + 4(y + 9)^2 = 76$
(2) $4(x - 3)^2 + 4(y + 9)^2 = 121$
(3) $4(x - 3)^2 + 4(y + 9)^2 = 166$
(2) $4(x - 3)^2 + 4(y + 9)^2 = 436$ 19 _____

20 There was a study done on oxygen consumption of snails as a function of pH, and the result was a degree 4 polynomial function whose graph is shown below.

Which statement about this function is *incorrect*?

(1) The degree of the polynomial is even.
(2) There is a positive leading coefficient.
(3) At two pH values, there is a relative maximum value.
(4) There are two intervals where the function is decreasing.

20 _____

21 Last year, the total revenue for Home Style, a national restaurant chain, increased 5.25% over the previous year. If this trend were to continue, which expression could the company's chief financial officer use to approximate their monthly percent increase in revenue? [Let m represent months.]

(1) $(1.0525)^m$

(3) $(1.00427)^m$

(2) $(1.0525)^{\frac{12}{m}}$

(4) $(1.00427)^{\frac{m}{12}}$

21 _____

Examination June 2016

22 Which value, to the *nearest tenth*, is *not* a solution of $p(x) = q(x)$ if $p(x) = x^3 + 3x^2 - 3x - 1$ and $q(x) = 3x + 8$?

(1) −3.9   (3) 2.1

(2) −1.1   (4) 4.7   22 _____

23 The population of Jamesburg for the years 2010–2013, respectively, was reported as follows:

250,000   250,937   251,878   252,822

How can this sequence be recursively modeled?

(1) $j_n = 250,000(1.00375)^{n-1}$

(2) $j_n = 250,000 + 937^{(n-1)}$

(3) $j_1 = 250,000$

$j_n = 1.00375 j_{n-1}$

(4) $j_1 = 250,000$

$j_n = j_{n-1} + 937$ 23 _____

24 The voltage used by most households can be modeled by a sine function. The maximum voltage is 120 volts, and there are 60 cycles *every second*. Which equation best represents the value of the voltage as it flows through the electric wires, where t is time in seconds?

(1) $V = 120 \sin(t)$ (3) $V = 120 \sin(60\pi t)$

(2) $V = 120 \sin(60t)$ (4) $V = 120 \sin(120\pi t)$ 24 _____

Examination June 2016

PART II

Answer all 8 questions in this part. Each correct answer will receive 2 credits. Clearly indicate the necessary steps, including appropriate formula substitutions, diagrams, graphs, charts, etc. For all questions in this part, a correct numerical answer with no work shown will receive only 1 credit.   [16 credits]

25 Solve for x: $\dfrac{1}{x} - \dfrac{1}{3} = -\dfrac{1}{3x}$

Examination June 2016

26 Describe how a controlled experiment can be created to examine the effect of ingredient X in a toothpaste.

27 Determine if $x - 5$ is a factor of $2x^3 - 4x^2 - 7x - 10$. Explain your answer.

28 On the axes below, graph *one* cycle of a cosine function with amplitude 3, period $\frac{\pi}{2}$, midline $y = -1$, and passing through the point (0, 2).

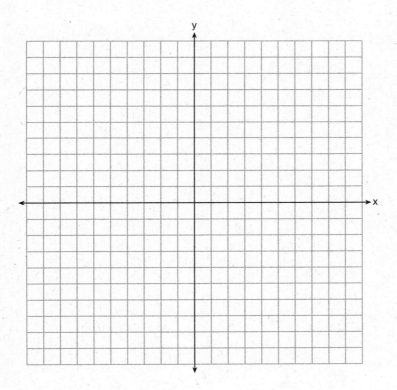

Examination June 2016

29 A suburban high school has a population of 1376 students. The number of students who participate in sports is 649. The number of students who participate in music is 433. If the probability that a student participates in either sports or music is $\frac{974}{1376}$, what is the probability that a student participates in both sports and music?

Examination June 2016

30 The directrix of the parabola $12(y + 3) = (x - 4)^2$ has the equation $y = -6$. Find the coordinates of the focus of the parabola.

31 Algebraically prove that $\dfrac{x^3 + 9}{x^3 + 8} = 1 + \dfrac{1}{x^3 + 8}$, where $x \neq -2$.

Examination June 2016

32 A house purchased 5 years ago for $100,000 was just sold for $135,000. Assuming exponential growth, approximate the annual growth rate, to the *nearest percent*.

Examination June 2016

PART III

Answer all 4 questions in this part. Each correct answer will receive 4 credits. Clearly indicate the necessary steps, including appropriate formula substitutions, diagrams, graphs, charts, etc. For all questions in this part, a correct numerical answer with no work shown will receive only 1 credit. [16 credits]

33 Solve the system of equations shown below algebraically.

$$(x - 3)^2 + (y + 2)^2 = 16$$
$$2x + 2y = 10$$

Examination June 2016

34 Alexa earns $33,000 in her first year of teaching and earns a 4% increase in each successive year.

Write a geometric series formula, S_n, for Alexa's total earnings over n years.

Use this formula to find Alexa's total earnings for her first 15 years of teaching, to the *nearest cent*.

35 Fifty-five students attending the prom were randomly selected to participate in a survey about the music choice at the prom. Sixty percent responded that a DJ would be preferred over a band. Members of the prom committee thought that the vote would have 50% for the DJ and 50% for the band.

A simulation was run 200 times, each of sample size 55, based on the premise that 60% of the students would prefer a DJ. The approximate normal simulation results are shown below.

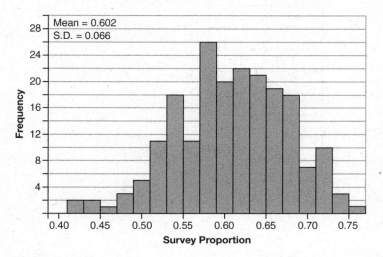

Using the results of the simulation, determine a plausible interval containing the middle 95% of the data. Round all values to the *nearest hundredth*.

Members of the prom committee are concerned that a vote of all students attending the prom may produce a 50%–50% split. Explain what statistical evidence supports this concern.

Examination June 2016

36 Which function shown below has a greater average rate of change on the interval [–2, 4]? Justify your answer.

x	f(x)
–4	0.3125
–3	0.625
–2	1.25
–1	2.5
0	5
1	10
2	20
3	40
4	80
5	160
6	320

$g(x) = 4x^3 - 5x^2 + 3$

Examination June 2016

PART IV

Answer the question in this part. A correct answer will receive 6 credits. Clearly indicate the necessary steps, including appropriate formula substitutions, diagrams, graphs, charts, etc. A correct numerical answer with no work shown will receive only 1 credit. [6 credits]

37 Drugs break down in the human body at different rates and therefore must be prescribed by doctors carefully to prevent complications, such as overdosing. The breakdown of a drug is represented by the function $N(t) = N_0(e)^{-rt}$, where $N(t)$ is the amount left in the body, N_0 is the initial dosage, r is the decay rate, and t is time in hours. Patient A, $A(t)$, is given 800 milligrams of a drug with a decay rate of 0.347. Patient B, $B(t)$, is given 400 milligrams of another drug with a decay rate of 0.231.

Write two functions, $A(t)$ and $B(t)$, to represent the breakdown of the respective drug given to each patient.

Question 37 is continued on the next page.

Question 37 continued.

Graph each function on the set of axes below.

To the *nearest hour*, t, when does the amount of the given drug remaining in patient B begin to exceed the amount of the given drug remaining in patient A?

The doctor will allow patient A to take another 800 milligram dose of the drug once only 15% of the original dose is left in the body. Determine, to the *nearest tenth of an hour*, how long patient A will have to wait to take another 800 milligram dose of the drug.

Answers
June 2016
Algebra II

Answer Key

PART I

1. (4)	5. (3)	9. (2)	13. (3)	17. (1)	21. (3)
2. (3)	6. (1)	10. (4)	14. (2)	18. (1)	22. (4)
3. (2)	7. (3)	11. (1)	15. (4)	19. (4)	23. (3)
4. (3)	8. (4)	12. (1)	16. (2)	20. (2)	24. (4)

PART II

25. 4
26. The experimenter must randomly split subjects into a control group and an experimental group.
27. No
28.

29. 108/1376
30. (4, 0)
31. Yes
32. 6%

PART III

33. (3, 2) and (7, –2)

34. $S_n = \dfrac{33{,}000 - 33{,}000 \cdot 1.04^n}{1 - 1.04}$, 660,778.39

35. Between .47 and .73

36. g

PART IV

37. $A(t) = 800(e)^{-0.347t}$,
    $B(t) = 400(e)^{-0.231t}$, 6, 5.5

In **Parts II–IV**, you are required to show how you arrived at your answers. For sample methods of solutions, see Barron's *Regents Exams and Answers* for Algebra II.

Examination
June 2017
Algebra II

HIGH SCHOOL MATH REFERENCE SHEET

Conversions

1 inch = 2.54 centimeters

1 meter = 39.37 inches

1 mile = 5280 feet

1 mile = 1760 yards

1 mile = 1.609 kilometers

1 kilometer = 0.62 mile

1 pound = 16 ounces

1 pound = 0.454 kilogram

1 kilogram = 2.2 pounds

1 ton = 2000 pounds

1 cup = 8 fluid ounces

1 pint = 2 cups

1 quart = 2 pints

1 gallon = 4 quarts

1 gallon = 3.785 liters

1 liter = 0.264 gallon

1 liter = 1000 cubic centimeters

Formulas	
Triangle	$A = \dfrac{1}{2}bh$
Parallelogram	$A = bh$
Circle	$A = \pi r^2$
Circle	$C = \pi d$ or $C = 2\pi r$
General Prisms	$V = Bh$

Examination June 2017

Cylinder	$V = \pi r^2 h$
Sphere	$V = \frac{4}{3}\pi r^3$
Cone	$V = \frac{1}{3}\pi r^2 h$
Pyramid	$V = \frac{1}{3}Bh$
Pythagorean Theorem	$a^2 + b^2 = c^2$
Quadratic Formula	$x = \dfrac{-b \pm \sqrt{b^2 - 4ac}}{2a}$
Arithmetic Sequence	$a_n = a_1 + (n-1)d$
Geometric Sequence	$a_n = a_1 r^{n-1}$
Geometric Series	$S_n = \dfrac{a_1 - a_1 r^n}{1 - r}$ where $r \neq 1$
Radians	1 radian = $\dfrac{180}{\pi}$ degrees
Degrees	1 degree = $\dfrac{\pi}{180}$ radians
Exponential Growth/Decay	$A = A_0 e^{k(t - t_0)} + B_0$

Examination June 2017

PART I

Answer all 24 questions in this part. Each correct answer will receive 2 credits. No partial credit will be allowed. For each statement or question, write in the space provided the numeral preceding the word or expression that best completes the statement or answers the question. [48 credits]

1 The graph of the function $p(x)$ is sketched below.

Which equation could represent $p(x)$?

(1) $p(x) = (x^2 - 9)(x - 2)$
(2) $p(x) = x^3 - 2x^2 + 9x + 18$
(3) $p(x) = (x^2 + 9)(x - 2)$
(4) $p(x) = x^3 - 2x^2 - 9x - 18$

1 _____

2 What is the solution to $8(2^{x + 3}) = 48$?

(1) $x = \dfrac{\ln 6}{\ln 2} - 3$

(2) $x = 0$

(3) $x = \dfrac{\ln 48}{\ln 16} - 3$

(4) $x = \ln 4 - 3$

2 _____

399

Examination June 2017

3 Cheap and Fast gas station is conducting a consumer satisfaction survey. Which method of collecting data would most likely lead to a biased sample?

(1) interviewing every 5th customer to come into the station

(2) interviewing customers chosen at random by a computer at the checkout

(3) interviewing customers who call an 800 number posted on the customers' receipts

(4) interviewing every customer who comes into the station on a day of the week chosen at random out of a hat

3 _____

4 The expression $6xi^3(-4xi + 5)$ is equivalent to

(1) $2x - 5i$ (3) $-24x^2 + 30x - i$

(2) $-24x^2 - 30xi$ (4) $26x - 24x^2i - 5i$

4 _____

5 If $f(x) = 3|x| - 1$ and $g(x) = 0.03x^3 - x + 1$, an approximate solution for the equation $f(x) = g(x)$ is

(1) 1.96 (3) (−0.99, 1.96)

(2) 11.29 (4) (11.29, 32.87)

5 _____

6 Given the parent function $p(x) = \cos x$, which phrase best describes the transformation used to obtain the graph of $g(x) = \cos(x + a) - b$, if a and b are positive constants?

(1) right a units, up b units

(2) right a units, down b units

(3) left a units, up b units

(4) left a units, down b units

6 _____

7 The solution to the equation $4x^2 + 98 = 0$ is

(1) ± 7

(2) $\pm 7i$

(3) $\pm \dfrac{7\sqrt{2}}{2}$

(4) $\pm \dfrac{7i\sqrt{2}}{2}$

7_____

8 Which equation is represented by the graph shown below?

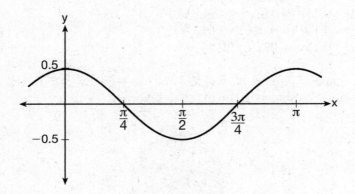

(1) $y = \dfrac{1}{2} \cos 2x$

(2) $y = \cos x$

(3) $y = \dfrac{1}{2} \cos x$

(4) $y = 2 \cos \dfrac{1}{2} x$

8_____

9 A manufacturing company has developed a cost model, $C(x) = 0.15x^3 + 0.01x^2 + 2x + 120$, where x is the number of items sold, in thousands. The sales price can be modeled by $S(x) = 30 - 0.01x$. Therefore, revenue is modeled by $R(x) = x \bullet S(x)$.

The company's profit, $P(x) = R(x) - C(x)$, could be modeled by

(1) $0.15x^3 + 0.02x^2 - 28x + 120$
(2) $-0.15x^3 - 0.02x^2 + 28x - 120$
(3) $-0.15x^3 + 0.01x^2 - 2.01x - 120$
(4) $-0.15x^3 + 32x + 120$

9 _____

10 A game spinner is divided into 6 equally sized regions, as shown in the diagram below.

For Miles to win, the spinner must land on the number 6. After spinning the spinner 10 times, and losing all 10 times, Miles complained that the spinner is unfair. At home, his dad ran 100 simulations of spinning the spinner 10 times, assuming the probability of winning each spin is $\frac{1}{6}$. The output of the simulation is shown in the following diagram.

Which explanation is appropriate for Miles and his dad to make?

(1) The spinner was likely unfair, since the number 6 failed to occur in about 20% of the simulations.
(2) The spinner was likely unfair, since the spinner should have landed on the number 6 by the sixth spin.
(3) The spinner was likely not unfair, since the number 6 failed to occur in about 20% of the simulations.
(4) The spinner was likely not unfair, since in the output the player wins once or twice in the majority of the simulations.

10 _____.

Examination June 2017

11 Which binomial is a factor of $x^4 - 4x^2 - 4x + 8$?

   (1) $x - 2$ (3) $x - 4$

   (2) $x + 2$ (4) $x + 4$ 11_____

12 Given that $\sin^2 \theta + \cos^2 \theta = 1$ and $\sin \theta = -\dfrac{\sqrt{2}}{5}$, what

   is a possible value of $\cos \theta$?

   (1) $\dfrac{5+\sqrt{2}}{5}$ (3) $\dfrac{3\sqrt{3}}{5}$

   (2) $\dfrac{\sqrt{23}}{5}$ (4) $\dfrac{\sqrt{35}}{5}$ 12_____

13 A student studying public policy created a model for the population of Detroit, where the population decreased 25% over a decade. He used the model $P = 714(0.75)^d$, where P is the population, in thousands, d decades after 2010. Another student, Suzanne, wants to use a model that would predict the population after y years. Suzanne's model is best represented by

   (1) $P = 714(0.6500)^y$ (3) $P = 714(0.9716)^y$

   (2) $P = 714(0.8500)^y$ (4) $P = 714(0.9750)^y$ 13_____

14 The probability that Gary and Jane have a child with blue eyes is 0.25, and the probability that they have a child with blond hair is 0.5. The probability that they have a child with both blue eyes and blond hair is 0.125. Given this information, the events blue eyes and blond hair are

               I: dependent

               II: independent

               III: mutually exclusive

   (1) I, only   (3) I and III

   (2) II, only   (4) II and III   14_____

Examination June 2017

15 Based on climate data that have been collected in Bar Harbor, Maine, the average monthly temperature, in degrees F, can be modeled by the equation $B(x) = 23.914\sin(0.508x - 2.116) + 55.300$. The same governmental agency collected average monthly temperature data for Phoenix, Arizona, and found the temperatures could be modeled by the equation $P(x) = 20.238\sin(0.525x - 2.148) + 86.729$.

Which statement can *not* be concluded based on the average monthly temperature models x months after starting data collection?

(1) The average monthly temperature variation is more in Bar Harbor than in Phoenix.

(2) The midline average monthly temperature for Bar Harbor is lower than the midline temperature for Phoenix.

(3) The maximum average monthly temperature for Bar Harbor is 79°F, to the nearest degree.

(4) The minimum average monthly temperature for Phoenix is 20°F, to the nearest degree.   15 _____

16 For $x \neq 0$, which expressions are equivalent to one divided by the sixth root of x?

$$\text{I. } \frac{\sqrt[6]{x}}{\sqrt[3]{x}} \qquad \text{II. } \frac{x^{\frac{1}{6}}}{x^{\frac{1}{3}}} \qquad \text{III. } x^{-\frac{1}{6}}$$

(1) I and II, only   (3) II and III, only

(2) I and III, only   (4) I, II, and III   16 _____

17 A parabola has its focus at $(1, 2)$ and its directrix is $y = -2$. The equation of this parabola could be

(1) $y = 8(x + 1)^2$ (3) $y = 8(x - 1)^2$

(2) $y = \frac{1}{8}(x + 1)^2$ (4) $y = \frac{1}{8}(x - 1)^2$ 17 _____

405

Examination June 2017

18 The function $p(t) = 110e^{0.03922t}$ models the population of a city, in millions, t years after 2010. As of today, consider the following two statements:

    I. The current population is 110 million.
   II. The population increases continuously by approximately 3.9% per year.

This model supports

(1) I, only   (3) both I and II
(2) II, only   (4) neither I nor II   18 _____

19 To solve $\dfrac{2x}{x-2} - \dfrac{11}{x} = \dfrac{8}{x^2 - 2x}$, Ren multiplied both

sides by the least common denominator. Which statement is true?

(1) 2 is an extraneous solution.

(2) $\dfrac{7}{2}$ is an extraneous solution.

(3) 0 and 2 are extraneous solutions.
(4) This equation does not contain any extraneous solutions.   19 _____

20 Given $f(9) = -2$, which function can be used to generate the sequence $-8, -7.25, -6.5, -5.75, \ldots$?

(1) $f(n) = -8 + 0.75n$
(2) $f(n) = -8 - 0.75(n - 1)$
(3) $f(n) = -8.75 + 0.75n$
(4) $f(n) = -0.75 + 8(n - 1)$ 20 _____

21 The function $f(x) = 2^{-0.25x} \cdot \sin\left(\dfrac{\pi}{2}x\right)$ represents a

damped sound wave function. What is the average rate of change for this function on the interval $[-7, 7]$, to the *nearest hundredth*?

(1) -3.66 (3) -0.26
(2) -0.30 (4) 3.36 21 _____

406

Examination June 2017

22 Mallory wants to buy a new window air conditioning unit. The cost for the unit is \$329.99. If she plans to run the unit three months out of the year for an annual operating cost of \$108.78, which function models the cost per year over the lifetime of the unit, $C(n)$, in terms of the number of years, n, that she owns the air conditioner?

(1) $C(n) = 329.99 + 108.78n$

(2) $C(n) = 329.99 + 326.34n$

(3) $C(n) = \dfrac{329.99 + 108.78n}{n}$

(4) $C(n) = \dfrac{329.99 + 326.34n}{n}$

22 _____

23 The expression $\dfrac{-3x^2 - 5x + 2}{x^3 + 2x^2}$ can be rewritten as

(1) $\dfrac{-3x - 3}{x^2 + 2x}$

(2) $\dfrac{-3x - 1}{x^2}$

(3) $-3x^{-1} + 1$

(4) $-3x^{-1} + x^{-2}$

23 _____

24 Jasmine decides to put \$100 in a savings account each month. The account pays 3% annual interest, compounded monthly. How much money, S, will Jasmine have after one year?

(1) $S = 100(1.03)^{12}$

(2) $S = \dfrac{100 - 100(1.0025)^{12}}{1 - 1.0025}$

(3) $S = 100(1.0025)^{12}$

(4) $S = \dfrac{100 - 100(1.03)^{12}}{1 - 1.03}$

24 _____

407

Examination June 2017

PART II

Answer all 8 questions in this part. Each correct answer will receive 2 credits. Clearly indicate the necessary steps, including appropriate formula substitutions, diagrams, graphs, charts, etc. For all questions in this part, a correct numerical answer with no work shown will receive only 1 credit. [16 credits]

25 Given $r(x) = x^3 - 4x^2 + 4x - 6$, find the value of $r(2)$.

What does your answer tell you about $x - 2$ as a factor of $r(x)$? Explain.

Examination June 2017

26 The weight of a bag of pears at the local market averages 8 pounds with a standard deviation of 0.5 pound. The weights of all the bags of pears at the market closely follow a normal distribution. Determine what percentage of bags, to the *nearest integer*, weighed *less* than 8.25 pounds.

27 Over the set of integers, factor the expression $4x^3 - x^2 + 16x - 4$ completely.

Examination June 2017

28 The graph below represents the height above the ground, h, in inches, of a point on a triathlete's bike wheel during a training ride in terms of time, t, in seconds.

Identify the period of the graph and describe what the period represents in this context.

29 Graph $y = 400(.85)^{2x} - 6$ on the set of axes below.

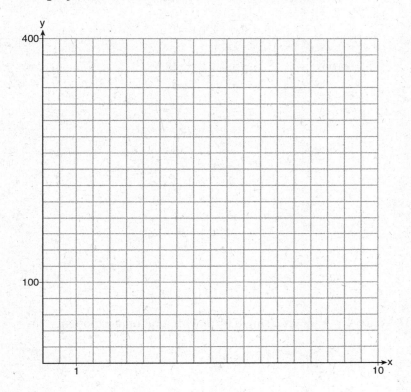

Examination June 2017

30 Solve algebraically for all values of x:

$$\sqrt{x-4} + x = 6$$

31 Write $\sqrt[3]{x} \cdot \sqrt{x}$ as a single term with a rational exponent.

Examination June 2017

32 Data collected about jogging from students with two older siblings are shown in the table below.

	Neither Sibling Jogs	One Sibling Jogs	Both Siblings Jog
Student Does Not Jog	1168	1823	1380
Student Jogs	188	416	400

Using these data, determine whether a student with two older siblings is more likely to jog if one sibling jogs or if both siblings jog. Justify your answer.

Examination June 2017

PART III

Answer all 4 questions in this part. Each correct answer will receive 4 credits. Clearly indicate the necessary steps, including appropriate formula substitutions, diagrams, graphs, charts, etc. For all questions in this part, a correct numerical answer with no work shown will receive only 1 credit. [16 credits]

33 Solve the following system of equations algebraically for all values of x, y, and z:

$$x + y + z = 1$$
$$2x + 4y + 6z = 2$$
$$-x + 3y - 5z = 11$$

Examination June 2017

34 Jim is looking to buy a vacation home for $172,600 near his favorite southern beach. The formula to compute a mortgage payment, M, is $M = P \cdot \dfrac{r(1+r)^N}{(1+r)^N - 1}$ where P is the principal amount of the loan, r is the monthly interest rate, and N is the number of monthly payments. Jim's bank offers a monthly interest rate of 0.305% for a 15-year mortgage.

With no down payment, determine Jim's mortgage payment, rounded to the *nearest dollar*.

Algebraically determine and state the down payment, rounded to the *nearest dollar*, that Jim needs to make in order for his mortgage payment to be $1100.

35 Graph $y = \log_2(x + 3) - 5$ on the set of axes below. Use an appropriate scale to include *both* intercepts.

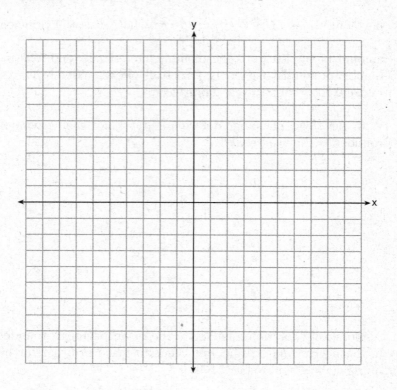

Describe the behavior of the given function as x approaches -3 and as x approaches positive infinity.

36 Charlie's Automotive Dealership is considering implementing a new check-in procedure for customers who are bringing their vehicles for routine maintenance. The dealership will launch the procedure if 50% or more of the customers give the new procedure a favorable rating when compared to the current procedure. The dealership devises a simulation based on the minimal requirement that 50% of the customers prefer the new procedure. Each dot on the graph below represents the proportion of the customers who preferred the new check-in procedure, each of sample size 40, simulated 100 times.

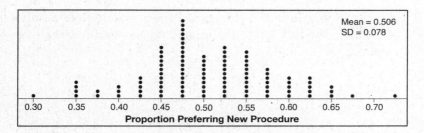

Assume the set of data is approximately normal and the dealership wants to be 95% confident of its results. Determine an interval containing the plausible sample values for which the dealership will launch the new procedure. Round your answer to the *nearest hundredth*.

Forty customers are selected randomly to undergo the new check-in procedure and the proportion of customers who prefer the new procedure is 32.5%. The dealership decides *not* to implement the new check-in procedure based on the results of the study. Use statistical evidence to explain this decision.

Examination June 2017

PART IV

Answer the question in this part. A correct answer will receive 6 credits. Clearly indicate the necessary steps, including appropriate formula substitutions, diagrams, graphs, charts, etc. A correct numerical answer with no work shown will receive only 1 credit. [6 credits]

37 A radioactive substance has a mass of 140 g at 3 p.m. and 100 g at 8 p.m. Write an equation in the form $A = A_0\left(\dfrac{1}{2}\right)^{\frac{t}{h}}$ that models this situation, where h is the constant representing the number of hours in the half-life, A_0 is the initial mass, and A is the mass t hours after 3 p.m.

Using this equation, solve for h, to the *nearest ten thousandth*.

Determine when the mass of the radioactive substance will be 40 g. Round your answer to the *nearest tenth of an hour*.

418

Answers June 2017
Algebra II
Answer Key

PART I

1. (1)
2. (1)
3. (3)
4. (2)
5. (2)
6. (4)
7. (4)
8. (1)
9. (2)
10. (3)
11. (1)
12. (2)
13. (3)
14. (2)
15. (4)
16. (4)
17. (4)
18. (2)
19. (1)
20. (3)
21. (3)
22. (3)
23. (4)
24. (2)

PART II

25. -6, $(x-2)$ is not a factor of $r(x)$
26. 69%
27. $(x^2+4)(4x-1)$
28. Period is $\frac{2}{3}$. The wheel takes $\frac{2}{3}$ seconds to make one rotation.
29.
30. 5
31. $x^{\frac{5}{6}}$
32. More likely if both siblings jog

PART III

33. $(0, 2, -1)$
34. 1247; $20,407
35. y approaches negative infinity as x approaches -3. y approaches positive infinity as x approaches positive infinity.
36. The confidence interval is between 0.35 and 0.66; 0.325 is not between 0.35 and 0.66.

PART IV

37. $100 = 140\left(\frac{1}{2}\right)^{\frac{5}{h}}$; $h = 10.3002$; $x = 18.6$

In **Parts II–IV**, you are required to show how you arrived at your answers. For sample methods of solutions, see Barron's *Regents Exams and Answers* for Algebra II.

Index

A
Addition of polynomials, 1
Algebra, identities, 12–13
Arithmetic sequences, finite, 274
Asymptotes, vertical and horizontal,
75–77
Average rate of change, 252

C
Calculator
composite function, 249–251
converting radian measure, 185–187
exponential equations, 94–95
graphing quadratic equations, 30–31
inverse function, 263
logarithms, 101–104, 106–108
normal distribution, 311–313
sin and cosine table of values, 202
sine and cosine functions, 191–192
standard deviation, 304–305
system of equations, 244–247
trigonometry equations, 178–179,
212–214
unit circle, 158
Complex numbers, simplifying,
130–135
Composite function
average rate of change, 252
end behavior, 252–253
key idea, 248
odd and even, 251
solving, 248–249
using calculator, 249–251
Compound interest, 120–122
Constants, multiplying polynomials, 1
Cubic equations, 42–44
Cubic expressions, factoring, 11–12

D
Directrix of a parabola, 31–35
Distributive property, for exponents,
86
Division, polynomials, 3–6
Double roots, 44

E
Elimination method, 230–235
Explicit formula sequences, 269–270
Exponential and logarithmic functions
horizontal shifts, 115–116
key idea, 111
real-world, 117–122
vertical shifts, 111–114
Exponential equations
and compound interest, 120–122
graphs of two-variable, 90–93
key idea, 88
and logarithm functions, 111–116
multistep, 104–106
solving, 88–90
using calculator, 94–95
Exponential expression, 82
Exponents
in complex numbers, 131
distributive property for, 86
dividing, 82–83
multiplying, 82
power to power, 83–86
properties of, 82–86

F
Factor theorem, 16–17
Factoring
cubic expressions, 11–12
defined, 8
perfect square, 10–11
quadratic trinomial, 9
Ferris wheel problem, 215–216
Finite arithmetic sequences, 273
Focus of a parabola, 31–35
Functions, matrix, 244

G
Geometric series sequences, 273
Graphing
complex numbers, 133–135
inverse function, 259–262
quadratic equations, 23–35, 30–31
radical functions, 144–147

421

Index

rational functions, 75–79
two-variable exponent equations,
 90–93
Graphing calculator
 equivalent rational expressions,
  77–78
 quadratic equations on, 30–31
 rational functions, 76–77
 solving rational equations, 78–79
Greatest common factor, 8–9
Growth rate, 119

H
Horizontal asymptotes, 75–77
Horizontal shifts, in exponential and
 logarithmic functions, 115–116

I
Identities, algebra, 12–13
Imaginary numbers, simplifying,
 130–131
Inferential statistics
 key idea, 318
 population proportion, 318–322
 randomization test, 323–324
 simulation, 323–325
Inverse functions
 basic, 256–257
 complicated, 264
 graphs of, 259–262
 key idea, 256
 of linear function, 257–258
 using calculator, 263

K
Key idea
 asymptotes, 75
 complex numbers, 130
 composite function, 248
 exponential and logarithmic
  .equations, 117
 exponential and logarithmic
  functions, 111
 exponential equations, 88
 exponential expression, 82
 factor theorem, 15
 inverse function, 256
 logarithms, 99
 polynomial equation, 18
 polynomial factoring, 8
 polynomial graph, 41
 polynomials, 1

quadratic graph, 23
radian measure, 183
radical equation, 137
radical functions, 144
rational equations, 68
rational expressions, 57
remainder theorem, 15
simplifying radicals, 126
trigonometry equations, 176
trigonometry identities, 221

L
Linear function, with inverse functions,
 257–259
Linear-quadratic system of equations,
 235
Logarithms
 calculating, 101–104, 106–108
 converting to exponential equations,
  100–101
 definition to solve, 99–100
 and exponential functions, 111–116
 key idea, 99
 using exponential equations, 104–106

M
Math, polynomial arithmetic, 1–6
Matrix functions, 244
Multiplication, polynomials, 1, 2–3

N
Normal distribution
 key idea, 309
 normal curve, 310–311
 using calculator, 311–313

P
Parabola, 23, 31–35
Perfect square, 10–11
Polynomial graphs
 basic cubic equations, 42–44
 key idea, 41
 polynomial functions with double
  roots, 44–50
Polynomial(s)
 addition of, 1
 arithmetic, 1–6, 176
 division in, 3–6
 equations, 18–21
 factoring, 8–13, 19
 multiplication, 1, 2–3
Population proportion, 318–322

422

Index

Practice
  answers, 342–362
  complex and imaginary numbers,
    135–136
  composite function, 253–255
  exponential and logarithmic
    functions, 123–125
  exponential equations, 96–99
  factoring, 14–15
  full, 375–395
  graphing rational equations, 79–81
  inferential statistics, 325–330
  linear function, 265–268
  logarithms, 109–111
  normal distribution, 313–315
  polynomial equations, 22–23
  polynomial graphs, 51–56
  polynomials, 7–8
  probability, 283–286, 290–292,
    297–300
  properties of exponents, 87–88
  quadratic graphs, 36–41
  radian measure, 187–189
  radical equation, 142–144
  radical functions, 147–156
  relational equations, 73–75
  relational expressions, 65–67
  remainder and factor theorem,
    17–18
  sequences, 274–275
  simplifying radicals, 128–130
  sine and cosine functions, 207–211
  solving trigonometry equations,
    214–215
  standard deviation, 308–309
  statistical studies, 317–318
  system of equations, 236–238, 243
  trigonometry equations, 180–182,
    214–215
  trigonometry identities, 225–227
  unit circle trigonometry, 171–175
Probability
  basic, 277–279
  with dependent events, 292–294
  dependent versus independent events,
    286–287
  events with tables or Venn diagrams,
    295–296
  questions, 288–289
  sample spaces, 276–277
  table sample spaces, 280–281
  as Venn diagrams, 281–282

Properties
  distributive property for exponents,
    86
  of exponents, 82–86

Q
Quadratic equations
  complex numbers in, 133
  defined, 18
  graphing, 30–31
  solving, 18–21
  standard form, 28–30
  using quadratic formula, 21
  vertex form, 25–28
Quadratic formula, 21
Quadratic graphs
  basic graph, 23–24
  a coefficient, 24–25
  parabola, 31–35
  standard form, 28–30
  using a calculator, 30–31
  vertex form, 25–28
Quadratic trinomial, 9

R
Radian measure, converting,
    183–187
Radical equations, solving, 137–142
Radical functions, graphs of square
    root function, 144–147
Radicals, simplifying, 126–128
Randomization test, 323–324
Rate, growth, 119
Rational equations
  with cross multiplication,
    68–69
  key idea, 68
  with multiple terms, 69–71
  rate word problems, 71–72
  simple, 68
Rational expressions
  adding, 60–64
  dividing, 60
  key idea, 57
  multiplying, 59–60
  reducing, 57–59
  subtracting, 64–65
Rational functions
  asymptotes, 75–77
  graphing, 76–79
  key idea, 75
Rational number, 57

423

Index

Real world scenarios
in exponential and logarithmic
functions, 117–122
ferris wheel problem, 215–216
temperature problem, 216–219
with trigonometric functions,
215–219
Recursive formula sequences, 270–271
Reference sheet, 340, 373–374,
397–398
Remainder theorem, 15–16
Roots, double, 44

S
Sample spaces, 276
Sequences
arithmetic and geometric, 271–273
explicit formula, 269–270
key idea, 269
recursive formula, 270–271
Simulation, 323–325
Sine and cosine functions
amplitude of, 192–194
finding the equation of, 203–207
frequency of, 197–201
graphs of, 189–192
table of values with calculator,
202
using calculator, 191–192
vertical translations of, 195–197
Square roots, 127–128
Standard deviation
key idea, 301
with repeated numbers, 305–308
two sets of data for, 301–304
using calculator, 304–305
Statistical studies, survey, 316–317
Strategies
calculator, 334–339
computational errors versus
conceptual errors, 333–334
partial credit, 332–333
points needed to pass, 341
question breakdown, 370
reference sheet, 340, 373–374,
397–398
scoring exam, 371–372
test-taking, 331
time management, 331–332
Substitution method, 228–229
Subtraction, polynomials, 1–2

Systems of equations
elimination method, 230–235
key idea, 228
linear-quadratic, 235
matrix functions, 244
substitution method, 228–229
three unknowns, 238–242
using calculator, 244–247

T
Temperature problems, 216–219
Terms, 363–369
Theorem(s)
factor, 16–17
number proofs, 13
remainder, 15–16
Trigonometry. *See* Unit circle
trigonometry
Trigonometry equations
with calculator, 178–179
graphical solving, 212–214
key idea, 176
solutions, 176–178
Trigonometry identities, five main,
222–225
Trinomial
perfect square, 10–11
quadratic, 9

U
Unit circle trigonometry
key idea, 157
points on a full unit circle,
165–170
points on a unit quarter circle,
164–165
right triangle, 157–164
using calculator, 158

V
Venn diagram and probability,
281–282, 295–296
Vertex, 23
Vertical asymptotes, 75–77
Vertical shifts, in exponential and
logarithmic functions, 111–114
Vertical translations, 195–197
Vocabulary, 363–369

W
Word problems, rate, 71–72

It's finally here—
online Regents exams from the experts!

- Take complete practice tests by date, or choose questions by topic
- All questions answered with detailed explanations
- Instant test results let you know where you need the most practice

Online Regents exams are available in the following subjects:

- Biology–The Living Environment
- Chemistry–The Physical Setting
- Earth Science–The Physical Setting
- English (Common Core)
- Global History and Geography
- Algebra I (Common Core)
- Geometry (Common Core)
- Algebra II (Common Core)
- Physics–The Physical Setting
- U.S. History and Government

Getting started is a point and a click away!
Only $24.99 for a subscription...includes all of the subjects listed above!

Teacher and classroom rates also available.

For more subscription information and to try our demo visit our site at *www.barronsregents.com*.

Prices subject to change without notice.

(#105) R8/16

**Barron's Let's Review—
Textbook companions to help
you ace your Regents Exams**

Let's Review: Algebra I
ISBN 978-1-4380-0604-8

Let's Review: Algebra II
ISBN 978-1-4380-0844-8

Let's Review: Biology—The Living
Environment, 6th Ed.
ISBN 978-1-4380-0216-3

Let's Review: Chemistry—The Physical
Setting, 6th Ed.
ISBN 978-1-4380-0959-9

Let's Review: Earth Science—The
Physical Setting, 5th Ed.
ISBN 978-1-4380-0911-7

Let's Review: English, 5th Ed.
ISBN 978-1-4380-0626-0

Let's Review: Geometry
ISBN 978-1-4380-0702-1

Let's Review: Global History and
Geography, 6th Ed.
ISBN 978-1-4380-1131-8

Let's Review: Physics, 5th Ed.
ISBN 978-1-4380-0630-7

Let's Review: U.S. History and
Government, 6th Ed.
ISBN 978-1-4380-0962-9

Each book: Paperback, $14.99, *Can18.50*

Available at your local bookstore or visit www.barronseduc.com

Barron's Educational Series, Inc.
250 Wireless Blvd.
Hauppauge, N.Y. 11788
Order toll-free: 1-800-645-3476
Order by fax: 1-631-434-3217

In Canada:
Georgetown Book Warehouse
34 Armstrong Ave.
Georgetown, Ontario L7G 4R9
Canadian orders: 1-800-247-7160
Order by fax: 1-800-887-1594

For extra Regents review visit our website: **www.barronsregents.com**

Prices subject to change without notice.

#225 R10/17